# Longman
# Chemistry
# for IGCSE

Jim Clark and
Ray Oliver

PEARSON
Longman

Pearson Education Limited,
Edinburgh Gate, Harlow,
Essex CM20 2JE, England
and Associated Companies throughout the world.

www.longman.co.uk

Fourth impression 2006
**ISBN-10: 1-4058-0208-1**
**ISBN-13: 978-1-4058-0208-5**

Produced by Cambridge Publishing Management Ltd, Cambridge
Cover design by Neil Straker Creative

The publisher's policy is to use paper manufactured from sustainable forests.

Printed in China
SWTC/04

### Acknowledgments

We are grateful to the following for permission to reproduce photographs:

l = left, r = right, c = centre, t = top, b = bottom

1l Getty Images; 1r Jacques Jangoux/Science Photo Library; 24l © William Whitehurst/Corbis; 24r Getty Images; 25 © Ray Juno/Corbis; 26 © Ed Young Corbis; 27 Charles D. Winter/ Science Photo Library; 31 Roberto de Gualiemo/Science Photo Library; 54 © Craig Aurness/Corbis; 57 Geoscience Features; 58 © Roger Wood/Corbis; 61 © David Muench/Corbis; 62, 68t Geoscience Features; 68b © Charles E. Rotkin/Corbis; 70 University of Cambridge; 79b Getty Images; 84t © Andrew Lambert/Leslie Garland Picture Library; 92t www.shoutpictures.com; 120 Paul Deakin; 124 © Firefly Productions/Corbis; 127 Johnson Mathey; 131 Thomas Eisner & Daniel Aneshanley, Cornell University; 134 Acestock; 137l NASA; 137r Acestock/Ronald Toms; 138 Diver Magazine, www.divernet.com; 139 © Tom Bean/Corbis; 151 Photolibrary.com/Oxford Scientific/ Niall Benvie; 166t Maximilian Stock Ltd/Science Photo Library; 166b Mazovian Museum, Plock, Poland/www.bridgeman.co.uk; 180tc Sinclair Stammers/Science Photo Library; 180tr Geoscience Features; 180bl © Layne Kennedy/Corbis; 182l © James L. Amos/ Corbis; 182r © Paul A. Souders/Corbis; 183 © Mark L. Stephenson/ Corbis; 184t Aluminium Federation; 184b Getty Images; 186 Corus; 187cr Getty Images; 187br © Michael John Keilty/Corbis; 190 The Natural History Museum, London; 191 Elizabeth Whiting & Associates/Alamy; 196 Novosti/Science Photo Library; 204bl Photolibrary.com/Oxford Scientific; 204bc Diver Magazine, www.divernet.com; 204br Robert Harding Picture Library; 204t Getty Images; 204c LASERIUM® Laser Show; 213 Bob Gibbons/Science Photo Library; 225 Alvey & Towers Picture Library; 229r © Asian Art & Archaeology Inc/Corbis; 231 Martin Bond/Science Photo Library; 239l Comstock/Getty Images; 239r www.shoutpictures.com; 241 Holt Studios/Paul Hobson; 246, 250 Andrew Lambert Photographs;

258 Staton R. Winter/Newsmakers/Camera Press; 266t & b Alvey & Towers Picture Library; 273 © Natalie Fobes/Corbis; 278b Maximilian Stock Ltd/Science Photo Library; 279t Tony Myers A.R.P.S/Firepix International; 279b © Mike Taylor/Leslie Garland Picture Library; 280 Martyn F. Chillmaid/Science Photo Library; 283 © Tony Arruza/Corbis; 288 Food Features.

All other photos by Jim Clark

Picture Research by Anne Lyons, with additional material for new edition by Sandie Huskinson-Rolfe of PHOTOSEEKERS.

Cover: Image Source/Alamy

Index: Indexing Specialists (UK) Ltd

Every effort has been made to trace and acknowledge ownership of copyright. If any have been overlooked, the publisher will be pleased to make the necessary changes at the earliest opportunity.

# Contents

# About this book

This book has several features to help you with IGCSE Chemistry.

## Introduction
Each chapter has a short introduction to help you start thinking about the topic and let you know what is in the chapter.

## End of chapter checklists
These lists summarise the material in the chapter. They could also help you to make revision notes because they form a list of things that you need to revise. (You need to check your specification to find out exactly what you need to know.)

## Margin boxes
The boxes in the margin give you extra help or information. They might explain something in a little more detail or guide you to linked topics in other parts of the book.

## Questions
There are short questions at the end of each chapter. These help you to test your understanding of the material from the chapter. Some of them may also be research questions – you will need to use the Internet and other books to answer these.

There are also questions at the end of each section. The end-of-section questions are written in an exam style and cover topics from all the chapters in the section.

---

### Section B: Chemical Calculations

#### Chapter 7: Relative Atomic Masses and Moles

**When you have completed this chapter, you will be able to:**
- define relative atomic mass ($A_r$) and relative formula mass ($M_r$)
- use the carbon-12 scale of masses
- understand the term 'weighted average'
- work out the relative formula mass of a compound
- understand how to calculate percentage compositions
- understand the mole and how we use the idea
- know the meaning of the Avogadro Constant
- understand the difference between empirical (simplest) and molecular formulae.

*Each gold bar contains almost $4 \times 10^{25}$ gold atoms. That's 4 followed by 25 noughts!*

You can make iron(II) sulphide by heating a mixture of iron and sulphur.

$$Fe(s) + S(l) \rightarrow FeS(s)$$

How do you know what proportions to mix them up in? You can't just mix equal masses of them because iron and sulphur atoms don't weigh the same. Iron atoms contain more protons and neutrons than sulphur atoms so that an iron atom is one and three-quarters times heavier than a sulphur atom. In this, or any other reaction, you can only get the proportions right if you know about the masses of the individual atoms taking part.

#### Relative atomic mass ($A_r$)

**Defining relative atomic mass**

Atoms are amazingly small. In order to get a gram of hydrogen, you would need to count out 602 204 500 000 000 000 000 000 atoms (to the nearest 100 000 000 000 000 000).

It would be silly to measure the masses of atoms in conventional mass units like grams. Instead, their masses are compared with the mass of an atom of the carbon-12 isotope, taken as a standard. We call this the 'carbon-12 scale'.

On this scale one atom of the carbon-12 isotope weighs *exactly* 12 units.

An atom of the commonest isotope of magnesium weighs twice as much as this and is therefore said to have a **relative isotopic mass** of 24.

An atom of the commonest isotope of hydrogen weighs only one twelfth as much as a carbon-12 atom, and so has a relative isotopic mass of 1.

*Remember that isotopes are atoms of the same element but with different masses.*

54

---

### End of Chapter Checklist
- one atom of carbon-12 weighs exactly 12 units. We call this the carbon-12 scale
- other kinds of atoms are compared with the mass of this isotope of carbon
- the relative formula mass ($M_r$) is the mass of a compound such as carbon dioxide (molecular) or sodium chloride (ionic)
- the percentage composition of a compound gives the percentage of each element
- the mole is the standard unit for the amount of a substance
- the empirical formula of a substance gives the simplest ratio of the atoms in a compound
- the molecular formula gives the actual numbers of atoms in one molecule.

### Questions

1. The relative formula mass of carbon dioxide is:

   A 28    C 56
   B 32    D 44      (1)

2. The empirical formula is:

   A the one found by experiment
   B the same as the molecular formula
   C the simplest ratio of atoms
   D none of these      (1)

3. a) What is the carbon-12 scale. Why don't we use kilograms instead? (3)
   b) Bromine atoms weigh 79 or 81 units. How many different diatomic molecules are there? (3)
   c) If the two isotopes of bromine are equally common, what is the weighted average? Explain why the relative atomic mass of bromine is given as 80 but there are no bromine atoms that weigh 80 units. (4)

4. a) Using hexane, $C_6H_{14}$, as your example, explain the meaning of empirical formula. (2)
   b) Calculate the percentage by mass of carbon and hydrogen in hexane. (4)
   c) What is the $M_r$ of hexane? (1)
   d) What is the mass of i) 1 mole of hexane, ii) 0.25 moles of hexane iii) 3.5 moles of hexane? (3)

5. a) What is the formula of hydrated calcium sulphate (2 molecules of water)? (2)
   b) What name is given to this water? (1)
   c) What is the $M_r$? (2)

   d) What is the mass of 0.1 mole? (2)
   e) When this compound is heated, the mass changes. Explain why this happens and write an equation assuming that half of the water is lost. (3)

6. a) Work out if the molecular formula of a hydrocarbon is $C_{12}H_{12}$ or $C_6H_6$. Explain your answer. The compound contains 7.7% hydrogen. Its $M_r$ is 78. (4)
   b) A compound contains 52.2% carbon, 13.1% hydrogen and the rest is oxygen. Find the empirical formula. If the $M_r$ is 46, find the molecular formula. (3)
   c) Find the empirical formula and the percentages of each element in benzene, $C_6H_6$. (3)

7. The common isotopes of chlorine weigh 35 and 37 units.
   a) How many different mass molecules are there for dichloromethane, $CH_2Cl_2$? (3)
   b) Draw and label each molecule. (3)
   c) Using a calculation, explain why the $A_r$ of chlorine is about 35.5 and not a whole number. (2)
   d) Lighter molecules diffuse more quickly. Which molecule of dichloromethane will diffuse quickest and why? (2)

8. Calculate the relative atomic mass of gallium given the percentage abundances:

   $^{69}Ga$ – 60.2%, $^{71}Ga$ – 39.8%. (6)

65

v

# Section A: Properties of Matter

## Chapter 1: Atomic Structure

This chapter explores the nature of atoms, and how they differ from element to element. The 100 or so elements are the building blocks from which everything is made – from the simplest substance, like carbon, to the most complex, like DNA.

**When you have completed this chapter, you will be able to:**

- explain the structure of atoms
- name and describe the particles in atoms
- explain atomic number and mass number
- understand the structure of the Periodic Table
- explain the link between the atomic number and the Periodic Table
- work out the arrangement of the electrons in atoms
- draw ring diagrams
- understand the importance of full energy levels of electrons
- explain the Laws of Conservation of Mass and Constant Composition.

Copper is an element. If you tried to chop it up into smaller and smaller bits, eventually you would end up with the smallest possible piece of copper. At that point you would have an individual copper atom. You can, of course, split that into still smaller pieces (protons, neutrons and electrons), but you would no longer have copper.

*Whether things are man-made...*

*...or natural, they are all made up of combinations of the same elements.*

Chemistry just rearranges existing atoms. For example, when propane burns in oxygen, existing carbon, hydrogen and oxygen atoms combine in new ways:

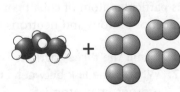 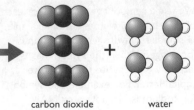

propane      oxygen      carbon dioxide      water

1

# The structure of the atom

Atoms are made of protons, neutrons and electrons.

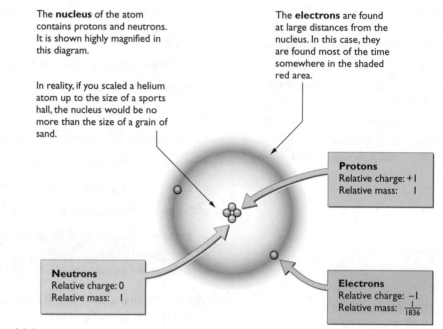

The **nucleus** of the atom contains protons and neutrons. It is shown highly magnified in this diagram.

In reality, if you scaled a helium atom up to the size of a sports hall, the nucleus would be no more than the size of a grain of sand.

The **electrons** are found at large distances from the nucleus. In this case, they are found most of the time somewhere in the shaded red area.

**Protons**
Relative charge: +1
Relative mass: 1

**Neutrons**
Relative charge: 0
Relative mass: 1

**Electrons**
Relative charge: −1
Relative mass: $\frac{1}{1836}$

*A helium atom.*

Virtually all the mass of the atom is concentrated in the nucleus, because the electrons weigh hardly anything.

The masses and charges are measured relative to each other because the actual values are incredibly small. For example, it would take about 600 000 000 000 000 000 000 000 protons to weigh 1 gram.

## Atomic number and mass number

The number of protons in an atom is called its **atomic number** or **proton number**. Each of the one hundred or so different elements has a different number of protons. For example, if an atom has 8 protons it must be an oxygen atom.

Atomic number = number of protons

The **mass number** is the total number of protons and neutrons in the nucleus of the atom.

Mass number = number of protons + number of neutrons

For any particular atom, this information can be shown simply (see the figure on the left).

This particular atom of cobalt contains 27 protons. To make the total number of protons and neutrons up to 59, there must also be 32 neutrons.

1. Name the three particles we find in atoms.
2. What is the link between atomic number and the number of protons in an atom?
3. What is meant by the mass number of an atom?

You may have come across diagrams of the atom in which the electrons are drawn orbiting the nucleus, rather like planets around the sun. This is misleading.

It is impossible to know exactly how the electrons are moving in an atom. All you can tell is that they have a particular energy and that they are likely to be found in a certain region of space at some particular distance from the nucleus. Electrons with different energies are found at different distances from the nucleus.

mass number counts
protons + neutrons

$^{59}_{27}\text{Co}$ ← symbol for element

atomic number counts
the number of protons

We use a symbol as shorthand for each of the chemical elements. For example, H stands for hydrogen and O stands for oxygen. When the symbol has two letters we just write the first letter as a capital letter. For example, the symbol for sodium is Na, not na or NA.

## Isotopes

The number of neutrons in an atom can vary slightly. For example, there are three kinds of carbon atom, called carbon-12, carbon-13 and carbon-14. They all have the same number of protons (because all carbon atoms have 6 protons – its atomic number), but the number of neutrons varies. These different atoms of carbon are called **isotopes**.

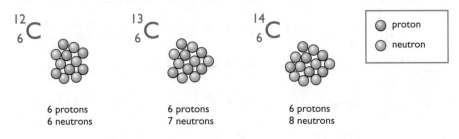

| | | | proton |
| | | | neutron |

6 protons    6 protons    6 protons
6 neutrons   7 neutrons   8 neutrons

*Nuclei of the three isotopes of carbon.*

Isotopes are atoms which have the same atomic number, but different mass numbers. They have the same number of protons, but different numbers of neutrons.

The fact that they have varying numbers of neutrons makes no difference whatsoever to their chemical reactions. The chemical properties are governed by the number and arrangement of the electrons and, as you will see shortly, that is identical for all three isotopes.

> **4.** Give one example of an element that is found as isotopes.
> **5.** Do isotopes behave in different ways in chemical reactions?

# The electrons

### Counting the number of electrons in an atom

Atoms are electrically neutral, and the positiveness of the protons is balanced by the negativeness of the electrons. In a neutral atom it follows that:

$$\text{Number of electrons} = \text{number of protons}$$

So, if an oxygen atom (atomic number = 8) has 8 protons, it must also have 8 electrons; if a chlorine atom (atomic number = 17) has 17 protons, it must also have 17 electrons.

You will see that the key feature in this is knowing the atomic number. You can find that from the Periodic Table.

> Remember that the number of protons is the same as the atomic number of the element.

### *Atomic number and the Periodic Table*

Atoms are arranged in the Periodic Table in order of increasing atomic number. You will find a full version of the Periodic Table on page 297. Most Periodic Tables have two numbers against each symbol – be careful to choose the right one. *The atomic number will always be the smaller number.* The other number will either be the mass number of the most common isotope of the element, or the relative atomic mass of the element. The Table will tell you which.

You use a Periodic Table to find out the atomic number of an element, and therefore how many protons and electrons there are in its atoms.

## The arrangement of the electrons

The electrons are found at considerable distances from the nucleus in a series of levels called **energy levels** or **shells**. Each energy level can only hold a certain number of electrons. Low energy levels are always filled before higher ones.

The diagram shows the maximum number of electrons that each energy level can hold.

The third level can expand to hold a total of 18 electrons.

increasing energy and distance from nucleus

third level — sometimes appears full with 8 electrons but can expand to hold a total of 18

second level — only room for 8 electrons

first level — only room for 2 electrons

(not to scale)   nucleus

### How to work out the electronic configurations

We will use chlorine as an example.

- Look up the atomic number in the Periodic Table. (Make sure that you don't use the wrong number if you have a choice. The atomic number will always be the *smaller* one.)

  *The Periodic Table tells you that chlorine's atomic number is 17.*

- This tells you the number of protons, and hence the number of electrons.

  *There are 17 protons, and so 17 electrons in a neutral chlorine atom.*

- Arrange the electrons in levels, always filling up an inner (lower energy) level before you go to an outer one.

  *These will be arranged as 2 in the first level, 8 in the second level, and 7 in the third level. This is written as **2,8,7**. When you have finished, always check to make sure that the electrons add up to the right number – in this case 17.*

Don't just accept this diagram! Use the Periodic Table on page 297 and work out each of these electronic structures for yourself (preferably in a random order to make it more difficult). Check your answers when you have finished.

### The electronic configurations of the first 20 elements in the Periodic Table

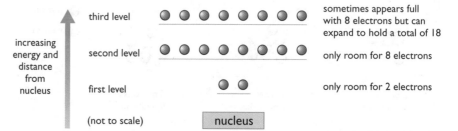

| group 1 | group 2 | | group 3 | group 4 | group 5 | group 6 | group 7 | group 0 |
|---|---|---|---|---|---|---|---|---|
| | | **H** 1 | | | | | | **He** 2 |
| **Li** 2,1 | **Be** 2,2 | 10 more elements | **B** 2,3 | **C** 2,4 | **N** 2,5 | **O** 2,6 | **F** 2,7 | **Ne** 2,8 |
| **Na** 2,8,1 | **Mg** 2,8,2 | | **Al** 2,8,3 | **Si** 2,8,4 | **P** 2,8,5 | **S** 2,8,6 | **Cl** 2,8,7 | **Ar** 2,8,8 |
| **K** 2,8,8,1 | **Ca** 2,8,8,2 | | | | | | | |

Vertical columns in the Periodic Table are called **groups**. Groups contain elements with similar properties.

There are two important generalisations you can make from this:

- *The number of electrons in the outer level is the same as the group number for groups 1 to 7.*

  This pattern extends right down the Periodic Table for these groups.

  So if you know that barium is in group 2, you know it has 2 electrons in its outer level; iodine (group 7) has 7 electrons in its outer level; lead (group 4) has 4 electrons in its outer level. Working out what is in the inner levels is much more difficult. The simple patterns we have described don't work beyond calcium.

- *The elements in group 0 have 8 electrons in their outer levels (apart from helium, which has 2).*

  These are often thought of as being 'full' levels. This is true for helium and neon, but not for the elements from argon downwards. For example, the third energy level will eventually contain 18 electrons.

  The group 0 elements are known as the **noble gases** because they are almost completely unreactive – in fact the three at the top of the group from helium to argon don't react with anything. This lack of reactivity is associated with their electronic structures – often described as **noble gas structures**.

### Drawing diagrams of electronic configurations

The electrons in their various energy levels can be shown by drawing circles with dots or crosses on them showing the electrons. It doesn't matter whether you draw dots or crosses.

Hydrogen has one electron and helium has two in the first level. The helium electrons are sometimes shown as a pair (as here), and sometimes as two separate electrons on opposite sides of the circle. Either form is acceptable.

The next four atoms are drawn like this:

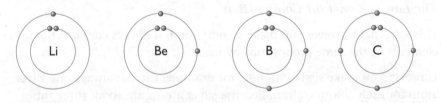

The electrons in the second energy level are drawn singly, up to a maximum of 4. After that, pair them up as necessary. It makes them much easier to count. More importantly, it gives a much better picture of the availability of the electrons in the atom for bonding purposes. This is explored in Chapter 3.

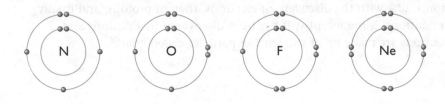

Drawing circles like this *does not* imply that the electrons are orbiting the nucleus along the circles. The circles represent energy levels. The further the level is from the nucleus, the higher its energy. For theoretical reasons it is impossible to work out exactly how an electron is moving in that energy level.

The atoms in the Periodic Table from sodium to argon fill the third level in exactly the same way, and potassium and calcium start to fill the fourth level.

6. How many electrons do the elements of Group 0 have in their outer shells?
7. Draw the electron arrangement in an atom of carbon.
8. Why are the noble gases so unreactive?

### The Law of Conservation of Mass

A simple version of this says that in a chemical reaction the total mass of everything at the end of the reaction is the same as the total mass at the beginning. This is bound to be the case if all you are doing is rearranging atoms which have got fixed masses.

The diagrams show the reaction between methane (natural gas) and chlorine in sunlight. The atoms have rearranged into new substances. The contents of both boxes must have the same mass because they are both made from exactly the same atoms – even if they are rearranged.

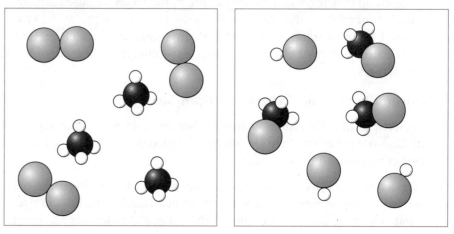

before                                    after

### The Law of Constant Composition

This says that however you make a compound, it always contains the same elements in the same proportions by mass.

However you make silver chloride, for example, there is always one silver atom for each chlorine atom. Because silver atoms are about three times heavier than chlorine atoms, the mass ratio is always approximately 3 g of silver to 1 g of chlorine.

### Refining the atomic theory

Scientists have gradually got closer and closer to the fine structure of atoms, first with the discovery of electrons, then of protons, and finally of neutrons. More recently, it has been discovered that protons and neutrons are made up of still smaller particles called quarks.

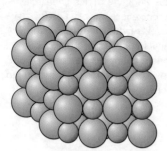

*A small part of a silver chloride crystal, AgCl.*

# End of Chapter Checklist

**In this chapter you have learnt that:**

- atoms contain smaller particles called protons (+ charge), electrons (− charge) and neutrons (no charge)

- atoms have a central nucleus containing protons and neutrons

- electrons in atoms are arranged in shells around the nucleus. The first ring can take up to two electrons, the second and third up to eight electrons

- the Periodic Table displays all the known elements in the order of their atomic numbers

- the arrangement of electrons in shells gives us information about groups in the Table

- atoms with full electron shells (2,8,8 for first, second and third) are very stable

- every element has its own atomic number and mass number

- the atomic number is the number of protons in the nucleus of an atom

- the mass number of an atom is the total number of protons and neutrons in the nucleus of an atom

- isotopes are atoms of the same chemical element with different numbers of neutrons.

# Questions

You will need to use the Periodic Table on page 297. More questions on atomic structure can be found at the end of Section A on page 43.

1.  Fluorine atoms have a mass number of 19.

    a)  Use the Periodic Table to find the atomic number of fluorine. (2)

    b)  Explain what *mass number* means. (1)

    c)  Write down the number of protons, neutrons and electrons in a fluorine atom. (3)

    d)  Draw a diagram to show the arrangement of the electrons in the fluorine atom. (2)

2.  Work out the numbers of protons, neutrons and electrons in each of the following atoms:

    a) $^{56}_{26}Fe$   b) $^{93}_{41}Nb$   c) $^{235}_{92}U$ (9)

3.  Chlorine has two isotopes, chlorine-35 and chlorine-37.

    a)  What are isotopes? (2)

    b)  Write down the numbers of protons, neutrons and electrons in the two isotopes. (2)

    c)  Write down the arrangement of the electrons in each of the two isotopes. (2)

4.  Draw diagrams to show the arrangement of the electrons in *a)* sodium, *b)* silicon, *c)* sulphur. (6)

5.  Find each of the following elements in the Periodic Table, and write down the number of electrons in their outer energy level.

    a) arsenic, As, b) bromine, Br,
    c) tin, Sn, d) xenon, Xe. (4)

6.  The questions refer to the electronic structures below. Don't worry if some of these are unfamiliar to you. All of these are the electronic structures of neutral atoms.

    **A** 2, 4              **B** 2, 8, 8

    **C** 2, 8, 18, 18, 7    **D** 2, 8, 18, 18, 8

    **E** 2, 8, 8, 2        **F** 2, 8, 18, 32, 18, 4

    a)  Which of these atoms are in group 4 of the Periodic Table? (2)

    b)  Which of these structures represents carbon? (2)

    c)  Which of these structures represents an element in group 7 of the Periodic Table? (2)

    d)  Which of these structures represent noble gases? (2)

    e)  Name element **E**. (2)

    f)  How many protons does element **F** have? Name the element. (2)

    g)  Element **G** has one more electron than element **B**. Draw a diagram to show how the electrons are arranged in an atom of **G**. (2)

## Chapter 2: States of Matter

> **When you have completed this chapter, you will be able to:**
>
> - explain changes of state and know the words for each change
> - understand the energy changes that accompany changes of state
> - understand how diffusion works.

### Particles everywhere

We believe that everything is made of particles. This chapter gives some of the evidence to support this idea.

There are three states of matter: **solid**, **liquid** and **gas**. When ice melts to water or water turns to steam there is a *change of state*. These are the special words to use when describing changes of state:

When ice becomes water it is melting (1), when water becomes ice it is freezing (2), when water becomes water vapour it is evaporating (3), when water vapour becomes water it is condensing (4). When a solid turns straight to a gas it is called sublimation (5). Only a few solids are capable of this change of state. An example is solid carbon dioxide.

Solid carbon dioxide is very cold, about –80°C. It sublimes as it warms up. The solid turns straight to carbon dioxide gas. This material is used in 'cloud machines' to produce special effects at rock concerts and discos. You can investigate sublimation like this:

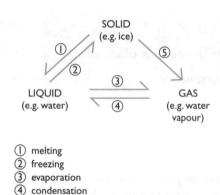

① melting
② freezing
③ evaporation
④ condensation
⑤ sublimation

### Activity 1: Investigating sublimation

**Investigative skills**

| P | O |
|---|---|

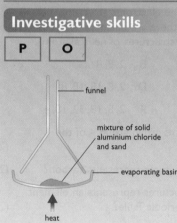

funnel

mixture of solid aluminium chloride and sand

evaporating basin

heat

**You will need:**

- evaporating basin and glass funnel to rest on top
- solid ammonium chloride and sand
- Bunsen burner, tripod and gauze

**Carry out the following:**

1. Mix together about 5 g of each solid.
2. Put the mixture in the evaporating basin and cover with the funnel.
3. Heat very gently.
4. Which solid sublimes? Where does it collect as it cools down again?
5. How could you use this idea to purify a solid which sublimes?

> 1. Explain the meaning of the following words: **a)** condense, **b)** sublime, **c)** melt.
> 2. What is the difference between freezing and melting?
> 3. Solid iodine sublimes on heating. How could you separate a mixture of iodine and sand?

# How are particles arranged?

In a solid like copper metal, the particles are arranged in a regular pattern. They are lined up in fixed positions, like bricks in a wall. When copper gets really hot, it melts. After this change of state the particles move around freely in the liquid copper. If the temperature goes up even more, the liquid will boil, like water in a kettle. After this change of state, the particles spread out everywhere. The copper has turned into a gas. How can we make a model of this?

There are forces of attraction between particles. In a solid the forces are strong. In a liquid the forces are weaker, allowing the particles to slide past each other. In a gas the forces are very weak indeed.

### Energy changes

We need to supply energy to make solids melt. A pure solid always melts at the same temperature, the melting point (m.p.). The melting point of ice is 0°C. More energy is needed to make liquids boil at their boiling point (b.p.). Pure water boils at 100°C at normal air pressure (one atmosphere pressure).

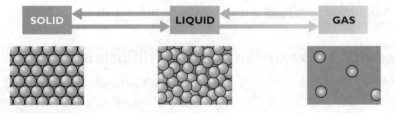

## Particles on the move

### Diffusion

At home or school you can soon tell if someone is cooking food with a strong smell. Even if there is no breeze or wind, the cooking smell still reaches you even if you are in another room. Gases spread out all by themselves. We call this **diffusion**. You can investigate the way diffusion works.

## Activity 2: Diffusion of ammonia and hydrogen chloride gases

### Investigative skills

O  A

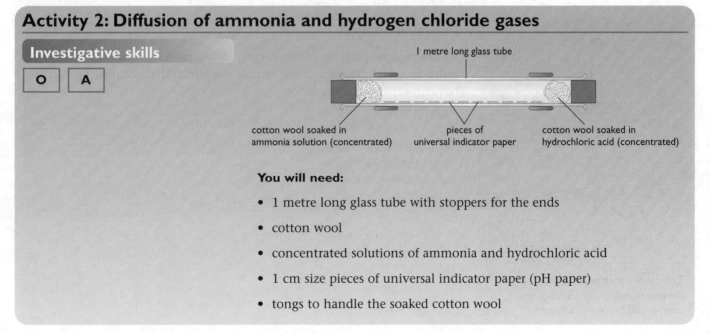

1 metre long glass tube

cotton wool soaked in ammonia solution (concentrated)

pieces of universal indicator paper

cotton wool soaked in hydrochloric acid (concentrated)

**You will need:**

- 1 metre long glass tube with stoppers for the ends
- cotton wool
- concentrated solutions of ammonia and hydrochloric acid
- 1 cm size pieces of universal indicator paper (pH paper)
- tongs to handle the soaked cotton wool

## Activity 2: Diffusion of ammonia and hydrogen chloride gases (continued)

**Carry out the following:**

1. Spread pieces of indicator paper along the inside of the tube.

2. Soak separate pieces of cotton wool in each solution.
   (**Warning**: harmful chemicals.)

3. Put each piece of soaked cotton wool inside the two ends of the tube at the same time and seal with a stopper (the ammonia is at one end, the acid at the other).

4. Watch for the appearance of a smoke ring inside the tube where the two gases meet and react.

5. How do you know that the particles of gases are moving?

6. If the particles move at the same speed, where will they meet? Is this what happened?

7. Ammonia particles are lighter than hydrogen chloride particles. Use this information to explain what you saw in the activity.

## Activity 3: Extension ideas on diffusion

### Investigative skills

| P | O | A |

1. What would happen if you dropped coloured sweets into a tall jar of water and left it for a few days?

2. Gelatine is mixed with pH indicator solution and left to set. How could you use this to investigate the movement of acids and alkalis within the solid gel?

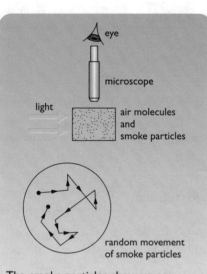

The smoke particles show up as moving points of light. This is one example of Brownian motion.

### Brownian motion

You cannot see particles moving, even with a powerful microscope. They are much too small. However, you can see the effects that the moving particles have on other, larger particles.

## Activity 4: Particles moving

**Investigative skills**

| A | E |
|---|---|

**You will need:**

- a microscope and a lamp

- a microscope slide and cover slip

- full-fat milk and a dropper

**Carry out the following:**

1. Place one drop of milk on the slide and cover with the cover slip.

2. Look at the fat globules in the milk and the way they move around.

3. Can you explain the movement? Here are some clues:
   Imagine a game where a team of players must move a giant balloon around a field. It needs lots of players to hold up the balloon from underneath and to push it along. Now imagine that you are in a helicopter far above. You cannot see the people, they are too small. But you can see that the large balloon is moving. Use this idea to explain Brownian motion.

### Dilution

When we add water to a concentrated solution, we say that it has been diluted. The solute particles in the concentrated solution are closer together than in the diluted solution. If the solute has a colour, it is possible to see how many times a solution can be diluted before the colour disappears.

## Activity 5: Dilution of a coloured solution

**Investigative skills**

| O | A |
|---|---|

**You will need:**

- test-tubes

- solid potassium manganate(VII) and spatula

- measuring cylinder or pipette

**Carry out the following:**

1. Place 10 cm³ water in a tube.

2. Add one crystal of the solid.

3. Shake carefully to dissolve.

4. Transfer 1 cm³ solution to a second tube and make up to a total volume of 10 cm³.

5. Repeat step 4 until you can no longer see the colour.

6. What evidence does this experiment give you about the size of the coloured particles in the first solution? Explain your answer.

# End of Chapter Checklist

**In this chapter you have learnt that:**

- the particles in solids, liquids and gases are arranged in different ways
- sublimation means that a solid changes straight into a gas when it is heated; it does not melt first
- diffusion means that particles spread out, like cooking smells in a house
- Brownian motion provides evidence that particles move, as with fat droplets in milk
- dilution gives evidence for the size of particles in solution.

# Questions

**1.** Particles are closest together in:

   A   solids
   B   gases
   C   solutions
   D   liquids       (1)

**2.** When a solid changes straight to a gas it is called:

   A   melting
   B   evaporation
   C   sublimation
   D   boiling       (1)

**3.** The opposite of evaporation is:

   A   freezing
   B   melting
   C   boiling
   D   condensing       (1)

**4.** Which of the following statements is/are true?

   I   in diffusion particles move
   II   gases diffuse faster than liquids
   III   the spread of cooking smells is an example of diffusion
   IV   evaporation and diffusion are really the same thing

   A   I and II
   B   I, II and IV
   C   I, II and III
   D   II and IV       (1)

**5.** In the following experiment:

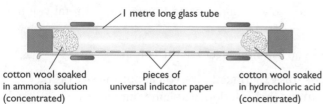

cotton wool soaked      pieces of      cotton wool soaked
in ammonia solution   universal indicator paper   in hydrochloric acid
(concentrated)                  (concentrated)

   A   the gas particles all move at the same speed
   B   the gas particles do not move at all
   C   the lighter gas particles move faster
   D   the heavier gas particles stay still   (1)

**6.** **a)** Explain what is meant by diffusion. Include a diagram in your answer.   (3)

   **b)** Describe what happens to the perfume particles from an aerosol can when it is used in a closed room.   (4)

   **c)** How could you measure the speed of diffusion of food dye in water?   (3)

**7.**

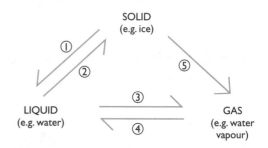

   **a)** Name each of the changes shown by numbers 1–5 in the diagram above.   (5)

   **b)** Explain what is happening to the particles at each stage.   (5)

**8.**

| Name | Formula | Mass | Order of diffusion |
|---|---|---|---|
| ammonia | $NH_3$ | 17 | |
| carbon dioxide | $CO_2$ | 44 | |
| hydrogen | $H_2$ | 2 | |

   **a)** Write 1st, 2nd and 3rd in the last column to show which gas will diffuse fastest, etc.   (2)

   **b)** If you had a bottle of each gas and removed the tops at the same time, what would happen? Explain your answer.   (4)

# Section A: Properties of Matter

## Chapter 3: Bonding

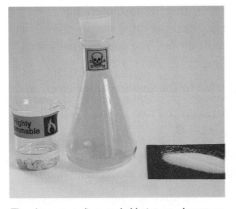

The elements sodium and chlorine are dangerous. The compound sodium chloride (salt) isn't.

Water has completely different properties from its elements, hydrogen and oxygen.

### When you have completed this chapter, you will be able to:

- describe covalent bonding in molecules
- understand what happens to electrons in bonding
- understand what is special about the noble gases
- understand how bonds break and new bonds form
- draw bonds in several different ways
- describe ionic (electrovalent) bonding
- explain the electrical charges in ions
- show how electrons are lost and gained
- describe metallic bonding
- describe intermolecular forces.

Sodium is a dangerously reactive metal. It is stored under oil to prevent it reacting with air or water. Chlorine is a very poisonous, reactive gas.

But salt, sodium chloride, is safe to eat in small quantities. Combining the elements to make salt obviously changes them significantly.

A mixture of hydrogen and oxygen gas would explode violently if you held a lighted match to it. Dropping a lighted match into water (a compound of hydrogen and oxygen) doesn't cause a literally earth-shattering explosion. Reacting the elements to make a compound has again made a huge difference to them.

## Covalent bonding

### What is a covalent bond?

In any bond, particles are held together by electrical attractions between something positively charged and something negatively charged. In a covalent bond, a pair of electrons is shared between two atoms. Each of the positively charged nuclei is attracted to the same negatively charged pair of electrons.

In most examples each atom in a covalent bond supplies one electron to the pair of electrons. That doesn't have to be the case. Both electrons may come from the same atom.

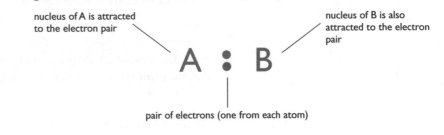

A and B are held together by this shared attraction.

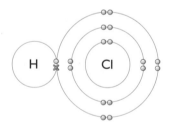

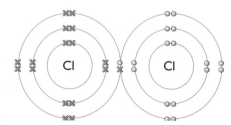

## Covalent bonding in a hydrogen molecule

Covalent bonds are often shown using 'dot-and-cross' diagrams. Although the electrons are drawn as dots or as crosses, there is absolutely no difference between them in reality. The dot and the cross simply show that the electrons have come from two different atoms. You could equally well use two different colour dots or two different colour crosses.

Both hydrogen nuclei are strongly attracted to the shared pair of electrons.

The bond can also be shown as a line between the two atoms. Each line represents one pair of shared electrons. H——H

The covalent bond between two hydrogen atoms is very strong. Hydrogen atoms therefore go around in pairs, called a hydrogen **molecule**, with the symbol $H_2$.

Molecules have a certain fixed number of atoms in them joined together by covalent bonds. Hydrogen molecules are said to be **diatomic** because they contain two atoms. Other sorts of molecules may have as many as thousands of atoms joined together.

### Why does hydrogen form molecules?

Whenever a bond is formed (of whatever kind), energy is released; that makes the things involved more stable than they were before. The more bonds an atom can form, the more energy is released and the more stable the system becomes.

In the hydrogen case, each hydrogen atom only has one electron to share, so it can only form one covalent bond. The $H_2$ molecule is still much more stable than two separate hydrogen atoms.

## Covalent bonding in a hydrogen chloride molecule

The chlorine atom has one unpaired electron in its outer level which it can share with the hydrogen atom to produce a covalent bond.

Notice that only the electrons in the outer energy level of the chlorine are used in bonding. The inner electrons are often left out of bonding diagrams.

### The significance of noble gas structures in covalent bonding

If you look at the arrangement of electrons around the chlorine atom in the covalently bonded molecule of HCl, you will see that its structure is now 2,8,8. That is the same as an argon atom. Similarly, the hydrogen now has two electrons in its outer level – the same as helium.

When atoms bond covalently, they often produce outer electronic structures the same as noble gases – in other words, with four pairs of electrons (or one pair in the case of hydrogen). There are, however, lots of examples where different numbers of pairs are formed, producing structures which are quite unlike noble gases.

## Covalent bonding in a chlorine molecule

Each chlorine has one unpaired electron in its outer energy level. These are shared between the two atoms to produce a covalent bond. Chlorine is another diatomic molecule, $Cl_2$.

## Covalent bonding in an oxygen molecule – double bonding

When atoms bond covalently, they do so in a way that forms the maximum number of bonds. That makes the final molecule more stable.

The top diagram shows that forming a single covalent bond between the two oxygen atoms still leaves unpaired electrons. If these are shared as well (as in the lower diagram), a more stable molecule is formed.

Oxygen is a diatomic molecule, $O_2$. The double bond can be shown as O=O.

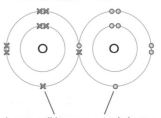

Each atom still has an unpaired electron. If these were shared as well, even more energy would be released.

## Covalent bonding in a nitrogen molecule

Each nitrogen atom has five electrons in its outer level. By sharing three more electrons, nitrogen forms a very stable molecule. The triple bond is very strong. This is why nitrogen is so unreactive.

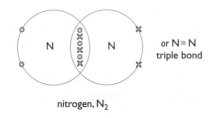

or N≡N
triple bond

nitrogen, $N_2$

## Covalent bonding in methane, ammonia, water and ethane

In methane, the carbon atom has four unpaired electrons. Each of these forms a covalent bond by sharing with the electron from a hydrogen atom. Methane has the formula $CH_4$.

In ammonia, the nitrogen only has three unpaired electrons and so can only form bonds with three hydrogen atoms to give $NH_3$.

In water, there are two unpaired electrons on the oxygen atom which can bond with hydrogen atoms to give $H_2O$.

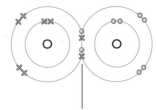

Sharing 2 electron pairs maximises the bonding and makes the system as stable as possible.

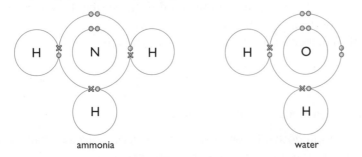

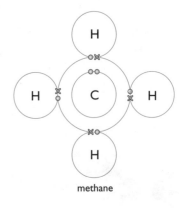

ammonia                    water                    methane

The structure of ethane is very similar to methane, natural gas. All of the bonds are single covalent bonds.

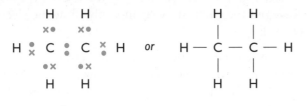

ethane

## Why is water more stable than a mixture of hydrogen and oxygen?

When hydrogen, $H_2$, and oxygen, $O_2$, combine to make water, covalent bonds have to be broken in the hydrogen and oxygen molecules, and new ones are made in water.

> It is important to realise that when most compounds are formed, you don't actually make them from atoms of their elements. For example, you never make a water molecule from individual hydrogen and oxygen atoms. Those atoms are already bonded in some way. A chemical reaction reorganises existing bonds.

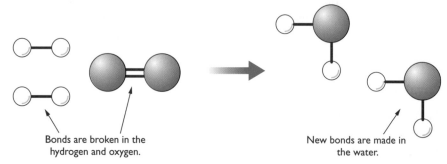

Bonds are broken in the hydrogen and oxygen.

New bonds are made in the water.

It costs energy to break bonds, but a lot of energy is released when new ones are made. The reaction between hydrogen and oxygen is explosive. This is because much more energy is released in making the new bonds than was used in breaking the old ones. Because so much energy is released, the water is much more stable than the original hydrogen and oxygen.

## Covalent bonding in carbon dioxide, $CO_2$

The carbon atom has four unpaired electrons and so can form four covalent bonds. Each oxygen has two unpaired electrons and can form two covalent bonds.

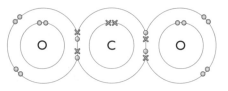

The number of bonds is maximised (and so the system becomes most stable) if the carbon forms two bonds with each oxygen. Using lines to represent pairs of shared electrons, you could draw carbon dioxide as:

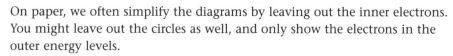

## Ways of representing covalent bonds

Apart from full dot-and-cross diagrams, covalent molecules can also be shown in other ways. In models, each link between the atoms represents a covalent bond – a pair of shared electrons.

On paper, we often simplify the diagrams by leaving out the inner electrons. You might leave out the circles as well, and only show the electrons in the outer energy levels.

Or you might draw each covalent bond as a straight line joining the atoms. Each line means a pair of shared electrons. In diagrams of this sort, sometimes you draw the non-bonding pairs of electrons in the outer level (called **lone pairs**); sometimes you leave them out.

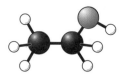

ethanol, $CH_3CH_2OH$

oxygen, $O_2$

ethene, $C_2H_4$

All these diagrams show the covalent bonding in ammonia, $NH_3$:

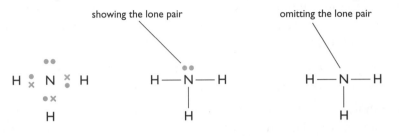

showing the lone pair      omitting the lone pair

1. How many electrons are there in: **a)** a single bond, **b)** a double bond?
2. What is special about the electrons in a noble gas? How does this affect the properties of noble gases?

## Ionic (electrovalent) bonding

In a covalent bond, the electrons are shared between two atoms. Both nuclei are attracted to the same electron pair.

Sometimes the electron pair is attracted to one atom much more strongly than the other one. The electron pair is then pulled very close to that atom, and away from the other one.

A ∶ B

pair of electrons shared between A and B in a covalent bond

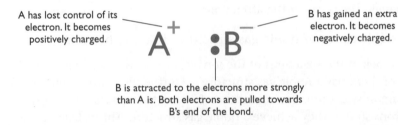

A has lost control of its electron. It becomes positively charged.

B has gained an extra electron. It becomes negatively charged.

B is attracted to the electrons more strongly than A is. Both electrons are pulled towards B's end of the bond.

A has become positively charged because it has effectively lost an electron. It still has the same number of positively charged protons, but now has one less electron to balance them. B is negatively charged because it has gained an extra negative electron.

Atom A has, in effect, given its electron to atom B.

The electrically charged particles are called **ions**. An ion is an atom (or group of atoms) which carries an electrical charge, either positive or negative.

- A positive ion is called a **cation**.

- A negative ion is called an **anion**.

**Ionic bonding** is bonding in which there has been a transfer of electrons from one atom to another to produce ions. The substance is held together by strong electrical attractions between positive and negative ions.

### Ionic bonding in sodium chloride

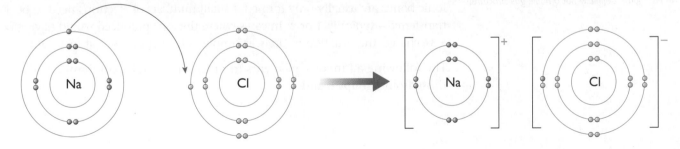

The single electron in the outer energy level of the sodium has been transferred to the chlorine. The sodium chloride is held together by the strong attraction between the sodium ion and the chloride ion. (Notice that it is called a **chloride** ion and not a chlorine ion.)

Overall, lots of energy is given out when this process happens – mainly due to the energy released when the strong bonding between the ions is set up.

You can draw dot-and-cross diagrams to show ionic bonding, but it is much quicker, and takes up much less space, to write electronic structures in the form 2,8,1 or 2,8,7.

### Ionic bonding in magnesium oxide

In this case, two electrons are transferred from the magnesium to the oxygen. The two electrons in the outer energy level of the magnesium are relatively easy to remove, and the oxygen has enough space in its outer level to receive them. More energy is given out this time, mainly due to the very, very strong attractions between the 2+ and 2– ions – the higher the number of charges, the stronger the attractions.

### The significance of noble gas structures in ionic bonding

If you look at the structures of the ions formed in the two examples above, each of them has a noble gas structure – 2,8 (the neon structure), or 2,8,8 (the argon structure). You might therefore say that atoms lose or gain electrons so that they achieve a noble gas structure. This is true of the elements in Groups 1 and 2 of the Periodic Table (forming 1+ and 2+ ions), and for those in Groups 6 and 7 when they form 2– and 1– ions – as in all these examples.

But there are lots of common ions which don't have noble gas structures. $Fe^{2+}$, $Fe^{3+}$, $Cu^{2+}$, $Zn^{2+}$, $Ag^+$ and $Pb^{2+}$ are all ions that you will come across during a chemistry course – although you won't have to write their electronic structures. Not one of them has a noble gas structure.

### Other examples of ionic bonding

Most examination questions restrict ionic bonding to elements from Groups 1, 2, 6 and 7 because that's where the simple examples are to be found.

Ionic bonds are usually only formed if small numbers of electrons need to be transferred – typically 1 or 2. In cases where the ions produced would have, say, a 3+ charge, the situation is rarely as simple as it might appear at first sight.

The following all involve elements from Groups 1 and 2 combining with those from Groups 6 and 7.

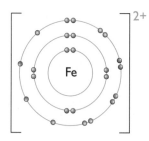

An $Fe^{2+}$ ion – definitely not a noble gas structure!

## Lithium fluoride

The lithium atom has one electron in its outer energy level that is easily lost, and the fluorine has space to receive one. Lithium fluoride is held together by the strong attractions between lithium and fluoride ions.

## Calcium chloride

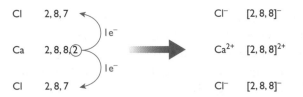

The calcium has two electrons in its outer energy level that are relatively easy to give away, but each chlorine atom only has room in its outer level to take one of them. You need two chlorines for every one calcium. The formula for calcium chloride is therefore $CaCl_2$. There will be very strong attractions holding the ions together because of the 2+ charge on the calcium ions.

## Potassium oxide

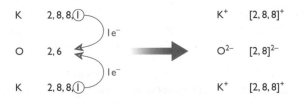

This time the oxygen has room for two electrons in its outer level, but each potassium can only supply one. Potassium oxide's formula is therefore $K_2O$. The presence of the double negative charge on the $O^{2-}$ ion will help to strengthen the attractions between the ions.

3. What is the main difference between covalent and ionic bonding?
4. Explain what is meant by: a) anion, b) cation.
5. Draw the electron arrangement in magnesium chloride.

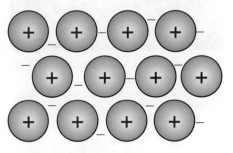

A model of metallic structure.

## Metallic bonding

Most metals are hard and have high melting points. That suggests that the forces which hold the particles in the metal together are very strong.

The diagram shows what happens when sodium atoms bond together to form the solid metal. The outer electron on each sodium atom becomes free to move throughout the whole structure. The electrons are said to be **delocalised**.

If a sodium atom loses its outer electron, that leaves behind a sodium ion.

The attraction of each positive ion to the delocalised electrons holds the structure together.

Metallic bonding is sometimes described as an array of positive ions in a '**sea of electrons**'.

In the case of sodium, only one electron per atom is delocalised, leaving ions with only one positive charge on them. The ions don't pack very efficiently either. The effect of all this is that the bonding in sodium is quite weak as metals go, which is why sodium is fairly soft with a melting point which is low for a metal.

By contrast, magnesium has two outer electrons, both of which are delocalised into the 'sea', leaving behind ions which carry a charge of 2+. It also packs more efficiently. There is a much stronger attraction between the more negative 'sea' and the doubly-charged ions, and so the bonding is stronger and the melting point is greater.

Metals like iron have even more outer electrons to delocalise and so the bonding is stronger still.

## Intermolecular forces

You will remember that water, $H_2O$, is a molecule with strong covalent bonds between the hydrogen and oxygen. In liquid water, or in ice, there must also be attractions between one molecule and its neighbours – otherwise they wouldn't stick together to make a liquid or a solid.

These forces of attraction between separate molecules are called **intermolecular forces** or **intermolecular attractions**. They are a lot weaker than covalent or ionic bonds, and vary in strength from substance to substance.

For example, the intermolecular forces between hydrogen molecules, $H_2$, are very, very weak. You have to cool hydrogen to –253°C before the molecules are travelling slowly enough for the intermolecular attractions to be able to hold them together as a liquid.

By contrast, sugar (also a covalent compound) is a solid which doesn't melt until 185°C. The intermolecular forces between sugar molecules must be quite strong.

Intermolecular forces arise from slight electrical distortions in molecules.

Some elements are very electronegative. This means that they attract electrons within covalent bonds. Instead of being evenly spread out, the electrons in the bond are closer to one end. Oxygen is very electronegative. This is why in a water molecule, we mark the slight positive (δ+) and negative (δ–) charges. Water is an example of a polar molecule – a molecule with slight charges caused by the uneven sharing of electrons in a bond. Each water molecule attracts others that are near to it. This follows the rule of electrostatics: opposite charges attract.

In melting, some but not all of the intermolecular forces are broken. In boiling, the attractions are totally disrupted and the molecules become free to move around as a gas. It is very important that you realise that melting or

*Breaking the intermolecular attractions in water to produce steam.*

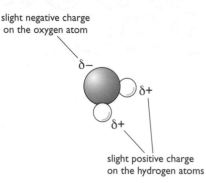

slight negative charge on the oxygen atom

δ–

δ+

δ+

slight positive charge on the hydrogen atoms

boiling a substance made of molecules breaks intermolecular forces – *not* covalent bonds. When you boil water, you get steam – not a mixture of hydrogen and oxygen atoms.

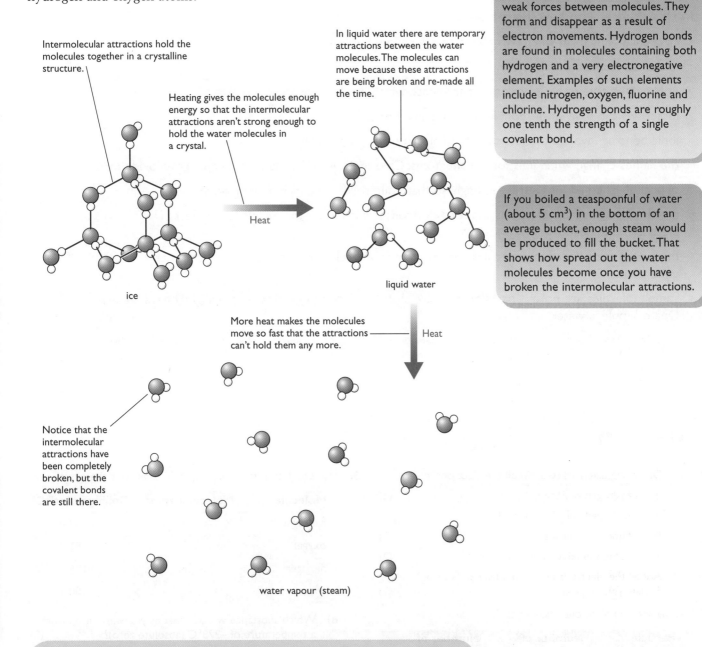

Intermolecular attractions hold the molecules together in a crystalline structure.

Heating gives the molecules enough energy so that the intermolecular attractions aren't strong enough to hold the water molecules in a crystal.

In liquid water there are temporary attractions between the water molecules. The molecules can move because these attractions are being broken and re-made all the time.

Heat

ice

liquid water

More heat makes the molecules move so fast that the attractions can't hold them any more.

Heat

Notice that the intermolecular attractions have been completely broken, but the covalent bonds are still there.

water vapour (steam)

If you boiled a teaspoonful of water (about 5 cm$^3$) in the bottom of an average bucket, enough steam would be produced to fill the bucket. That shows how spread out the water molecules become once you have broken the intermolecular attractions.

6. How do metals conduct electricity?
7. Why does magnesium have a higher melting point than sodium?
8. What are intermolecular forces? Give one example.

# End of Chapter Checklist

**In this chapter you have learnt that:**

- covalent bonds form when two electrons are shared between atoms
- molecules contain covalent bonds
- noble gases have stable electron arrangements, which is why they are unreactive
- molecules can have single, double or even triple bonds
- ionic bonding involves the loss or gain of electrons between atoms
- the names change when ions form from atoms. Chlorine gives chloride ions, oxygen gives oxide ions
- oppositely charged ions attract strongly so that ionic solids have high melting points
- metals contain mobile electrons, a 'sea of electrons', which allow the metals to conduct electricity and heat
- there are forces between molecules known as intermolecular attractions — these forces are quite weak
- some molecules are polar — parts of them are slightly positive (+) and parts are slightly negative (−). One example is water.

# Questions

1. *a)* Draw ring diagrams to show all the electrons in:

   *i)* a hydrogen molecule (2)

   *ii)* a hydrogen chloride molecule (2)

   *iii)* a fluorine molecule (2)

   *iv)* a carbon dioxide molecule. (2)

   *b)* Are all the electrons in an atom used in bonding? Explain your answer. (2)

2. Copy and complete the following table.

| Particle | Number of protons | Number of electrons |
|----------|-------------------|---------------------|
| hydrogen atom | | |
| hydrogen ion | | |
| chlorine atom | | |
| chloride ion | | |
| sodium ion | | |

(10)

3. Use the data in the following table to answer the questions.

| Molecule | Molecular mass | Boiling point/°C |
|----------|----------------|------------------|
| hydrogen | 2 | −253 |
| oxygen | 32 | −183 |
| nitrogen | 28 | −196 |
| water | 18 | +100 |

   *a)* Which substance will boil first as you warm it up from a temperature of −273°C (absolute zero)? (2)

   *b)* Plot a graph of molecular mass against boiling point. (3)

   *i)* Is there a pattern? (1)

   *ii)* Is there an exception? (2)

   *c)* *i)* Which molecules are likely to have stronger intermolecular attractions? (1)

   *ii)* What is the evidence for this? (1)

4.  **a)**  What is meant by a *covalent bond*? How does this bond hold two atoms together? (2)

    **b)**  Draw dot-and-cross diagrams to show the covalent bonding in *i)* methane, $CH_4$, *ii)* hydrogen sulphide, $H_2S$, *iii)* phosphine, $PH_3$, *iv)* silicon tetrachloride, $SiCl_4$. (4)

5.  Draw dot-and-cross diagrams to show the covalent bonding in **a)** ethane, $C_2H_6$, **b)** ethene, $C_2H_4$ (4) and **c)** ethanol, $CH_3CH_2OH$. You will find models of ethene and ethanol on page 19 which might help you. (2)

6.  **a)**  What is meant by *i)* an ion, *ii)* an ionic bond? (2)

    **b)**  In each of the following cases, write down the electronic structures of the original atoms and then explain (in words or diagrams) what happens when:

    *i)*  sodium bonds with chlorine to make sodium chloride

    *ii)*  lithium bonds with oxygen to make lithium oxide

    *iii)*  magnesium bonds with fluorine to make magnesium fluoride. (6)

7.  **a)**  A solid metal is often described as having 'an array of positive ions in a sea of electrons'. Write down the electronic structure of a magnesium atom and use it to explain what this phrase means. (4)

    **b)**  Metallic bonds are not fully broken until the metal has first melted and then boiled. The boiling points of sodium, magnesium and aluminium are 890°C, 1110°C and 2470°C respectively. What does this suggest about the strengths of the metallic bonds in these three elements? (2)

    **c)**  Find these three metals in the Periodic Table, and suggest why the boiling points show this pattern. (2)

    **d)**  Assuming that an electric current is simply a flow of electrons, suggest why all these elements are good conductors of electricity. (2)

8.  The table gives details of the boiling temperatures of some substances made up of covalent molecules. Arrange these substances in increasing order of the strength of their intermolecular attractions.

| | Boiling point (°C) |
|---|---|
| ammonia | −33 |
| ethanamide | 221 |
| ethanol | 78.5 |
| hydrogen | −253 |
| phosphorus trifluoride | −101 |
| water | 100 |

(6)

Don't panic if you don't recognise some of the names. The substances could just as well have been described as A, B, C, D, E and F.

9.  Boron and aluminium are both in Group 3 of the Periodic Table. Both form compounds with fluorine ($BF_3$ and $AlF_3$). Unusually for elements found in the same group of the Periodic Table, their compounds are bonded differently. $BF_3$ is covalent whereas $AlF_3$ is a straightforward ionic compound.

    **a)**  Draw a diagram to show the covalent bonding in $BF_3$. (2)

    **b)**  Explain, using diagrams or otherwise, the origin of the ionic bonding in $AlF_3$. (2)

    **c)**  $BF_3$ is described as an *electron deficient* compound. What do you think that might mean? (1)

# Section A: Properties of Matter

# Chapter 4: Structure

**When you have completed this chapter, you will be able to:**

- describe examples of giant structures
- understand the structures of metals and alloys
- explain the properties of giant ionic structures
- understand how to measure melting and boiling points
- investigate the solubility of materials
- describe some examples of allotropes.

Diamond (a form of carbon) is obviously crystalline, and is the hardest naturally occurring substance.

Ice is also crystalline, but melts easily to form water.

The photographs show some substances with quite different physical properties – hardness, melting point and solubility, for example. This chapter explores some of the reasons for these differences, based on the bonding in the substances.

## Giant structures

You can divide substances into two quite different types – giant structures and molecular structures.

You will remember that molecules are made up of *fixed* numbers of atoms joined together by covalent bonds. The number of atoms per molecule is usually fairly small, but can run into thousands in the case of big molecules like plastics, proteins or DNA.

By contrast, giant structures contain huge numbers of either atoms or ions arranged in some regular way, but the number of particles isn't fixed.

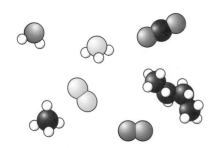

Some simple molecules.

Examples will make this clear.

# Giant metallic structures

Remember that metals consist of a regular array of positive ions in a 'sea of electrons'. The metal is held together by the attractions between the positive ions and the delocalised electrons.

## The simple physical properties of metals

*Metals tend to be strong, with high melting and boiling points* because of the powerful attractions involved.

*Metals conduct electricity*. This is because the delocalised electrons are free to move throughout the structure. Imagine what happens if a piece of metal is attached to an electrical power source.

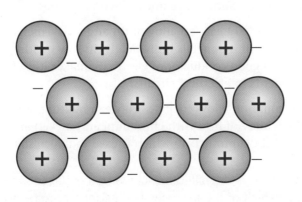

A model of metallic structure.

*Metals are good conductors of heat*. This is again due to the mobile delocalised electrons. If you heat one end of a piece of metal, the energy is picked up by the electrons. As the electrons move around in the metal, the heat energy is transferred throughout the structure.

## The workability of metals

If a metal is subjected to just a small force, it will stretch and then return to its original shape when the force is released. The metal is described as being **elastic**.

But if a large force is applied, the particles slide over each other and stay in their new positions.

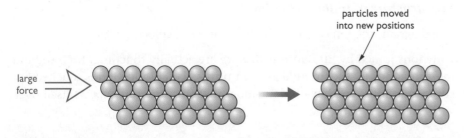

large force

particles moved into new positions

Metals are usually easy to shape because their regular packing makes it simple for the atoms to slide over each other. Metals are said to be **malleable** and **ductile**. Malleable means that it is easily beaten into shape. Ductile means that it is easily pulled out into wires.

*Steel being rolled into strips.*

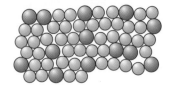

## Alloys

Metals can be made harder by **alloying** them with other metals. An alloy is a mixture of metals – for example, **brass** is a mixture of copper and zinc.

In an alloy the different metals have slightly differently sized atoms. This breaks up the regular arrangement and makes it more difficult for the layers to slide.

The diagram shows how mixing atoms of only slightly different sizes disrupts the regular packing, and makes it much more difficult for particles to slide over each other when a force is applied. This tends to make alloys harder than the individual metals which make them up.

In some cases, alloys have unexpected properties.

For example, solder – an alloy of tin and lead – melts at a lower temperature than either of the metals individually. Its low melting point and the fact that it is a good conductor of electricity makes it useful for joining components in electrical circuits.

Other common alloys include bronze (a mixture of copper and tin), stainless steel (an alloy of iron with chromium and nickel) and the mixture of copper and nickel ('cupronickel') which is used to make 'silver' coins.

*Solder is an alloy of lead and tin. Solder has a melting point lower than that of either of the individual metals.*

1. What is the difference between a metal and an alloy?
2. Which of the following has a giant structure? **a)** oxygen, **b)** diamond, **c)** carbon dioxide.
3. Explain why solder is more useful than either tin or lead alone in joining wires together.

## Giant ionic structures

An ion is an atom or group of atoms which carries an electrical charge – either positive or negative. All ionic compounds consist of huge lattices of positive and negative ions packed together in a regular way. A **lattice** is a regular array of particles. The lattice is held together by the strong attractions between the positively and negatively charged ions.

### The structure of sodium chloride

In a diagram, the ions are usually drawn in an 'exploded' view.

Only ions joined by lines in the diagram are actually touching. Each sodium ion is touched by six chloride ions. In turn, each chloride ion is touched by six sodium ions. You have to remember that this structure repeats itself over vast numbers of ions.

### The structure of magnesium oxide

Magnesium oxide, MgO, contains magnesium ions, $Mg^{2+}$, and oxide ions, $O^{2-}$. It has exactly the same structure as sodium chloride.

The only difference is that the magnesium oxide lattice is held together by stronger forces of attraction. This is because in magnesium oxide, 2+ ions are attracting 2– ions. In sodium chloride, the attractions are weaker because they are only between 1+ and 1– ions.

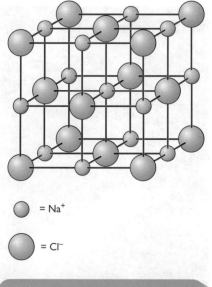

= $Na^+$

= $Cl^-$

**Warning!** The lines in this diagram are *not* covalent bonds. They are just there to help to show the arrangement of the ions. Those ions joined by lines are touching each other.

## The simple physical properties of ionic substances

*Ionic compounds have high melting points and boiling points* because of the strong forces holding the lattices together. Magnesium oxide has much higher melting and boiling points than sodium chloride because the attractive forces are much stronger.

Comparison of magnesium oxide and sodium chloride.

| Formula | MgO | NaCl |
|---|---|---|
| **Melting point/Kelvin** | 3173 | 1081 |
| **Solubility in water** | almost insoluble | very soluble |
| **Charge on metal ion** | +2 | +1 |
| **Charge on non-metal ion** | –2 | –1 |
| **Name** | magnesium oxide | sodium chloride |

The ions in magnesium oxide have double the charges on those in sodium chloride. This explains the differences in their properties. The higher the charge, the stronger the attraction between the oppositely charged ions.

*Ionic compounds tend to be crystalline.* This reflects the regular arrangement of ions in the lattice. Sometimes the crystals are too small to be seen except under powerful microscopes. Magnesium oxide, for example, is always seen as a white powder because the individual crystals are too small to be seen with the naked eye.

*Ionic crystals tend to be brittle.* This is because any small distortion of a crystal will bring ions with the same charge alongside each other. Like charges repel and so the crystal splits itself apart.

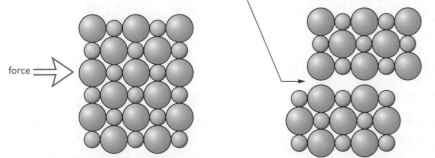

*The shape of the sodium chloride crystal reflects the arrangement of the ions.*

*Ionic substances tend to be soluble in water.* Although water is a covalent molecule, the electrons in the bonds are attracted towards the oxygen end of the bond. This makes the oxygen slightly negative. It leaves the hydrogen slightly short of electrons and therefore slightly positive.

Because of this electrical distortion, water is described as a **polar** molecule. There are quite strong attractions between the polar water molecules and the ions in the lattice.

The slightly positive hydrogens in the water molecules cluster around the negative ions, and the slightly negative oxygens are attracted to the positive ions.

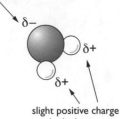

slight negative charge on the oxygen atom

slight positive charge on the hydrogen atoms

Chapter 4: Structure

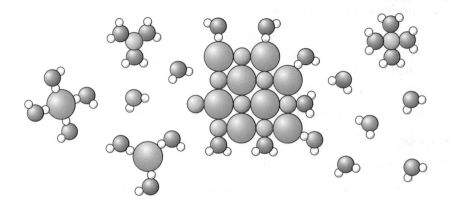

The water molecules then literally pull the crystal apart.

Magnesium oxide isn't soluble in water because the attractions between the water molecules and the ions aren't strong enough to break the very powerful ionic bonds between magnesium and oxide ions.

*Ionic compounds tend to be insoluble in organic solvents.* Organic solvents contain molecules which have much less electrical distortion than there is in water – their molecules are less polar. There isn't enough attraction between these molecules and the ions in the crystal to break the strong forces holding the lattice together.

Organic solvents include things like alcohol (ethanol) and hydrocarbons, such as are found in petrol. If you are interested in these, you could explore the organic chemistry section of this book.

## Activity 1: Testing the solubility of materials

### Investigative skills

| O | A |
|---|---|

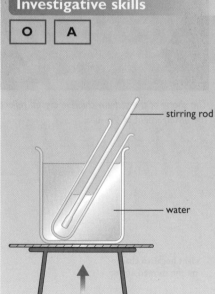

stirring rod

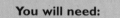
water

heat

**You will need:**

- Bunsen burner
- gauze
- stirring rod
- 10 cm³ measuring cylinder
- thermometer with 0–100°C range
- hexane
- a selection of solids, e.g. sodium chloride (salt), copper sulphate, iodine crystals (**Warning**: handle with care.), stearic acid.
- tripod
- spatula
- beaker of water
- test-tubes

**Carry out the following:**

1. Measure 5 cm³ cold water and pour into a test-tube.

2. Add one spatula full of salt and stir to see if it dissolves.

3. Repeat steps 1 and 2 with the other solids.

4. Measure 5 cm³ cold hexane into a clean test-tube.

5. Add 1 spatula full of salt and stir to see if it dissolves.

6. Repeat steps 4 and 5 using solid iodine crystals (just a couple of crystals).

7. Tabulate your results to show which solids (solutes) are soluble.

8. Try one solid at a range of temperatures using the apparatus shown.

## Activity 2: Testing the conductivity of solutions

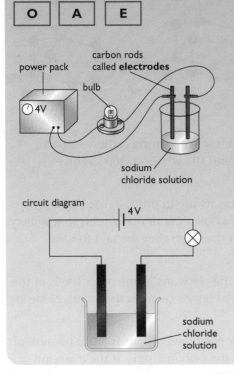

power pack

carbon rods called **electrodes**

bulb

4V

sodium chloride solution

circuit diagram

4V

sodium chloride solution

**You will need:**

- a power supply (4–6 volts d.c.)
- a lamp or ammeter
- carbon rod electrodes
- crocodile clips
- various liquids in beakers, e.g. distilled water, salt water, alcohol, sugar water, cooking oil, copper sulphate solution, dilute sulphuric acid

**Carry out the following:**

1. Set up the circuit (see diagram).

2. Test each liquid in turn to see if it conducts electricity. How will you know if it does?

3. Liquids containing molecules do not conduct electricity, but liquids containing ions (charged particles) can conduct electricity.

4. Tabulate your results.

5. Which material/s contain ions?

---

4. What is meant by solubility?
5. How can you measure the melting point of a solid?
6. How could you show that a solution contains ions and not just molecules?

---

### The electrical behaviour of ionic substances

Ionic compounds don't conduct electricity when they are solid, because they don't contain any mobile electrons. They do, however, conduct electricity when they melt, or if they are dissolved in water. This happens because the ions then become free to move around. How this enables the compound to conduct electricity is explained in Chapter 16: Electrolysis.

### Giant covalent structures

### Diamond

Diamond is a form of pure carbon. Each carbon atom has four unpaired electrons in its outer energy level (shell) and it uses these to form four covalent bonds. In diamond, each carbon bonds strongly to four other carbon atoms. The diagram shows enough of the structure to see what is happening.

This is a giant covalent structure – it continues on and on in three dimensions. It is not a molecule, because the number of atoms joined up in a real diamond is completely variable, depending on the size of the crystal. Molecules always contain fixed numbers of atoms joined by covalent bonds.

In the diagram some carbon atoms only seem to be forming two bonds (or even one bond), but that's not really the case. We are only showing a small bit of the whole structure. The lines in this diagram each represents a covalent bond.

Draw this structure in the following stages:

This is a very easy structure to draw as long as you practise it. You should be able to produce a reasonable sketch in 30 seconds.

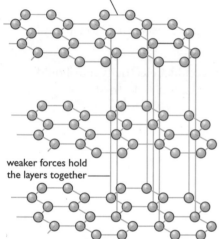

strong covalent bonds within each layer

weaker forces hold the layers together

*The structure of graphite.*

*Layers of graphite will flake off in the same way that an individual card separates easily from the pack.*

*Rubbing layers of graphite off on paper.*

*Diamond is very hard.* This is because of the very strong carbon–carbon covalent bonds which extend throughout the whole crystal in three dimensions.

The great hardness of diamond explains its use in drills, as used by dentists, and to cut stone or concrete. The best known use of diamond is in jewellery. Cut and polished diamonds sparkle in the light: they are said to have brilliance.

*Diamond doesn't conduct electricity.* All the electrons in the outer levels of the carbon atoms are tightly held in covalent bonds between the atoms. None are free to move around.

*Diamond doesn't dissolve in water or any other solvent.* This is again because of the powerful covalent bonds between the carbon atoms. If the diamond dissolved, these would have to be broken.

### Graphite

Graphite is also a form of carbon, but the atoms are arranged differently. Graphite has a layer structure rather like a pack of cards. In a pack of cards each card is strong, but the individual cards are easily separated. The same is true in graphite.

*Graphite is a soft material with a slimy feel.* Although the forces holding the atoms together in each layer are very strong, the attractions between the layers are much weaker. Layers can easily be flaked off.

Graphite (mixed with clay to make it harder) is used in pencils. When you write with a pencil, you are leaving a trail of graphite layers behind on the paper.

Pure graphite is so slippery that it is used as a dry lubricant – for example, to lubricate locks.

*Graphite is insoluble in any solvents.* To dissolve the graphite, you don't just have to break the layers apart – you have to break up the whole structure, including the covalent bonds. That needs very large amounts of energy because the bonds are so strong.

*Graphite is less dense than diamond*, because the layers in graphite are relatively far apart. The distance between the graphite layers is more than twice the distance between atoms in each layer. In a sense, a graphite crystal contains a lot of wasted space, which isn't there in a diamond crystal.

*Graphite conducts electricity.* If you look at the diagram of the arrangement of the atoms in each layer in the graphite, you will see that each carbon atom is only joined to three others.

Each carbon atom uses three of its electrons to form these simple covalent bonds. The fourth electron in the outer layer of each atom is free to move around throughout the whole of the layer. The movement of these electrons allows the graphite to conduct electricity.

### Heating diamond and graphite

Neither diamond nor graphite melt when heated. At a temperature well above 3000°C they both turn straight into gases. We call this sublimation. It takes a lot of heat energy to make them sublime since so many covalent bonds must be broken.

### What is allotropy?

Some elements occur in more than one form. For example carbon exists naturally as diamond and graphite. The carbon atoms are arranged in different patterns inside the two materials. This gives them different properties. Diamond is the hardest natural material whereas its **allotrope** graphite is one of the softest.

Oxygen, $O_2$, and ozone, $O_3$, are also allotropes. They are both gases. Phosphorus is another element which is allotropic. Compare the two allotropes of phosphorus:

*Quartz crystals.*

| Property | White phosphorus | Red phosphorus |
|---|---|---|
| melting point/°C | 44 | sublimes at about 400°C |
| appearance | yellow, waxy | red, powdery |
| solubility | soluble in benzene | insoluble |
| structure | simple molecular | giant structure |

Sulphur is another element which exists as allotropes. It can form needle-shaped crystals called monoclinic sulphur, or crystals which are like two pyramids joined base to base, called rhombic sulphur.

> 7. Give two uses each for diamond and graphite.
> 8. Describe two differences between white and red phosphorus.
> 9. Explain why the structure of graphite makes it so soft that we can use it in pencils.

## Simple molecular structures

Remember that molecules contain fixed numbers of atoms joined by strong covalent bonds. The forces of attraction between one molecule and its neighbours are much weaker than the covalent bonds, and vary in strength from compound to compound.

Because these intermolecular attractions are relatively weak, *simple molecular compounds tend to be gases, liquids or low melting point solids.*

Try making models of the salt (sodium chloride) and graphite structures. You could use straws or sticks for the bonds. Use clay or plasticine for the particles. Work out how many particles are attached to one central particle in each of the structures.

Where is the weak point in the graphite structure?

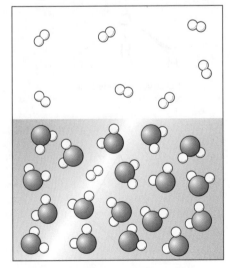

*Hydrogen gas is almost insoluble in water.*

*Molecular substances tend to be insoluble in water unless they react with it.* In order for a substance to dissolve, quite strong attractions between water molecules have to be broken so that the dissolving molecules can fit between them. Any new attractions between water molecules and the covalent molecules are not usually big enough to compensate for this.

On the other hand, *molecular substances are often soluble in organic solvents*. In this case, the intermolecular attractions between the two different types of molecules are much the same as in the pure substances.

*Molecular substances don't conduct electricity*, because the molecules don't have any overall electrical charge and there are no electrons mobile enough to move from molecule to molecule.

### Shapes of simple molecules

If you join together three atoms, there are only two possible shapes. In carbon dioxide, the three atoms are in a straight line. It is a linear molecule. When frozen solid, carbon dioxide forms simple molecular crystals, as does ice.

carbon dioxide

with double bonds shown

In water, the three atoms are not in a straight line. We call this a non-linear molecule.

water

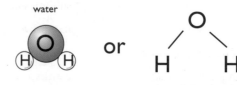

with single bonds shown

methane

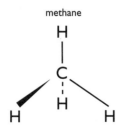

3-D structure of methane is tetrahedral.

When we have four atoms in a single molecule, such as ammonia, a new shape is possible. The nitrogen is at the top with the three hydrogen atoms below, a bit like the top of a pyramid. This is known as a pyramidal shape.

ammonia

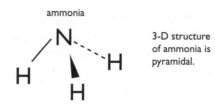

3-D structure of ammonia is pyramidal.

Adding one more atom gives a total of five, as in methane, natural gas. Methane has a symmetrical structure. The carbon atom is in the centre with four hydrogen atoms spaced out around it. This is called a tetrahedral shape. The four hydrogen atoms are as far apart as they can get. The shapes of molecules are related to the ways electrons are arranged. All electrons have the same negative charge, and like charges repel.

When cooled sufficiently until they freeze solid, both ammonia and methane also form simple molecular crystals. The element iodine is a solid

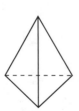

at room temperature. It sublimes when heated to give a purple gas, at about 400°C. Compare this with the giant structures of diamond or graphite. It shows that solid iodine must have a simple molecular structure since it vaporises so easily.

# End of Chapter Checklist

**In this chapter you have learnt that:**

- metals have giant metallic structures with positive ions and a 'sea' of electrons which can move around
- alloys are mixtures of metals, for example brass contains copper and zinc
- sodium chloride has a giant ionic structure, it dissolves in water and the solution conducts electricity
- solids change into liquids at their melting points
- liquids change into gases at their boiling points
- ionic substances are soluble in water and some molecular substances dissolve in hexane
- diamond and graphite are different forms of carbon called allotropes. Oxygen, sulphur and phosphorus are also allotropic
- molecules have special shapes.

# Questions

1. Which of the following does not have a giant structure?

   A   graphite
   B   oxygen
   C   sodium chloride
   D   copper                                    (1)

2. An alloy is:

   A   a type of element
   B   a compound of metals
   C   the same as aluminium
   D   a mixture of metals                        (1)

3. Which of the following statements is/are true?

   I     salt is sodium chloride
   II    salt has a giant structure
   III   salt is made of ions
   IV    salt is soluble in water

   A   I and II
   B   I, II, III and IV
   C   II and III
   D   II    (1)                                   (1)

4.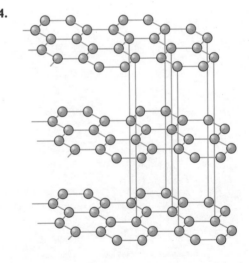

   The diagram shows:

   A   the structure of graphite
   B   the structure of diamond
   C   the structure of sodium chloride
   D   a metallic structure                        (1)

5. Water is an example of:

    A    a mixture of hydrogen and oxygen
    B    a giant structure
    C    a compound of hydrogen and oxygen    (1)
    D    a non-polar liquid

6. *a)* Draw a sketch to show the structure of a metal    (3)

   *b)* Explain why metals conduct electricity.    (2)

   *c)* What is an alloy? Give *one* example of an alloy.    (2)

   *d)* Alloys are usually harder than pure metals. Explain why.    (3)

7. *a)* Name *three* elements which form allotropes.    (3)

   *b)* Choose one element and sketch the structures of *two* of its allotropes.    (4)

   *c)* Describe how the properties of these allotropes are linked to their structures.    (3)

8. *a)* Draw simple diagrams to show the structures of diamond and graphite.    (2)

   *b)* Choose any *one* physical property where diamond and graphite have similar characteristics, and any *two* physical properties where they are different. Use your diagrams to explain the similarity and the differences.    (4)

## Chapter 5: Separating Mixtures

> **When you have completed this chapter, you will be able to:**
> - explain the differences between elements, mixtures and compounds
> - identify different kinds of solutions
> - explain what is meant by a saturated solution
> - understand how to measure solubility
> - understand how to use filtration and distillation
> - understand how we separate crude oil
> - explain how chromatography works.

## Elements, compounds and mixtures

### Elements

Elements are substances which can't be split into anything more simple by chemical means. All the atoms in an element have the same atomic number. You can recognise them in models or diagrams because they consist of atoms of a single colour or size.

> It isn't quite true to say that elements consist of only one type of atom. Most elements consist of mixtures of isotopes – with the same atomic number but different numbers of neutrons. When we draw diagrams or make models we aren't usually interested in the differences between the isotopes.

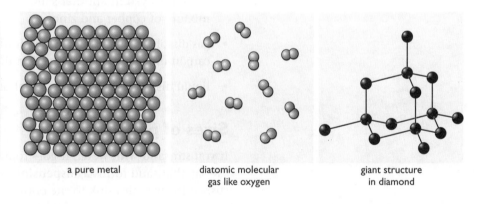

a pure metal          diatomic molecular          giant structure
                      gas like oxygen             in diamond

### Compounds

All compounds are made from combinations of two or more elements in fixed proportions, and joined by strong bonds.

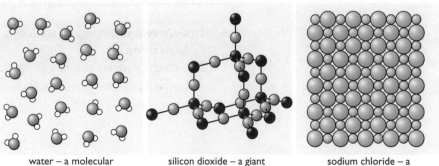

water – a molecular          silicon dioxide – a giant          sodium chloride – a
compound                     covalent compound                  giant ionic structure

### Mixtures

In a mixture, the various components can be in any proportions. An alloy is a mixture rather than a compound because of the totally variable proportions.

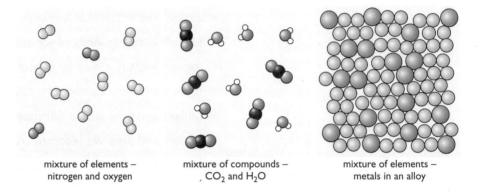

| mixture of elements – nitrogen and oxygen | mixture of compounds – $CO_2$ and $H_2O$ | mixture of elements – metals in an alloy |

We can purify mixtures by separating them into their different parts. First we need to understand some technical words.

A **solution** is formed when a substance **dissolves** in a liquid. The liquid is often water, but there are others we can use. Here are some examples of solutions.

- solid dissolved in a liquid – salt in sea water

- solid dissolved in another solid – alloys are mixtures of metals; brass is a mixture of copper and zinc

- gas dissolved in a liquid – fizzy drinks are also called carbonated drinks; carbon dioxide gas is dissolved in the drink

- liquid dissolved in a liquid – alcohol in water, for example rum.

## Sizes of particles

If you stir sand with water, it goes round and round but does not dissolve. We say that sand forms a **suspension** with water. Sooner or later the suspended particles sink to the bottom. Particles in suspensions are large.

Salt dissolves easily in water. The particles are very small. Solutions contain small particles.

Oil and water do not mix. But if you add some soap to oil and water and then shake it, the oil and water stay mixed. We call this an **emulsion**. Salad dressings are often emulsions and so is milk.

1. Give two examples of solutions: **a)** found at home, **b)** found in the laboratory.
2. Which contains the larger particles, a suspension or a solution?

# The solubility of solids

## Measuring solubility

The solubility of a solid at a particular temperature is usually defined as 'the mass of solute which will saturate 100 g of water at that temperature'.

A **saturated solution** contains as much dissolved solute as possible at a particular temperature. There must be some undissolved solute present.

To measure the solubility of potassium nitrate at exactly 40°C, you start by making a saturated solution of potassium nitrate in water at a temperature just over 40°C.

> The **solute** is the substance which dissolves in the solvent. In this case the solvent is water. It is possible to get 'supersaturated' solutions with some solutes. These contain more dissolved substance than you would expect at a particular temperature. If you add even one tiny crystal of solid to these, all the extra solute will crystallise out, and you are left with a normal saturated solution. Having undissolved solid present when you make a saturated solution prevents supersaturated solutions forming.

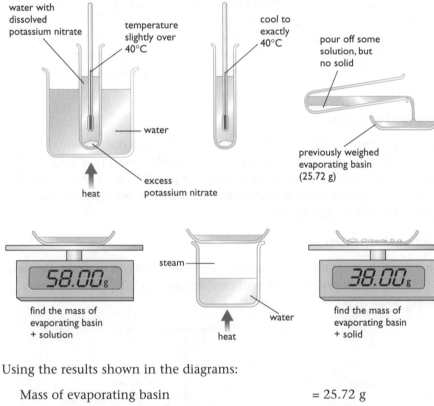

Using the results shown in the diagrams:

| | |
|---|---|
| Mass of evaporating basin | = 25.72 g |
| Mass of evaporating basin + solution | = 58.00 g |
| Mass of evaporating basin + dry crystals | = 38.00 g |
| Mass of crystals | = 38.00 – 25.72 g |
| | = 12.28 g |
| Mass of water | = 58.00 – 38.00 g |
| | = 20.00 g |

If 12.28 g of crystals saturate 20.00 g of water, you would need five times as much to saturate 100 g. That works out as 61.4 g. Therefore, the solubility of potassium nitrate at 40°C is 61.4 g per 100 g of water.

## Solubility curves

The solubility of solids changes with temperature, and you can plot this on a **solubility curve**. Most solids have solubility curves like A or B in the sketch. Their solubility increases with temperature – either dramatically, or just a bit.

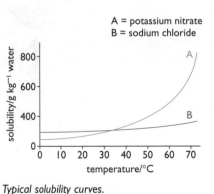

*Typical solubility curves.*

# Separation methods

## Crystallisation

Impure solids can sometimes be separated by crystallisation. This is possible when the pure substance and the impurity have different solubilities in a solvent. The impure mixture is first dissolved in the warm solvent. As the temperature falls, the material becomes less soluble and the more abundant material then crystallises out. The impurity remains in solution. If the impurity is insoluble, it can be removed by filtration of the solution.

## Filtration

If you pour salt water into a filter paper, it all goes through. You can prove it by letting the water that has passed through (the filtrate) evaporate. Salt crystals are left behind. If you do the same with a mixture of sand and water the result is different. The sand particles are too big to squeeze through the holes in the filter paper. You can therefore filter sand from a mixture of sand and water.

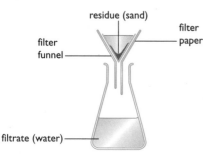

## Simple distillation

If you try filtering a mixture of water and liquid ink, it doesn't work. Everything goes through the filter paper. We need to use a new idea, **distillation**. Distillation separates mixtures of liquids. Water boils at 100°C but ink boils at a different temperature, much higher. We can distil pure water from inky water like this.

## Fractional distillation

Some mixtures contain lots of liquids. These are harder to separate. We need to separate each liquid at a different temperature. Each liquid has its own boiling point (b.p.).

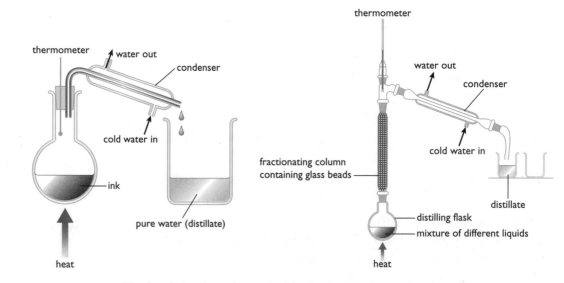

The beads in the column let the hot vapours condense and evaporate many times. This gives a better separation than simple distillation.

Ethanol and water have different boiling points and so can be separated.

## Separating water and ethanol

**You will need:**

- a mixture of water and ethanol (alcohol)
- a fractionating column (see diagram)
- an evaporating basin to test the distillate, the first liquid to distil

**Carry out the following:**

1. The apparatus is set up with the flask half full of the mixture.
2. Heat carefully. (**Warning**: ethanol is flammable.) Observe what happens.
3. Note the temperature when the first drops of liquid distil.
4. Pour 5 cm$^3$ of this liquid into an evaporating basin. Carefully try to light the liquid using a splint.
5. Is the distillate water or ethanol? What evidence supports this?

---

3. Imagine you are stranded on a desert island and need water to drink. How could you separate a sand and water mixture? Will the water be pure?
4. At what temperature does pure water boil? Does ink boil at the same temperature?
5. You are given a mixture of alcohol and water. Describe how you would separate them. Explain why your chosen method works.

---

### Separating crude oil by fractional distillation

We use this method to separate crude oil into petrol, diesel, paraffin and other liquids, each of which has a different boiling point. The crude oil is heated and passed into a **fractionating column** which is cooler at the top and hotter at the bottom. The crude oil is split into various **fractions**.

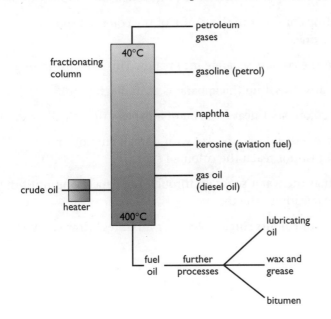

Things boil at lower temperatures if you reduce the pressure. Distillation under reduced pressure prevents the large molecules breaking up as a result of the high temperatures.

How far up the column a particular hydrocarbon gets depends on its boiling point. Suppose a hydrocarbon boils at 120°C. At the bottom of the column, the temperature is much higher than 120°C and so the hydrocarbon remains as a gas. As it travels up through the column, the temperature gets lower. When the temperature falls to 120°C, that hydrocarbon will start to turn to a liquid. It condenses and can be removed.

The hydrocarbons in the petroleum gases have boiling points which are so low that the temperature of the column never falls low enough for them to condense to liquids.

The temperature of the column isn't hot enough to boil the large hydrocarbons found in the fuel oil and this remains as a liquid. Some of the fuel oil is fractionally distilled under reduced pressure. The residue at the end of all this is **bitumen**, which is used in road making.

### Paper chromatography

We need a special way to separate small amounts of mixed colours such as those in coloured inks. It is called **chromatography**.

## Activity 1: Using chromatography to separate colours

### Investigative skills

(a) Horizontal

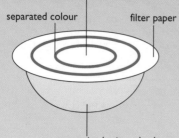

centre where spot was dropped and where solvent (e.g. water) is applied

separated colour

filter paper

evaporating basin or beaker to support paper

(b) Vertical

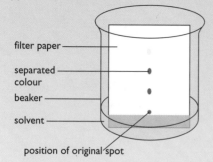

filter paper

separated colour

beaker

solvent

position of original spot

*Different methods of paper chromatography.*

**You will need:**

- chromatography paper or filter paper, both circles and strips
- droppers
- evaporating basin
- beaker
- food dyes or coloured sweets or marker pens (water-soluble type)

**Carry out the following:**

1. Set up the apparatus as in diagram (a).
2. Put a colour spot in the centre of the circle and allow it to dry.
3. Add drops of water to the centre of the coloured spot, very slowly, one by one.
4. Watch the colours separate into bands. This is a **chromatogram**.
5. Alternatively, set up the apparatus as in diagram (b).
6. Put a colour spot near one end of a paper strip and allow to dry.
7. Lower the strip into a beaker with a small amount of water. The water level must not reach the coloured spot.
8. Watch as the water soaks up through the paper. Wait until the water reaches nearly to the top.
9. Were your colours single colours or mixtures? How can you tell?

6. How can you test water and ethanol to find out which is which?
7. What do you observe when water-soluble black ink is used to make a chromatogram? Explain your observations.
8. Ballpoint pen ink is soluble in alcohol but not in water. How could you make a chromatogram of this ink?

## Separating liquids that don't mix

Oil floats on water. It does not mix or dissolve in water. We can use a special separating funnel to separate the water from the oil.

If we pour a mixture of oil and water into a separating funnel, the water will settle below the oil, so it will come out first when the tap is opened. This allows us to separate the two liquids.

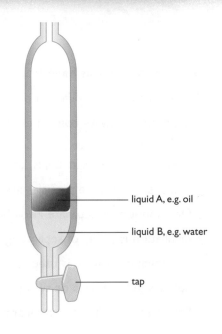

liquid A, e.g. oil

liquid B, e.g. water

tap

## Activity 2: Investigating charcoal as a colour absorber

### Investigative skills

P

Design an experiment to find out whether charcoal lumps or powder are better at removing food colour from water.

# End of Chapter Checklist

**In this chapter you have learnt that:**

- mixtures can have different proportions

- compounds always contain elements combined in the same proportions, for example water or carbon dioxide

- solutions can contain solids, liquids or gases

- in a suspension, the particles are large and separate out

- filtration lets us separate insoluble solids, such as sand, from water

- distillation is a way of separating liquid mixtures

- fractional distillation is used to separate complex mixtures such as crude oil

- colours in food dyes or sweets can be separated by chromatography

- liquids that don't mix are separated using a separating funnel.

# Questions

1. *a)* Describe the differences between mixtures and compounds using hydrogen, oxygen and water as examples. (5)

   *b)* Name *two* examples of simple molecular compounds. (2)

   *c)* Name *two* examples of giant covalent structures. (2)

   *d)* Are alloys mixtures or compounds? (1)

2. Which of the following is *not* a method of separation?

   A   chromatography      B   filtration
   C   melting             D   distillation      (1)

3. Fractional distillation separates liquids which have different:

   A   colours             B   boiling points
   C   melting points      D   conductivities    (1)

4.

liquid A, e.g. oil

liquid B, e.g. water

tap

*a)* Which liquid in the separating funnel, X or Y, is more dense? Give a reason. (2)

*b)* Name two liquids that could be separated by this method. (2)

*c)* A third liquid, Z, is added to the funnel. It is insoluble in X but soluble in Y. Describe what happens when the mixture is shaken. (4)

*d)* How else can you separate liquid mixtures? (2)

5. *a)* Describe how you would make a chromatogram of water-soluble black ink. (4)

   *b)* Draw the apparatus you would need and and label it. (4)

   *c)* What changes would you make if the ink was insoluble in water but soluble in alcohol? (2)

6. *a)* Describe *three* different examples of filtration at home or school. (6)

   *b)* A student says that using a teabag is an example of filtration. Is this true? Explain your answer. (2)

   *c)* Name *two* products distilled from crude oil and give a use for each. (2)

7. The table gives the solubility data for sodium sulphate.

   | Temp/°C | 0 | 10 | 20 | 30 | 40 | 50 | 60 | 70 | 80 |
   |---|---|---|---|---|---|---|---|---|---|
   | Solubility/ g kg$^{-1}$ water | 0.45 | 90 | 200 | 400 | 480 | 465 | 450 | 440 | 435 |

   *a)* Plot a solubility curve for sodium sulphate. (5)

   *b)* What is unusual about the curve? (2)

   *c)* What would you see happening when a saturated solution of sodium sulphate was warmed from 40°C to 80°C? Explain your answer. (3)

# End of Section Questions

You may need to refer to the Periodic Table on page 297.

**1.** The diagrams show an atom and an ion.

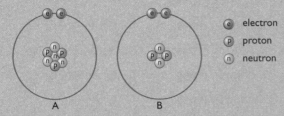

e) electron
p) proton
n) neutron

A    B

**a)** Which of the two structures represents an atom? Explain your choice. *(2 marks)*

**b)** Use the Periodic Table to help you to write the symbol (including the charge) for the structure representing an ion. *(2 marks)*

**c)** Complete the following table showing the relative masses and charges of the various particles.

| Particle | Relative charge | Relative mass |
| --- | --- | --- |
| proton | +1 | |
| neutron | 0 | |
| electron | | 1/1836 |

*(3 marks)*

**d)** Strontium combines with bromine to form strontium bromide. What happens to the electrons in the outer levels when strontium atoms combine with bromine atoms? *(2 marks)*

**e)** What is the formula for strontium bromide? *(1 mark)*

***Total 10 marks***

**2. a)** Draw dot-and-cross diagrams to show the arrangement of the electrons in *i)* a chlorine atom, *ii)* a chloride ion, *iii)* a chlorine molecule. *(4 marks)*

**b)** Dichloromethane, $CH_2Cl_2$, is a liquid with a low boiling point used in paint strippers. Draw a dot-and-cross diagram to show the bonding in dichloromethane. You need only show the electrons in the outer levels of the atoms. *(3 marks)*

**c)** Dichloromethane contains strong carbon–hydrogen and carbon–chlorine bonds. Despite the presence of these strong bonds, dichloromethane is a liquid. Explain why. *(3 marks)*

***Total 10 marks***

**3.**

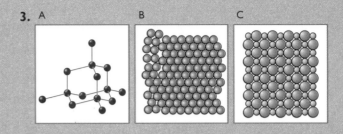

A    B    C

**a)** Which of the diagrams represents the arrangement of the particles in *i)* magnesium metal, *ii)* solid sodium chloride, *iii)* diamond? *(2 marks)*

**b)** Explain why:

*i)* magnesium conducts electricity *(2 marks)*

*ii)* diamond is extremely hard. *(2 marks)*

**c)** *i)* State any *one* physical property of graphite which is different from diamond. *(1 mark)*

*ii)* Explain how the difference arises from the arrangement of the atoms in the two substances. *(3 marks)*

***Total 10 marks***

**4.** Here is some information about four substances named P, Q, R and S.

| Substance | State | Soluble in water? | Miscible with water? |
| --- | --- | --- | --- |
| P | solid | yes | – |
| Q | solid | no | – |
| R | liquid | yes | yes |
| S | liquid | no | no |

Describe how you would separate the following mixtures. Include labelled diagrams for each one and explain any assumptions you need to make.

**a)** a mixture of P + Q. *(3 marks)*

**b)** a mixture of R + S. *(3 marks)*

**c)** a mixture of Q + S. *(4 marks)*

***Total 10 marks***

## Chapter 6: Formulae and Equations

> **When you have completed this chapter, you will be able to:**
>
> - write chemical formulae for molecules
> - understand how to balance plus and minus charges for ionic formulae
> - understand the different endings of names such as sulphate and sulphide
> - work out the formulae of ionic compounds using a data chart
> - describe experiments to find formulae
> - understand the way we use brackets in formulae
> - write and balance equations
> - use state symbols in equations.

## Writing formulae

Formulae for covalent substances

*Common everyday examples*

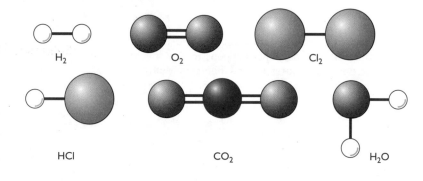

Remember that each line represents a covalent bond – a pair of shared electrons. The formula simply counts the number of atoms combined to make the compound. You won't usually have to work out the formula of a covalent compound – most of the ones you will meet are so simple and so common that you just remember them.

*Suppose you had to work one out*

Suppose you had to find the formula for phosphine – a simple compound of phosphorus and hydrogen.

Find phosphorus in the Periodic Table on page 297. Its atomic number is 15, and so the atom has 15 protons and 15 electrons. The electrons would be arranged 2,8,5. All you are interested in are the electrons in the outer level.

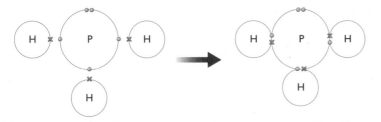

The phosphorus can create three covalent bonds by sharing the single electrons with three hydrogen atoms. That means that the formula for phosphine must be $PH_3$.

1. How many atoms are there in one molecule of **a)** hydrogen, **b)** carbon dioxide?
2. Name two molecules that contain double bonds.
3. Diatomic molecules contain only two atoms. Name three diatomic molecules.

## Formulae for ionic compounds

There are so many different ionic compounds which you might come across, that it would be impossible to learn all their formulae. You need a simple way to work them out. You could work a few of them out from first principles using their electronic structures, but that would take ages. Others would be too difficult. You need a simple short-cut method.

### The need for equal numbers of 'pluses' and 'minuses'

Ions are atoms or groups of atoms which carry electrical charges, either positive or negative. Compounds are electrically neutral. Therefore in an ionic compound there must be the right number of each sort of ion so that the total positive charge exactly cancels out the total negative charge. Obviously, then, if you are going to work out a formula, you need to know the charges on the ions.

### (1) Cases where you can work out the number of charges on an ion

Any element in Group 2 has two outer electrons which it will lose to form a 2+ ion. Any element in Group 6 has six outer electrons, and it has room to gain two more. This leads to a 2– ion. Similar arguments apply in the other groups shown in the table below.

| Element in Periodic Table Group | Charge on ion |
|---|---|
| 1 | +1 |
| 2 | +2 |
| 3 | +3 |
| 6 | –2 |
| 7 | –1 |

## (2) Cases where the name tells you the charge

All metals form positive ions. Names like lead(II) oxide, or iron(III) chloride or copper(II) sulphate tell you directly about the charge on the metal ion. The number after the metal tells you the number of charges. So...

Lead(II) oxide contains a $Pb^{2+}$ ion. Iron(III) chloride contains an $Fe^{3+}$ ion. Copper(II) sulphate contains a $Cu^{2+}$ ion. Find the symbols for the metals from a Periodic Table if you need to.

### Charges on some ions

| Positive ions | | | Negative ions | | |
|---|---|---|---|---|---|
| **Charge** | **Name of ion** | **Formula** | **Charge** | **Name of ion** | **Formula** |
| 1+ | ammonium | $NH_4^+$ | 1– | bromide | $Br^-$ |
| | copper(I) | $Cu^+$ | | chloride | $Cl^-$ |
| | hydrogen | $H^+$ | | hydroxide | $OH^-$ |
| | lithium | $Li^+$ | | fluoride | $F^-$ |
| | potassium | $K^+$ | | iodide | $I^-$ |
| | silver | $Ag^+$ | | nitrate | $NO_3^-$ |
| | sodium | $Na^+$ | | | |
| 2+ | barium | $Ba^{2+}$ | 2– | carbonate | $CO_3^{2-}$ |
| | calcium | $Ca^{2+}$ | | oxide | $O^{2-}$ |
| | copper(II) | $Cu^{2+}$ | | sulphate | $SO_4^{2-}$ |
| | iron(II) | $Fe^{2+}$ | | sulphide | $S^{2-}$ |
| | lead(II) | $Pb^{2+}$ | | | |
| | magnesium | $Mg^{2+}$ | | | |
| | nickel(II) | $Ni^{2+}$ | | | |
| | strontium | $Sr^{2+}$ | | | |
| | zinc | $Zn^{2+}$ | | | |
| 3+ | aluminium | $Al^{3+}$ | 3– | nitride | $N^{3-}$ |
| | iron(III) | $Fe^{3+}$ | | phosphate | $PO_4^{3-}$ |

You will come across other ions during the course, but these are the important ones for now. The ions in this list are the tricky ones – be sure to learn both the formula and the charge for each ion.

4. What are the charges on these metal ions?
   **a)** potassium **b)** zinc **c)** aluminium
5. What are the charges on these non-metal ions?
   **a)** chloride **b)** oxide **c)** sulphate
6. What pattern do you notice in the signs of the charges on metallic and non-metallic ions?

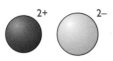

copper(II) sulphide

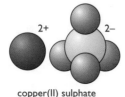

copper(II) sulphate

### Confusing endings!

Don't confuse ions like sulph**ate** with sulph**ide**. A name like copper(II) sulphide means that it contains copper and sulphur *only*. Any 'ide' ending means that there isn't anything complicated there. Sodium chloride, for example, is just sodium and chlorine combined together.

Once you have an 'ate' ending, it means that there is something else there as well – often, but not always, oxygen. So, for example, copper(II) sulphate contains copper, sulphur and oxygen.

Not looking carefully at word endings is one of the commonest mistakes students make when they start to write formulae.

## Working out the formula for an ionic compound

**Example 1: To find the formula for sodium oxide**

Sodium is in Group 1, so the ion is $Na^+$.

Oxygen is in Group 6, so the ion is $O^{2-}$.

To have equal numbers of positive and negative charges, you would need two sodium ions to provide the two positive charges to cancel the two negative charges on one oxide ion. In other words, you need:

$Na^+$  $Na^+$  $O^{2-}$

The formula is therefore $Na_2O$, which means two ions $Na^+$ and one ion $O^{2-}$.

**Example 2: To find the formula for barium nitrate**

Barium is in Group 2, so the ion is $Ba^{2+}$.

Nitrate ions are $NO_3^-$. You will have to remember this.

To have equal numbers of positive and negative charges, you would need two nitrate ions for each barium ion.

| positive ions | negative ions |
|---|---|
| $Ba^{2+}$ | $NO_3^-$ |
| | $NO_3^-$ |
| total charges          2+ | 2– |

The formula is $Ba(NO_3)_2$, which means one ion $Ba^{2+}$ and two ions $NO_3^-$.

Notice the brackets around the nitrate group. *Brackets must be written if you have more than one of these polyatomic ions* (ions containing more than one type of atom). In any other situation, they are completely unnecessary.

**Example 3: To find the formula for iron(III) sulphate**

Iron(III) tells you that the metal ion is $Fe^{3+}$.

Sulphate ions are $SO_4^{2-}$.

To have equal numbers of positive and negative charges, you would need two iron(III) ions for every three sulphate ions – to give 6+ and 6– in total.

The formula is $Fe_2(SO_4)_3$, which means that there are two ions $Fe^{3+}$ and three ions $SO_4^{2-}$ making 6+ charges and 6– charges overall.

*worked*
examples

### Subscripts and superscripts

Subscripts are small numbers written below the line. This is how we use them:

- $O_2$ means two oxygen atoms joined together in one molecule.

- $CO_2$ means one carbon atom joined to two oxygen atoms in one molecule.

Superscripts are small numbers written above the line. This is how we use them:

- $Na^+$ is a sodium ion with a charge of +1

- $Ca^{2+}$ is a calcium ion with a charge of +2

- $SO_4^{2-}$ is a sulphate ion with a charge of 2–, containing one sulphur atom and four oxygen atoms.

If you didn't write the brackets, the formula for barium nitrate would look like this: $BaNO_{32}$. That would read as 1 barium, 1 nitrogen and 32 oxygens.

## Why aren't ion charges shown in formulae?

Actually, they can be shown. For example, the formula for sodium chloride is NaCl. It is sometimes written $Na^+Cl^-$ if you are trying to make a particular point, but for most purposes the charges are left out. In an ionic compound, the charges are there – whether you write them or not.

### Teacher demonstration 1: Finding the formula of copper oxide

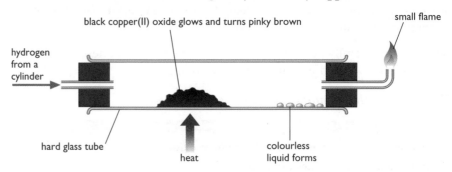

black copper(II) oxide glows and turns pinky brown

small flame

hydrogen from a cylinder

hard glass tube

heat

colourless liquid forms

We can find the percentages of copper and oxygen in copper oxide by experiment. A weighed amount of copper oxide is heated in a stream of hydrogen gas. The copper oxide is reduced to copper metal which can then be weighed. Here are some sample results:

Mass of copper oxide  = 2.50 g

Mass of copper metal  = 2.00 g

Mass of oxygen in oxide  = 0.50 g

Percentage of copper  = $\frac{2.00}{2.50} \times 100 = 80\%$

Percentage of oxygen  = $\frac{0.50}{2.50} \times 100 = 20\%$

These results allow us to work out the chemical formula of copper oxide (fully explained in chapter 7).

Number of moles of copper  = $\frac{2.00}{64}$ = 0.031 moles

Number of moles of oxygen  = $\frac{0.50}{16}$ = 0.031

Ratio of copper to oxygen  = 1:1

Formula of copper oxide  = CuO

The formula of magnesium oxide could be found by burning a weighed amount of magnesium metal and then weighing the magnesium oxide produced.

### Teacher demonstration 2: Finding the formula of water

Everyone knows that water has the formula $H_2O$, but why not $H_3O$, or $H_4O$? We can use electrolysis to split water into its elements and compare the amounts of hydrogen and oxygen produced.

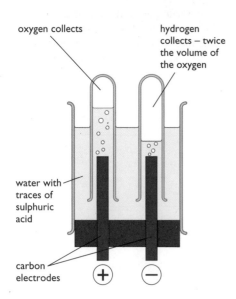

oxygen collects

hydrogen collects – twice the volume of the oxygen

water with traces of sulphuric acid

carbon electrodes

The diagram shows the apparatus needed. The trace of sulphuric acid is added to improve the electrical conductivity of the water. The acid is not changed by the electrolysis.

The results show that twice as much hydrogen is produced. The volume ratio is 2:1 for hydrogen:oxygen. This shows us that in water the ratio of hydrogen to oxygen is 2:1, giving the formula $H_2O$.

# Writing equations

## What all the numbers mean

When you write equations it is important to be able to count up how many of each sort of atom you have got. In particular you must understand the difference between big numbers written in front of formulae such as the **2** in **2HCl**, and the smaller, subscripted (written slightly lower on the line) numbers such as the **4** in $CH_4$.

What, for example, is the difference between $2Cl$ and $Cl_2$? The position of the 2 shows whether or not the atoms are joined together.

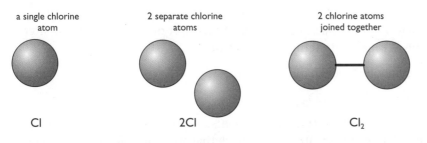

| a single chlorine atom | 2 separate chlorine atoms | 2 chlorine atoms joined together |
|---|---|---|
| Cl | 2Cl | $Cl_2$ |

Look at the way the numbers work in $2H_2SO_4$. The big number in front tells you that you have two sulphuric acid molecules. The little 4, for example, tells you that you have four oxygen atoms in each molecule. A small number in a formula only applies to the atom immediately before it in the formula.

If you count the atoms in $2H_2SO_4$, you will find four hydrogens, two sulphurs and eight oxygens.

If you have brackets in a formula, a small number refers to everything inside the brackets. For example, in the formula $Ca(OH)_2$, the 2 applies to both the oxygen and the hydrogen. The formula shows one calcium, two oxygens and two hydrogens.

## Balancing equations

Chemical reactions involve taking elements or compounds and shuffling their atoms around into new combinations. It follows that you must always end up with the same number of atoms that you started with.

Suppose you had to write an equation for the reaction between methane, $CH_4$, and oxygen, $O_2$. Methane burns in oxygen to form carbon dioxide and water. Think of this in terms of rearranging the atoms in some models.

| methane | oxygen | carbon dioxide | water |

This can't be right! During the rearrangement you seem to have gained an oxygen atom and lost two hydrogens. The reaction must be more complicated than this. Since the substances are all correct, the numbers must be wrong.

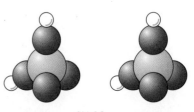

$2H_2SO_4$

Here are some simple rules about balancing equations:

1. Number and kind of atoms on left-hand side = number and kind of atoms on right-hand side. In other words when chemicals react, nothing is lost. The particles (atoms, molecules or ions) just join together in new ways.

2. Big numbers multiply everything that follows, for example 3C means three carbon atoms and 3NaCl means three sodium atoms and also three chlorine atoms.

3. Small numbers multiply the atom or brackets just before them. For example:

$CO_2$ means one carbon atom and two oxygen atoms; $CaCO_3$ means one calcium atom, one carbon atom and three oxygen atoms; $NO_3$ means one nitrogen atom and three oxygen atoms. $(NO_3)_2$ means that everything inside the brackets is doubled. This means there are two nitrogen atoms and six oxygen atoms. We use brackets for numbers of atoms that are found together as a permanent group. Another example is the sulphate group – $Al_2(SO_4)_3$ means two aluminium atoms, three sulphur atoms and twelve oxygen atoms.

Try again:

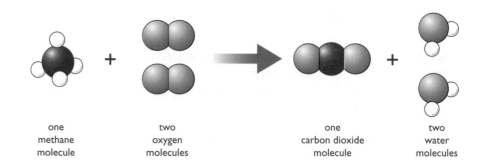

| one | two | one | two |
|-----|-----|-----|-----|
| methane molecule | oxygen molecules | carbon dioxide molecule | water molecules |

There are now the same number of each sort of atom before and after. This is called **balancing the equation**.

In symbols, this equation would be:

$$CH_4 + 2O_2 \rightarrow CO_2 + 2H_2O$$

Each symbol (C or H or O) represents one atom of that element. Count them up in the equation, and check that there are the same numbers on both sides.

### How to balance equations

Follow the rules and you can balance equations.

- Work across the equation from left to right, checking one element after another, *except* if an element appears in several places in the equation. In that case, leave it until the end – you will often find that the numbers will then balance.

- If you have a group of atoms (like a sulphate group, $SO_4$) which is unchanged from one side of the equation to the other, count that up as a whole group. Don't bother to count individual sulphurs and oxygens.

- Check everything at the end.

***worked*** examples

**Example 1: Zinc + hydrochloric acid → zinc chloride + hydrogen**

Balance the equation:

$$Zn + HCl \rightarrow ZnCl_2 + H_2$$

Work from left to right. Count the zinc atoms. One on each side – this is OK.

Count the hydrogen atoms. One on the left; two on the right. If you end up with two you must have started with two. The only way of achieving this is to have 2HCl. (You *must not* change the formula to $H_2Cl$ – there's no such substance.)

$$Zn + 2HCl \rightarrow ZnCl_2 + H_2$$

Now count the chlorines. There are two on each side. Finally check everything again to make sure.

This is really important! You must *never, never* change a correct formula in balancing an equation. All you are allowed to do is to write big numbers in front of the formula.

**Example 2: Silver nitrate solution + calcium chloride solution → calcium nitrate solution + silver chloride**

Balance the equation:

$$AgNO_3 + CaCl_2 \rightarrow Ca(NO_3)_2 + AgCl$$

Working from left to right: one silver atom on both sides. That's OK.

The nitrate group is unchanged as it goes from left to right, so save time by counting it as a whole. There's one $NO_3$ group on the left, two on the right. That needs correcting. You must have started with $2AgNO_3$.

$$2AgNO_3 + CaCl_2 \rightarrow Ca(NO_3)_2 + AgCl$$

Now check the calcium. One on each side.

Now the chlorine. There are two on the left, but only one on the right. You need $2AgCl$.

$$2AgNO_3 + CaCl_2 \rightarrow Ca(NO_3)_2 + 2AgCl$$

Finally, check everything again. It's all OK. You actually changed the numbers of silver atoms on the left-hand side after you checked them at the beginning. It so happens that the problem corrected itself when you put the two in front of the AgCl on the right – but that won't always happen.

**Example 3: Ethane + oxygen → carbon dioxide + water**

Balance the equation:

$$C_2H_6 + O_2 \rightarrow CO_2 + H_2O$$

Starting from the left, balance the carbons:

$$C_2H_6 + O_2 \rightarrow 2CO_2 + H_2O$$

Now the hydrogens:

$$C_2H_6 + O_2 \rightarrow 2CO_2 + 3H_2O$$

And finally the oxygens. There are seven oxygens (4 + 3) on the right-hand side, but only two on the left. The problem is that the oxygens have to go around in pairs – so how can you get an odd number (7) of them on the left-hand side?

The trick with this is to allow yourself to have halves in your equation; seven oxygen atoms, O, is the same as three and a half oxygen molecules, $O_2$.

$$C_2H_6 + 3\tfrac{1}{2}O_2 \rightarrow 2CO_2 + 3H_2O$$

You might think that you can't have half an oxygen molecule, but to get rid of that problem you only have to double everything.

$$2C_2H_6 + 7O_2 \rightarrow 4CO_2 + 6H_2O$$

> Don't continue until you are sure you understand why there are seven oxygen atoms on the right-hand side.

## State symbols

State symbols are often written after the formulae of the various substances in an equation to show what physical state everything is in. You need to know four different state symbols:

- (s)     solid
- (l)     liquid
- (g)     gas
- (aq)   in aqueous solution (solution in water)

If those were written into one of the equations we've just worked out it would look like this:

$$Zn(s) + 2HCl(aq) \rightarrow ZnCl_2(aq) + H_2(g)$$

This means that solid zinc reacts with an aqueous solution of hydrochloric acid. The products are an aqueous solution of zinc chloride and hydrogen gas.

> **State symbols**
>
> Water can be solid (s) ice, liquid water (l) or a gas (g) that we call water vapour. The small letters in brackets are the state symbols and we add them to equations. There is one other state symbol we use. For example, a solution of salt in water would be written with the symbol (aq) which is short for aqueous. An aqueous solution is a solution in water. State symbols give us extra information about materials reacting in a chemical equation.
>
> Adding state symbols to your equations will gain you extra marks in the examination.

**7.** What are the four state symbols?

**8.** What is the difference between liquid water and an aqueous solution?

**9.** How would you write sulphur, symbol S, as a solid, a liquid and a gas?

# End of Chapter Checklist

**In this chapter you have learnt that:**

● the shorthand for chemicals is called a chemical formula

● chemical formulae show the numbers of each type of element present in the compound and are found by experiment

● ionic formulae must have equal numbers of plus (+) and minus (−) charges

● the endings of some ions can sound very similar, such as sulphate and sulphide, but they are different

● big numbers, small numbers and brackets are all useful in writing equations

● when you balance an equation, there will be equal numbers of the same kinds of atoms on each side but they will be arranged differently

● the state symbols are (s) solid, (l) liquid, (g) gas and (aq) for aqueous solutions.

# Questions

You will need to use the Periodic Table on page 297.

**1. a)** Write the formula and charge for sulphate, calcium and sodium ions. (3)

**b)** What charges (+ or −) are carried by *i)* metal ions, *ii)* non-metal ions? Name an exception. (3)

**c)** Write the formulae for *i)* calcium sulphate, *ii)* sodium sulphate, *iii)* barium chloride, *iv)* iron(III) oxide. (2)

**d)** Explain, using examples, the way we use small numbers in chemical formulae. (2)

**2. a)** Write the formulae for the oxides of sodium, calcium and aluminium. (3)

**b)** How many atoms are there in each of the molecules listed in *a)*? (3)

**c)** Explain the use of brackets in formulae, giving *two* examples. (4)

**3. a)** Using water as an example, explain the four state symbols. (4)

**b)** Write the following formulae with state symbols: *i)* solid sodium chloride, *ii)* molten sodium chloride, *iii)* sodium chloride solution. (3)

**c)** When water vapour condenses on a cold mirror, what state symbols would you use in the equation? (2)

**d)** What state symbol would you use for a cola drink? (1)

**4. a)** Write the correct formula for each item in this word equation:

silver nitrate solution + barium chloride solution → barium nitrate solution + silver chloride solid (4)

**b)** Balance the equation in *a)*. (4)

**c)** Silver chloride is insoluble in water. How do we show this? (2)

**5.** Ethane gas, $C_2H_6$, burns in oxygen to give carbon dioxide and water vapour.

**a)** Write a word equation for the reaction. (2)

**b)** Write the correct formula for each item in the equation. (3)

**c)** Balance the equation. (4)

**d)** What change would you notice if the water vapour were cooled to 50°C? (1)

6. Work out the formulae of the following compounds:

lead(II) oxide
magnesium sulphate
potassium carbonate
calcium nitrate
iron(II) sulphate
aluminium sulphate
cobalt(II) chloride
silver nitrate
ammonium nitrate
sodium sulphate

sodium bromide
zinc chloride
ammonium sulphide
iron(III) hydroxide
copper(II) carbonate
calcium hydroxide
calcium oxide
iron(III) fluoride
rubidium iodide
chromium(III) oxide.          (20)

7. a) Hydrogen sulphide is a simple covalent compound of hydrogen and sulphur.

   i) Write down the electronic structures of hydrogen and sulphur.          (2)

   ii) Draw a dot-and-cross diagram to show the bonding in hydrogen sulphide.          (2)

   iii) What is the formula for hydrogen sulphide?          (1)

   b) Silane is the simplest compound of silicon and hydrogen. Work out the formula of silane by drawing a dot-and-cross diagram of it.          (2)

8. Balance the following equations:

   a) $Ca + H_2O \rightarrow Ca(OH)_2 + H_2$

   b) $Al + Cr_2O_3 \rightarrow Al_2O_3 + Cr$

   c) $Fe_2O_3 + CO \rightarrow Fe + CO_2$

   d) $NaHCO_3 + H_2SO_4 \rightarrow Na_2SO_4 + CO_2 + H_2O$

   e) $C_8H_{18} + O_2 \rightarrow CO_2 + H_2O$

   f) $Fe + HCl \rightarrow FeCl_2 + H_2$

   g) $Zn + H_2SO_4 \rightarrow ZnSO_4 + H_2$

   h) $Fe_3O_4 + H_2 \rightarrow Fe + H_2O$

   i) $Mg + O_2 \rightarrow MgO$

   j) $Pb + AgNO_3 \rightarrow Pb(NO_3)_2 + Ag$

   k) $AgNO_3 + MgCl_2 \rightarrow Mg(NO_3)_2 + AgCl$

   l) $C_3H_8 + O_2 \rightarrow CO_2 + H_2O$

   m) $Fe_2O_3 + C \rightarrow Fe + CO$          (13)

9. Rewrite the following equations as balanced symbol equations:

   a) sodium carbonate + hydrochloric acid (HCl) → sodium chloride + carbon dioxide + water          (2)

   b) sodium hydroxide + sulphuric acid (H₂SO₄) → sodium sulphate + water          (2)

   c) sodium + water → sodium hydroxide + hydrogen (H₂) (2)

   d) sodium + chlorine (Cl₂) → sodium chloride          (2)

   e) iron(III) oxide + nitric acid (HNO₃) → iron(III) nitrate + water          (2)

   f) zinc + oxygen (O₂) → zinc oxide          (2)

   g) copper(II) oxide + hydrochloric acid → copper(II) chloride + water          (2)

   h) barium chloride + sodium sulphate → barium sulphate + sodium chloride          (2)

   i) zinc + lead(II) nitrate → lead + zinc nitrate          (2)

   j) copper(II) sulphate + potassium hydroxide → copper(II) hydroxide + potassium sulphate          (2)

   k) magnesium + copper(II) oxide → magnesium oxide + copper          (2)

   l) sodium + oxygen (O₂) → sodium oxide          (2)

   m) iron + chlorine (Cl₂) → iron(III) chloride          (2)

10. Write balanced symbol equations from the following descriptions. Everything must have a state symbol attached. (2)

   a) Solid calcium carbonate reacts with a dilute solution of hydrochloric acid (HCl) to give a solution of calcium chloride and carbon dioxide gas. Water is also formed. (2)

   b) Zinc metal reacts with copper(II) sulphate solution to give solid copper and a solution of zinc sulphate.          (2)

   c) Magnesium reacts with dilute sulphuric acid to give magnesium sulphate solution and hydrogen.          (2)

   d) Iron(III) sulphate solution and sodium hydroxide solution react to give solid iron(III) hydroxide and a solution of sodium sulphate.          (2)

   e) Solid aluminium reacts with a dilute solution of hydrochloric acid (HCl) to give a solution of aluminium chloride and hydrogen (H₂) gas.          (2)

   f) Solid iron(III) oxide reacts with a dilute solution of sulphuric acid to give iron(III) sulphate solution and water.          (2)

   g) Solid lead(II) carbonate reacts with a dilute solution of nitric acid (HNO₃) to give a solution of lead(II) nitrate, carbon dioxide and water.          (2)

   h) Magnesium reacts if heated in steam to produce white solid magnesium oxide and hydrogen (H₂).          (2)

   i) A mixture of carbon and copper(II) oxide heated together produces copper and carbon dioxide.          (2)

## Chapter 7: Relative Atomic Masses and Moles

*Each gold bar contains almost $4 \times 10^{25}$ gold atoms. That's 4 followed by 25 noughts!*

**When you have completed this chapter, you will be able to:**

- define relative atomic mass ($A_r$) and relative formula mass ($M_r$)
- use the carbon-12 scale of masses
- understand the term 'weighted average'
- work out the relative formula mass of a compound
- understand how to calculate percentage compositions
- understand the mole and how we use the idea
- know the meaning of the Avogadro Constant
- understand the difference between empirical (simplest) and molecular formulae.

You can make iron(II) sulphide by heating a mixture of iron and sulphur.

$$Fe(s) + S(l) \rightarrow FeS(s)$$

How do you know what proportions to mix them up in? You can't just mix equal masses of them because iron and sulphur atoms don't weigh the same. Iron atoms contain more protons and neutrons than sulphur atoms so that an iron atom is one and three-quarters times heavier than a sulphur atom. In this, or any other reaction, you can only get the proportions right if you know about the masses of the individual atoms taking part.

## Relative atomic mass ($A_r$)

### Defining relative atomic mass

Atoms are amazingly small. In order to get a gram of hydrogen, you would need to count out 602 204 500 000 000 000 000 000 atoms (to the nearest 100 000 000 000 000 000).

It would be silly to measure the masses of atoms in conventional mass units like grams. Instead, their masses are compared with the mass of an atom of the carbon-12 isotope, taken as a standard. We call this the **'carbon-12 scale'**.

On this scale one atom of the carbon-12 isotope weighs *exactly* 12 units.

An atom of the commonest isotope of magnesium weighs twice as much as this and is therefore said to have a **relative isotopic mass** of 24.

An atom of the commonest isotope of hydrogen weighs only one twelfth as much as a carbon-12 atom, and so has a relative isotopic mass of 1.

Remember that isotopes are atoms of the same element but with different masses.

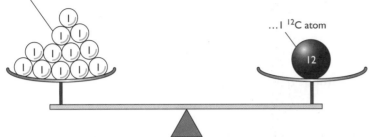

12 $^1$H atoms have the same mass as…

…1 $^{12}$C atom

12

This is a slight approximation – to be accurate, each of these hydrogen atoms has a mass of 1.008 on the carbon-12 scale. We take it as being exactly 1.

The basic unit on the carbon-12 scale is one twelfth of the mass of a $^{12}$C atom – approximately the mass of the commonest hydrogen atom. For example, a fluorine-19 atom has a relative isotopic mass of 19 because its atoms have a mass 19 times that basic unit.

The **relative atomic mass** of an element (as opposed to one of its isotopes) is given the symbol $A_r$ and is defined like this:

> **The relative atomic mass of an element is the weighted average mass of the isotopes of the element. It is measured on a scale on which a carbon-12 atom has a mass of exactly 12 units**.

## *Explaining the term 'weighted average'*

In any sample of chlorine, some atoms have a relative mass of 35; others a relative mass of 37. A simple average of 35 and 37 is, of course, 36 – but that isn't the relative atomic mass of chlorine. The problem is that there aren't equal numbers of $^{35}$Cl and $^{37}$Cl atoms.

A typical sample of chlorine has:  $^{35}$Cl   75%
$^{37}$Cl   25%

35   35

35   37

*The masses of four typical chlorine atoms. The average of these isn't 36.*

If you had 100 typical atoms of chlorine, 75 would be $^{35}$Cl, and 25 would be $^{37}$Cl.

The total mass of the 100 atoms would be    $(75 \times 35) + (25 \times 37)$
=    3550

The average mass of 1 atom would be    $\dfrac{3550}{100}$
=    35.5

The weighted average is closer to 35 than to 37, because there are more $^{35}$Cl atoms than $^{37}$Cl atoms. A weighted average allows for the unequal proportions.

35.5 is the relative atomic mass ($A_r$) of chlorine.

## More examples of calculating relative atomic masses

### *Magnesium*

The isotopes of magnesium and their percentage abundances are:

$^{24}$Mg    78.6%
$^{25}$Mg    10.1%
$^{26}$Mg    11.3%

Again, assume that you have 100 typical atoms.

The total mass would be $= \frac{(78.6 \times 24) + (10.1 \times 25) + (11.3 \times 26)}{}$

$= 2432.7$

The $A_r$ would be $= \frac{2432.7}{100}$

$= 24.3$ (to 3 significant figures)

The relative atomic mass of magnesium is 24.3.

### Lithium

The abundance data might be given in a different form. You might get a graph with the most common isotope being given a relative abundance of 100, and the others quoted relative to that.

In this case, there would be 8.0 atoms of $^6$Li for every 100 of $^7$Li.

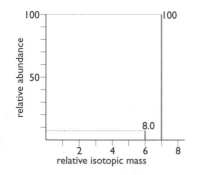

Relative abundance of lithium isotopes.

The total mass of 108 atoms would be $= (8.0 \times 6) + (100 \times 7)$

$= 748$

The average mass of 1 atom (the $A_r$) $= \frac{748}{108}$

$= 6.9$

The relative atomic mass of lithium is 6.9.

1. Why isn't the mass of a chlorine molecule always the same?
2. What is a weighted average? Give one example.
3. Explain the meaning of the graph showing the abundance data for lithium.

## Relative formula mass (M$_r$)

You can measure the masses of compounds on the same carbon-12 scale. For example, it turns out that a water molecule, $H_2O$, has a mass of 18 on the $^{12}$C scale. Where you are talking about compounds, you use the term **relative formula mass**. Relative formula mass is sometimes called relative molecular mass.

Relative formula mass is given the symbol **M$_r$**.

### Working out some relative formula masses

#### To find the M$_r$ of magnesium carbonate, MgCO$_3$

Relative atomic masses: C = 12; O = 16; Mg = 24

All you have to do is to add up the relative atomic masses to give you the relative formula mass of the whole compound. In this case, you need to add up the masses of 1 × Mg, 1 × C and 3 × O.

$M_r = 24 + 12 + (3 \times 16)$

$= 84$

*To find the $M_r$ of calcium hydroxide, $Ca(OH)_2$*

Relative atomic masses: H = 1; O = 16; Ca = 40

$$M_r \qquad = 40 + (16 + 1) \times 2$$
$$= 74$$

*To find the $M_r$ of copper(II) sulphate crystals, $CuSO_4 \cdot 5H_2O$*

Relative atomic masses: H = 1; O = 16; S = 32; Cu = 64

$$M_r \qquad = 64 + 32 + (4 \times 16) + 5 \times [(2 \times 1) + 16]$$
$$= 250$$

Most people have no difficulty with $M_r$ sums until they get to an example involving water of crystallisation (the $5H_2O$ in this example). $5H_2O$ means 5 molecules of water – so to get the total mass of this, work out the $M_r$ of water (18) and then multiply it by 5. It is dangerous to do the hydrogens and oxygens separately. The common mistake is to work out 10 hydrogens (quite correctly!), but then only count one oxygen rather than five.

4. Work out the $M_r$ for: **a)** magnesium sulphate, **b)** hydrated magnesium sulphate (each molecule of magnesium sulphate has seven molecules of water).
5. Find the percentage by mass of carbon in carbon dioxide.
6. Which has a higher percentage by mass of carbon, methane or ethane?

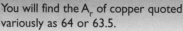

You will find the $A_r$ of copper quoted variously as 64 or 63.5.

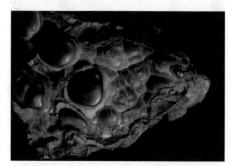

*Malachite is a copper ore. You can think of it as behaving like a mixture of copper(II) carbonate and copper(II) hydroxide.*

### Using relative formula mass to find percentage composition

Having found the relative formula mass of a compound, it is then easy to work out the percentage by mass of any part of it. Examples make this clear.

*To find the percentage by mass of copper in copper(II) oxide, CuO*

Relative atomic masses: O = 16; Cu = 64

$$M_r \text{ of CuO} \qquad = 64 + 16$$
$$= 80$$

Of this, 64 is copper.

$$\text{Percentage of copper} \qquad = \frac{64}{80} \times 100$$
$$= 80\%$$

## To find the percentage by mass of copper in malachite, $CuCO_3 \cdot Cu(OH)_2$

Relative atomic masses: H = 1; C = 12; O = 16; Cu = 64

$M_r$ of $CuCO_3 \cdot Cu(OH)_2$ = 64 + 12 + (3 × 16) + 64 + 2 × (16 + 1)
= 222

Of this, (2 × 64) is copper.

$$\text{Percentage of copper} = \frac{(2 \times 64)}{222} \times 100$$
$$= 57.7\%$$

## To find the percentage of water in alabaster, $CaSO_4 \cdot 2H_2O$

Relative atomic masses: H = 1; O = 16; S = 32; Ca = 40

$M_r$ of alabaster = 40 + 32 + (4 × 16) + 2 × [(2 × 1) + 16]
= 172

Of this, 2 × [(2 × 1) + 16] g is water. That's 36 g.

$$\text{Percentage of water} = \frac{36}{172} \times 100$$
$$= 20.9\%$$

### Percentage yield

You can neutralise an acid by adding an alkali.

$$\text{acid} + \text{alkali} \rightarrow \text{salt} + \text{water}$$

Even if all of the acid and alkali react and you evaporate the solution carefully, you will be lucky to collect all of the solid salt that was formed. Some may be left on the apparatus or some might be spilled. If your calculation shows that you should get 10 g of salt but you only recover 9 g, then your yield is 9 g out of a possible 10 g. As a fraction it is $\frac{9}{10}$ or as a percentage yield it is $\frac{9}{10} \times 100 = 90\%$ yield.

It is difficult to measure out exactly the right amounts of chemicals that will react together and leave nothing over. Often we use slightly too much of one chemical, called using an **excess**. This ensures that the other chemical has reacted completely. Here is an example.

$$\text{copper oxide} + \text{sulphuric acid} \rightarrow \text{copper sulphate solution} + \text{water}$$

In reactions like this, we usually add a little too much of the solid, copper oxide, to the acid. This makes sure that all of the acid has been changed to the product, the salt copper sulphate. It is easy to remove the excess copper oxide by filtering. The solution can be evaporated to give blue crystals of copper sulphate. Some of the solution will be lost during the filtration stage. The yield of copper sulphate will be less than 100%.

The mass of copper sulphate crystals we could obtain depends on the amount of acid at the start. The acid is present in a lesser amount. It is not in excess. The rule is: the yield is limited by the chemical reagent present in lesser amount.

Alabaster, $CaSO_4 \cdot 2H_2O$, is a soft mineral which is easily carved.

## Activity 1:

### Investigative skills

P

How could you find the percentage yield of copper sulphate in the reaction of copper oxide with dilute sulphuric acid?

It does not matter how much extra copper oxide we use. We filter off the excess and remove it.

When we burn a fuel such as natural gas, it is the oxygen in the air which is present in excess. The amount of carbon dioxide produced is limited by the amount of natural gas burned, not by the amount of oxygen.

# The mole

In Chemistry, the mole is a measure of **amount of substance**. A mole is a particular mass of that substance. You can use such expressions as:

- a mole of copper(II) sulphate crystals, $CuSO_4 \cdot 5H_2O$

- a mole of oxygen gas, $O_2$

- 0.1 mole of zinc oxide, ZnO

- 3 moles of magnesium, Mg

The abbreviation for mole is **mol**. The mass of one mole is called the **molar mass**.

You find the mass of 1 mole of a substance in the following way:

**Work out the relative formula mass, and attach the units 'grams'.**

## Working out the masses of a mole of substance

### 1 mole of oxygen gas, $O_2$

Relative atomic mass: O = 16

$$M_r \text{ of oxygen, } O_2 = 2 \times 16$$
$$= 32$$

1 mole of oxygen, $O_2$, weighs 32 g.

### 1 mole of iron(II) sulphate crystals, $FeSO_4 \cdot 7H_2O$

Relative atomic masses: H = 1; O = 16; S = 32; Fe = 56

$$M_r \text{ of crystals} = 56 + 32 + (4 \times 16) + 7 \times [(2 \times 1) + 16]$$
$$= 278$$

1 mole of iron(II) sulphate crystals weighs 278 g.

## The importance of quoting the formula

Whenever you talk about a mole of something, you **must** quote its formula, otherwise there is the risk of confusion.

For example, if you talk about 1 mole of oxygen, this could mean:

- 1 mole of oxygen atoms, O, weighing 16 g

- 1 mole of oxygen molecules, $O_2$, weighing 32 g.

> One mole of molecules or of atoms of a material has a mass in grams that is the same as the $M_r$, relative formula mass, or the $A_r$, relative atomic mass.
>
> For example, carbon dioxide ($CO_2$) has a $M_r$ of (12 + 16 + 16) = 44, so one mole of carbon dioxide is 44 g.

*1 mole of copper(II) sulphate crystals, $CuSO_4 \cdot 5H_2O$*

Or if you were talking about 1 mole of copper(II) sulphate, this could mean:

- 1 mole of anhydrous copper(II) sulphate, $CuSO_4$ (160 g)
- 1 mole of copper(II) sulphate crystals, $CuSO_4 \cdot 5H_2O$ (250 g)

## Simple calculations with moles

You need to be able to interconvert between a mass in grams and a number of moles for a given substance. There is a simple formula that you can learn:

$$\text{number of moles} = \frac{\text{mass (g)}}{\text{mass of 1 mole (g)}}$$

You can rearrange that to find whatever you want. If rearranging this expression causes you problems, you can learn a simple triangular arrangement which does the whole thing for you.

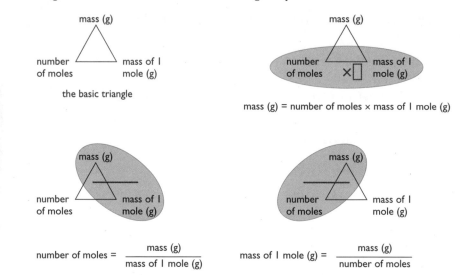

mass (g) = number of moles × mass of 1 mole (g)

$$\text{number of moles} = \frac{\text{mass (g)}}{\text{mass of 1 mole (g)}} \qquad \text{mass of 1 mole (g)} = \frac{\text{mass (g)}}{\text{number of moles}}$$

Look at this carefully and make sure that you understand how you can use it to work out the three equations that you might need.

### Finding the mass of 0.2 moles of calcium carbonate, $CaCO_3$

Relative atomic masses: C = 12; O = 16; Ca = 40

First find the relative formula mass of calcium carbonate.

$$M_r \text{ of } CaCO_3 = 40 + 12 + (3 \times 16)$$
$$= 100$$

1 mol of $CaCO_3$ weighs 100 g.

$$\text{mass (g)} = \text{no. of moles} \times \text{mass of 1 mole (g)}$$
$$= 0.2 \times 100 \text{ g}$$
$$= 20 \text{ g}$$

0.2 moles of $CaCO_3$ has a mass of 20 g.

*Finding the number of moles in 54 g of water, H₂O*

Relative atomic masses: H = 1; O = 16

1 mol of $H_2O$ weighs 18 g.

$$\text{number of moles} = \frac{\text{mass (g)}}{\text{mass of 1 mole (g)}}$$

$$= \frac{54}{18}$$

$$= 3 \text{ mol}$$

54 g of water is 3 moles.

## Moles and the Avogadro Constant

Suppose you had 1 mole of $^{12}C$. It would have a mass of 12 g, and contain a huge number of carbon atoms – in fact, about $6 \times 10^{23}$ carbon atoms – that's 6 followed by 23 noughts. This number of atoms in 12 g of $^{12}C$ is called the Avogadro Constant.

*1 mole of anything else contains this same number of particles.* For example:

- 1 mole of magnesium contains $6 \times 10^{23}$ magnesium atoms, Mg, and has a mass of 24 g.

- 1 mole of water contains $6 \times 10^{23}$ water molecules, $H_2O$, and has a mass of 18 g.

- 1 mole of sodium chloride contains $6 \times 10^{23}$ formula units, NaCl, and has a mass of 58.5 g. (You can't say 'molecules of NaCl' because sodium chloride is ionic.)

- 1 mole of oxide ions contains $6 \times 10^{23}$ $O^{2-}$ ions and has a mass of 16 g.

- 1 mole of electrons contains $6 \times 10^{23}$ electrons.

*1 × 10³² water molecules go over Niagara Falls every second during the summer. That's 100 million million million million million water molecules per second.*

> **7.** What is the mass of five moles of: **a)** water, **b)** ethane?
> **8.** What is the rule that links together moles, mass and the mass of one mole of a substance?

## Using moles to find formulae

### Interpreting symbols in terms of moles

Assume that you know the formula for something like copper(II) oxide, for example. The formula is CuO.

When you are doing sums, it is often useful to interpret a symbol as meaning more than just 'an atom of copper' or 'an atom of oxygen'. For calculation purposes we take the symbol Cu to mean **1 mole of copper atoms.**

In a formula, 'Cu' means '64 g of copper'. 'O' means '16 g of oxygen'. ($A_r$s: O = 16; Cu = 64.) So in copper(II) oxide, the copper and oxygen are combined in the ratio of 64 g of copper to 16 g of oxygen.

You can read a formula like $H_2O$ as meaning that two moles of hydrogen atoms are combined with one mole of oxygen atoms. In other words, 2 g of hydrogen are combined with 16 g of oxygen. ($A_r$s: H = 1; O = 16)

## Working out formulae

### The formula for magnesium oxide

Suppose you did an experiment to find out how much magnesium and oxygen reacted together to form magnesium oxide. Suppose 2.4 g of magnesium combined with 1.6 g of oxygen. You can use these figures to find the formula of magnesium oxide. ($A_r$s: O = 16; Mg = 24)

| | **Mg** | **O** |
|---|---|---|
| Combining masses: | 2.4 g | 1.6 g |
| No. of moles of atoms: | $\dfrac{2.4}{24}$ | $\dfrac{1.6}{16}$ |
| | = 0.10 | = 0.10 |
| Ratio of moles: | 1 : | 1 |
| Simplest formula: | MgO | |

This simplest formula is called the **empirical formula**. The empirical formula just tells you the **ratio** of the various atoms. Without more information it isn't possible to work out the 'true' or 'molecular' formula which could be $Mg_2O_2$, $Mg_3O_3$, etc. For ionic substances, the formula quoted is always the empirical formula.

### The formula for red copper oxide

You might get the data in a more complicated form. This example is about a less common form of copper oxide. If hydrogen is passed over the hot oxide, it is reduced to metallic copper. These figures could be obtained using the apparatus shown on page 48.

| | |
|---|---|
| Mass of empty tube | = 52.2 g |
| Mass of tube + copper oxide (before experiment) | = 66.6 g |
| Mass of tube + copper (after experiment) | = 65.0 g |

The tube loses mass because oxygen has been removed from the copper oxide, leaving metallic copper.

| | | |
|---|---|---|
| Mass of oxygen | = 66.6 – 65.0 g | = 1.6 g |
| Mass of copper | = 65.0 – 52.2 g | = 12.8 g |

Now you have all the information to find the empirical formula. ($A_r$s: O = 16; Cu = 64)

| | **Cu** | **O** |
|---|---|---|
| Combining masses: | 12.8 g | 1.6 g |
| No. of moles of atoms: | $\dfrac{12.8}{64}$ | $\dfrac{1.6}{16}$ |
| | = 0.20 | = 0.10 |
| Ratio of moles: | 2 : | 1 |
| Simplest (empirical) formula: | $Cu_2O$ | |

Crystals of the copper-containing mineral cuprite, $Cu_2O$.

## Working out formulae using percentage composition figures

Often, figures for the compound are given as percentages by mass. For example: Find the empirical formula of a compound containing 85.7% C, 14.3% H by mass. ($A_r$s: H = 1; C = 12)

This isn't a problem! Those percentage figures apply to any amount of substance you choose – so choose 100 g. In which case, the percentages convert simply into masses. 85.7% of 100 g is 85.7 g.

| | C | H |
|---|---|---|
| Given percentages | 85.7% | 14.3% |
| Combining masses in 100 g | 85.7 g | 14.3 g |
| No. of moles of atoms: | $\dfrac{85.7}{12}$ | $\dfrac{14.3}{1}$ |
| | = 7.14 | = 14.3 |
| Ratio of moles: | 1 : | 2 |
| Simplest (empirical) formula: | $CH_2$ | |

> Usually the ratio will be fairly obvious, but if you can't spot it at once, divide everything by the smallest number and see if that helps. Sometimes, you may find that a ratio comes out as, for example, 1:1.5. In that case, all you have to do is double both numbers to give the true whole number ratio of 2:3.

> **9.** What is the empirical formula of a compound that contains 53.8% by mass of iron and 46.2% sulphur?

### Converting empirical formulae into molecular formulae

In the example we have just looked at, $CH_2$ can't possibly be the real formula of the hydrocarbon – the carbon would have spare unbonded electrons. The **molecular formula** (the true formula) would have to be some multiple of $CH_2$, like $C_2H_4$ or $C_3H_6$ or whatever – as long as the ratio is still 1 carbon to 2 hydrogens.

You could find the molecular formula if you knew the relative formula mass of the compound (or the mass of 1 mole – which is just the $M_r$ expressed in grams).

In the previous example, suppose you knew that the relative formula mass of the compound was 56. $CH_2$ has a relative formula mass of only 14. ($A_r$s: H = 1; C = 12)

All you need to find out is how many times 14 goes into 56.

$\dfrac{56}{14}$ = 4 and so you need 4 lots of $CH_2$ – in other words, $C_4H_8$.

The molecular formula is $C_4H_8$.

### Empirical formula calculations involving water of crystallisation

When some substances crystallise from solution, water becomes chemically bound up with the salt. This is called **water of crystallisation**. The salt is said to be **hydrated**. Examples include $CuSO_4 \cdot 5H_2O$ and $MgCl_2 \cdot 6H_2O$.

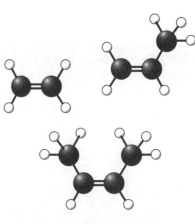

All of these have an empirical formula $CH_2$.

> Even having got the molecular formula, you still don't know the structure. There are several isomers of $C_4H_8$ which have different shapes.

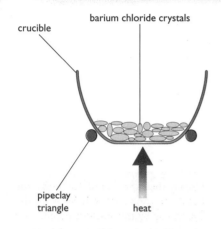

crucible

barium chloride crystals

pipeclay triangle

heat

### Finding the 'n' in BaCl₂·nH₂O

When you heat a salt which contains water of crystallisation, the water is driven off, leaving the anhydrous salt behind. Hydrated barium chloride is a commonly used example because the barium chloride itself doesn't decompose even on quite strong heating. If you heated barium chloride crystals in a crucible you might end up with these results:

| | |
|---|---|
| Mass of crucible | = 30.00 g |
| Mass of crucible + barium chloride crystals, $BaCl_2 \cdot nH_2O$ | = 32.44 g |
| Mass of crucible + anhydrous barium chloride, $BaCl_2$ | = 32.08 g |

To find 'n' you need to find the ratio of the number of moles of $BaCl_2$ to the number of moles of water. It's just another empirical formula sum. ($A_r$s: H = 1; O = 16; Cl = 35.5; Ba = 137)

| | | |
|---|---|---|
| Mass of $BaCl_2$ | = 32.08 − 30.00 g | = 2.08 g |
| Mass of water | = 32.44 − 32.08 g | = 0.36 g |

| | **$BaCl_2$** | **$H_2O$** |
|---|---|---|
| Combining masses: | 2.08 g | 0.36 g |
| No. of moles: | $\dfrac{2.08}{208}$ | $\dfrac{0.36}{18}$ |
| | = 0.01 | = 0.02 |
| Ratio of moles: | 1 : | 2 |
| Empirical formula: | $BaCl_2 \cdot 2H_2O$ | |

208 is the $M_r$ of $BaCl_2$. 18 is the $M_r$ of water. Check them if you aren't sure.

# End of Chapter Checklist

- one atom of carbon-12 weighs exactly 12 units. We call this the carbon-12 scale

- other kinds of atoms are compared with the mass of this isotope of carbon

- the relative formula mass ($M_r$) is the mass of a compound such as carbon dioxide (molecular) or sodium chloride (ionic)

- the percentage composition of a compound gives the percentage of each element

- the mole is the standard unit for the amount of a substance

- the empirical formula of a substance gives the simplest ratio of the atoms in a compound

- the molecular formula gives the actual numbers of atoms in one molecule.

# Questions

1. The relative formula mass of carbon dioxide is:

   A  28      C  56
   B  32      D  44     (1)

2. The empirical formula is:

   A  the one found by experiment
   B  the same as the molecular formula
   C  the simplest ratio of atoms
   D  none of these     (1)

3. a) What is the carbon-12 scale. Why don't we use kilograms instead?     (3)

   b) Bromine atoms weigh 79 or 81 units. How many different diatomic molecules are there?     (3)

   c) If the two isotopes of bromine are equally common, what is the weighted average? Explain why the relative atomic mass of bromine is given as 80 but there are no bromine atoms that weigh 80 units.     (4)

4. a) Using hexane, $C_6H_{14}$, as your example, explain the meaning of empirical formula.     (2)

   b) Calculate the percentage by mass of carbon and hydrogen in hexane.     (4)

   c) What is the $M_r$ of hexane?     (1)

   d) What is the mass of i) 1 mole of hexane, ii) 0.25 moles of hexane iii) 3.5 moles of hexane?     (3)

5. a) What is the formula of hydrated calcium sulphate (2 molecules of water)?     (2)

   b) What name is given to this water?     (1)

   c) What is the $M_r$?     (2)

   d) What is the mass of 0.1 mole?     (2)

   e) When this compound is heated, the mass changes. Explain why this happens and write an equation assuming that half of the water is lost.     (3)

6. a) Work out if the molecular formula of a hydrocarbon is $C_{12}H_{12}$ or $C_6H_6$. Explain your answer. The compound contains 7.7% hydrogen. Its $M_r$ is 78.     (4)

   b) A compound contains 52.2% carbon, 13.1% hydrogen and the rest is oxygen. Find the empirical formula. If the $M_r$ is 46, find the molecular formula.     (3)

   c) Find the empirical formula and the percentages of each element in benzene, $C_6H_6$.     (3)

7. The common isotopes of chlorine weigh 35 and 37 units.

   a) How many different mass molecules are there for dichloromethane, $CH_2Cl_2$?     (3)

   b) Draw and label each molecule.     (3)

   c) Using a calculation, explain why the $A_r$ of chlorine is about 35.5 and not a whole number.     (2)

   d) Lighter molecules diffuse more quickly. Which molecule of dichloromethane will diffuse quickest and why?     (2)

8. Calculate the relative atomic mass of gallium given the percentage abundances:

   $^{69}$Ga – 60.2%, $^{71}$Ga – 39.8%.     (6)

9. Calculate the relative atomic mass of silicon given:

| Relative isotopic mass | Relative abundance |
|---|---|
| 28 | 100 |
| 29 | 5.10 |
| 30 | 3.36 |

(4)

10. a) Define relative atomic mass. (12)

b) Calculate the relative atomic masses of copper and sulphur from the percentage abundances of their isotopes. Use your answers to find the relative formula mass of copper(II) sulphide, CuS.

$^{63}Cu - 69.1\%$; $^{65}Cu - 30.9\%$.
$^{32}S - 95.0\%$; $^{33}S - 0.76\%$; $^{34}S - 4.22\%$; $^{36}S - 0.020\%$. (6)

11. Calculate the relative formula masses of the following compounds.

($A_r$s: H = 1; C = 12; N = 14; O = 16; Na = 23; S = 32; Ca = 40; Cr = 52; Fe = 56)

a) $CO_2$ (2)

b) $(NH_4)_2SO_4$ (2)

c) $Na_2CO_3 \cdot 10H_2O$ (2)

d) $Cr_2(SO_4)_3$ (2)

e) $(NH_4)_2SO_4 \cdot FeSO_4 \cdot 6H_2O$ (2)

12. Find the percentage of the named substance in each of the following examples.

($A_r$s: H = 1; C = 12; O = 16; Mg = 24; S = 32)

a) Carbon in propane, $C_3H_8$. (3)

b) Water in magnesium sulphate crystals, $MgSO_4 \cdot 7H_2O$. (5)

13. Work out the percentage of nitrogen in each of the following substances (all used as nitrogen fertilisers).

($A_r$s: H = 1; C = 12; N = 14; O = 16; S = 32; K = 39)

a) urea, $CO(NH_2)_2$ (2)

b) potassium nitrate, $KNO_3$ (2)

c) ammonium nitrate, $NH_4NO_3$ (2)

d) ammonium sulphate, $(NH_4)_2SO_4$. (2)

14. Work out the mass of the following:

($A_r$s: H = 1; C = 12; N = 14; O = 16; Na = 23; Pb = 207)

a) 1 mole of lead(II) nitrate, $Pb(NO_3)_2$ (2)

b) 4.30 moles of methane, $CH_4$ (2)

c) 0.24 moles of sodium carbonate crystals, $Na_2CO_3 \cdot 10H_2O$. (2)

15. How many moles are represented by each of the following:

($A_r$s: H = 1; O = 16; S = 32; Fe = 56; Cu = 64)

a) 50 g of copper(II) sulphate crystals, $CuSO_4 \cdot 5H_2O$ (2)

b) 1 tonne of iron, Fe (1 tonne is 1000 kg) (2)

c) 0.032 g of sulphur dioxide, $SO_2$. (2)

16. Some more questions about converting between moles and grams of a substance:

($A_r$s: H = 1; O = 16; Na = 23; Cl = 35.5; Ca = 40; Cu = 64)

a) What is the mass of 4 mol of sodium chloride, NaCl? (2)

b) How many moles is 37 g of calcium hydroxide, $Ca(OH)_2$? (2)

c) How many moles is 1 kg (1000 g) of calcium, Ca? (2)

d) What is the mass of 0.125 mol of copper(II) oxide, CuO? (2)

e) 0.1 mol of a substance weighed 4 g. What is the weight of 1 mole? (2)

f) 0.004 mol of a substance weighed 1 g. What is the relative formula mass of the compound? (2)

17. ($A_r$s: H = 1; C = 12; N = 14; O = 16; Na = 23; S = 32; K = 39; Br = 80)

Find the empirical formulae of the following compounds which contained:

a) 5.85 g K; 2.10 g N; 4.80 g O (4)

b) 3.22 g Na; 4.48 g S; 3.36 g O (4)

c) 22.0% C; 4.6% H; 73.4% Br (by mass). (4)

18. 1.24 g of phosphorus was burnt completely in oxygen to give 2.84 g of phosphorus oxide. Find a) the empirical formula of the oxide, b) the molecular formula of the oxide, given that 1 mole of the oxide weighs 284 g. ($A_r$s: O = 16; P = 31) (8)

19. An organic compound contained C – 66.7%, H – 11.1%, O – 22.2% by mass. Its relative formula mass was 72. Find a) the empirical formula of the compound, b) the molecular formula of the compound. ($A_r$s: H = 1; C = 12; O = 16) (8)

20. In an experiment to find the number of molecules of water of crystallisation in sodium sulphate crystals, $Na_2SO_4 \cdot nH_2O$, 3.22 g of sodium sulphate crystals were heated gently. When all the water of crystallisation had been driven off, 1.42 g of anhydrous sodium sulphate was left. Find the value for 'n' in the formula. ($A_r$s: H = 1; O = 16; Na = 23; S = 32) (8)

21. Gypsum is hydrated calcium sulphate, $CaSO_4 \cdot nH_2O$. A sample of gypsum was heated in a crucible until all the water of crystallisation had been driven off. The following results were obtained:

Mass of crucible = 37.34 g
Mass of crucible + gypsum, $CaSO_4 \cdot nH_2O$ = 45.94 g
Mass of crucible + anhydrous calcium sulphate, $CaSO_4$ = 44.14 g

Calculate the value of 'n' in the formula $CaSO_4 \cdot nH_2O$. (10)

($A_r$s: H = 1; O = 16; S = 32; Ca = 40)

## Chapter 8: Calculations from Equations

> **When you have completed this chapter, you will be able to:**
>
> - use moles to work out the masses of substances in chemical equations
> - calculate the volumes of gases in equations
> - understand Avogadro's Law
> - use the molar volume of a gas in calculations
> - change between moles of gases and volumes of gases
> - understand everyday examples of chemical calculations.

## Calculations involving only masses

Calculations usually give you a mass of starting material and ask you to calculate how much product you are likely to get. You will also meet examples done the other way around. You are told the mass of the product and are asked to find out how much of the starting material you would need. In almost all the cases you will be given the equation for the reaction.

### A problem involving heating limestone

When limestone, $CaCO_3$, is heated, calcium oxide is formed. Suppose you wanted to calculate the mass of calcium oxide produced by heating 25 g of limestone. (Relative atomic masses: C = 12; O = 16; Ca = 40)

### *The calculation*

*First write the equation:*

$$CaCO_3(s) \rightarrow CaO(s) + CO_2(g)$$

*Interpret the equation in terms of moles:*

Remember that each formula represents 1 mole of that substance.

   1 mol $CaCO_3$ produces 1 mol CaO (and 1 mol $CO_2$)

*Substitute masses where needed:*

   100 g (1 mol) $CaCO_3$ produces 56 g (1 mol) CaO

(Notice that we haven't calculated the mass of carbon dioxide. In this question you aren't interested in it.)

*Do the simple proportion sum:*

If   100 g of calcium carbonate gives 56 g of calcium oxide
    1 g of calcium carbonate gives $\frac{56}{100}$ g of calcium oxide = 0.56 g
    25 g of calcium carbonate gives 25 × 0.56 g of calcium oxide
    = 14 g of calcium oxide

> **Equations**
>
> Remember that you need to write equations linked to your calculations.

> The $M_r$ of $CaCO_3$ is 100, and the $M_r$ of CaO is 56. Work them out!

> Your maths may be good enough that you don't need to take all these steps to get to the answer. You must, however, show all your working.

Galena, PbS.

## A problem involving the manufacture of lead

Lead is extracted from galena, PbS. The ore is heated in air to produce lead(II) oxide, PbO.

$$2PbS(s) + 3O_2(g) \rightarrow 2PbO(s) + 2SO_2(g)$$

The lead(II) oxide is reduced to lead by heating it with carbon in a blast furnace.

$$PbO(s) + C(s) \rightarrow Pb(l) + CO(g)$$

The molten lead is tapped from the bottom of the furnace.

Calculate:

(a) The mass of sulphur dioxide produced when 1 tonne of galena is heated.

(b) The mass of lead that would be produced from 1 tonne of galena.

(Relative atomic masses: O = 16; S = 32; Pb = 207)

### Calculation part (a)

*First write the equation:*

$$2PbS(s) + 3O_2(g) \rightarrow 2PbO(s) + 2SO_2(g)$$

*Interpret the equation in terms of moles:*

2 mol PbS produces 2 mol $SO_2$ (the others aren't important for this calculation)

*Substitute masses where relevant:*

$2 \times 239$ g PbS produces $2 \times 64$ g $SO_2$
478 g PbS produces 128 g $SO_2$

Now there seems to be a problem. The question is asking about tonnes and not grams. You could work out how many grams there are in a tonne and then do hard sums with large numbers. However, it's much easier to think a bit, and realise that the ratio is always going to be the same whatever the units – so that...

if        478 g PbS produces 128 g $SO_2$
then     478 tonnes PbS produces 128 tonnes $SO_2$

*Do the simple proportion sum:*

If        478 tonnes PbS produces 128 tonnes $SO_2$

1 tonne PbS gives $\dfrac{128}{478}$ tonnes $SO_2$ = 0.267 tonnes $SO_2$

### Calculation part (b)

*First write the equation:*

There are two equations to think about:

$$2PbS(s) + 3O_2(g) \rightarrow 2PbO(s) + 2SO_2(g)$$
$$PbO(s) + C(s) \rightarrow Pb(l) + CO(g)$$

*Interpret the equation in terms of moles:*

Lead ingots.

Trace the lead through the equations:

2 mol PbS produces 2 mol PbO
2 mol PbO produces 2 mol Pb

We've doubled the second equation so that we can trace what happens to all the 2PbO from the first one.

In other words, every 2 moles of PbS produces 2 moles of lead.

*Substitute masses where needed:*

In this case, the masses are only the PbS and the final lead.

2 × 239 g PbS produces 2 × 207 g Pb
478 g PbS produces 414 g Pb

So:    478 tonnes PbS produces 414 tonnes Pb

*Do the simple proportion sum:*

If    478 tonnes PbS produces 414 tonnes Pb

1 tonne PbS gives $\frac{414}{478}$ tonnes Pb = 0.866 tonnes

0.866 tonnes of lead are produced from 1 tonne of galena.

You could save a bit of time here by realising that if 2 mol PbS produces 2 mol Pb, that's exactly the same as 1 mol PbS producing 1 mol Pb. That would save you having to multiply two numbers by 2.

---

**1.** What is the difference between a product and a reactant in a chemical equation?
**2.** Find the equation for heating lead sulphide (galena) in air (see page 68). How many moles are there of each substance?
**3.** Explain the state symbols in the equation in question 2.

---

# Calculations involving gas volumes

### Units of volume

Volumes (of gases or liquids) are measured in:    cubic centimetres (cm³)

or    cubic decimetres (dm³)

or    litres (l)

**1 litre = 1 dm³ = 1000 cm³**

### Avogadro's Law

**Equal volumes of gases at the same temperature and pressure contain equal numbers of molecules.**

This means that if you have 100 cm³ of hydrogen at some temperature and pressure, it contains exactly the same number of molecules as there are in 100 cm³ of carbon dioxide or any other gas under those conditions – irrespective of the size of the molecules.

Avogadro's Law can be used for some very simple calculations involving gases.

If you want to talk about 1000 cm³, the cubic decimetre is the preferred unit rather than the litre.

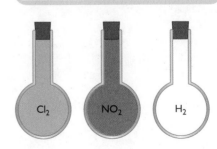

*Three identical flasks containing different gases at the same temperature and pressure all contain equal numbers of molecules.*

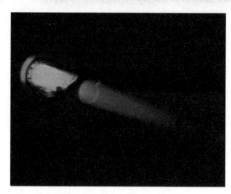
*Burning methane in air.*

## *Calculating the volume of oxygen needed to burn 100 cm³ of methane*

Methane (natural gas) burns in air according to the equation:

$$CH_4(g) + 2O_2(g) \rightarrow CO_2(g) + 2H_2O(l)$$

The equation says that you need twice as many molecules of oxygen as you do of methane. According to Avogadro's Law, this means that you will need twice the volume of oxygen as you have of methane.

So, if you have to burn 100 cm³ of methane, you will need 200 cm³ of oxygen.

Similarly, for every molecule of methane you burn, you will get a molecule of carbon dioxide formed. Therefore every 100 cm³ of methane will give 100 cm³ of carbon dioxide, because 100 cm³ of methane and 100 cm³ of $CO_2$ contain the same number of molecules.

Avogadro's Law only applies to gases. If the water is formed as a liquid, you can't say anything about the volume of water produced. However, if it was formed as steam, then you could say that every 100 cm³ of methane will produce 200 cm³ of steam.

## *Calculating the volume of air needed to burn 100 cm³ of methane*

When you do a sum of this sort, be careful to notice whether you are being asked about air or oxygen. Air is only approximately one fifth oxygen – and so you need five times more air than you would need oxygen.

In this example, if you need 200 cm³ of oxygen, you will need five times as much air – in other words, 1000 cm³.

## The molar volume of a gas

1 mole of any gas contains the same number of molecules and so occupies the same volume as 1 mole of any other gas at the same temperature and pressure.

The volume occupied by 1 mole of a gas is called the **molar volume**. At room temperature and pressure, the molar volume is approximately 24 dm³ (24 000 cm³).

**I mole of any gas occupies 24 dm³ (24 000 cm³) at rtp.**

> The abbreviation **rtp** is commonly used for 'room temperature and pressure'.

We sometimes need to work out the volume of a gas under standard conditions of temperature and pressure (stp). This means a temperature of 0°C (or 273 K) and a pressure of one atmosphere (or 760 mm mercury).

**I mole of any gas occupies 22.4 dm³ (or 22 400 cm³) at stp.**

## Simple calculations with the molar volume

### *Calculating the volume of a given mass of gas*

Calculate the volume of 0.01 g of hydrogen at rtp.

First we need to know how many moles of hydrogen we have.

($A_r$: H = 1)

    1 mol $H_2$ has a mass of 2 g

0.01 g of hydrogen is $\frac{0.01}{2}$ moles = 0.005 moles

1 mole of hydrogen occupies 24 000 cm$^3$

0.005 mol of hydrogen occupies 0.005 × 24 000 cm$^3$

= 120 cm$^3$

### Calculating the mass of a given volume of gas

Early airships were filled with flammable hydrogen. Suppose an airship contained 180 000 cubic metres of hydrogen. We can calculate the mass of this hydrogen.

180 000 cubic metres is 180 000 000 dm$^3$. Assuming the gas was at room temperature and pressure, each 24 dm$^3$ represents one mole of hydrogen.

$$\text{Number of moles of } H_2 = \frac{180\,000\,000}{24} = 7\,500\,000 \text{ mol}$$

Each mole of hydrogen, H$_2$, weighs 2 g. (A$_r$: H = 1)

Mass of hydrogen = 7 500 000 × 2 g
= 15 000 000 g (= 15 000 kg)

If you were to convert that into more sensible sized units the mass of hydrogen is 15 tonnes.

### Using the molar volume in calculations from equations

### Calculating the molar volume from densities

$$\text{Density} = \frac{\text{mass}}{\text{volume}}$$

We can measure the density of a gas like this:

- Fit a tube and pressure valve to a flask and use a vacuum pump to remove all the air.

- Weigh the empty flask.

- Fill the flask with a gas such as nitrogen.

- Weigh the flask again and find the increase in mass.

The volume of gas = the volume of the flask

The mass of gas = increase in mass

From this we can calculate the density of the gas. Here is some sample data

| Gas | M$_r$ | Density at stp in g/dm³ | Molar volume in dm³ |
|---|---|---|---|
| Nitrogen | 28 | 1.250 | $\frac{28}{1.25}$ = 22.40 |
| Oxygen | 32 | 1.43 | $\frac{32}{1.43}$ = 22.37 |
| Carbon dioxide | 44 | 1.96 | $\frac{44}{1.96}$ = 22.44 |
| Ammonia | 17 | 0.76 | ? work it out |

The molar volume of a gas is constant, within the limits of experimental error.

## Activity 1: Gas volumes

**Investigative skills**

P

How could you measure the volume of carbon dioxide produced on burning 1 g of a candle in air?

The calculation shows that 1 g of limpet shells, CaCO$_3$, would react with hydrochloric acid to give 0.24 dm$^3$ of carbon dioxide.

## Working out the volume of gas produced during a reaction

Calculate the volume of carbon dioxide produced at room temperature and pressure when an excess of dilute hydrochloric acid is added to 1.00 g of calcium carbonate. (You use an excess of acid to make sure that all the calcium carbonate reacts.)

($A_r$s: C = 12; O = 16; Ca = 40. Molar volume = 24 dm$^3$ at rtp)

*First write the equation:*

$$CaCO_3(s) + 2HCl(aq) \rightarrow CaCl_2(aq) + CO_2(g) + H_2O(l)$$

*Interpret the equation in terms of moles:*

1 mol $CaCO_3$ gives 1 mol $CO_2$

*Substitute masses and volumes where appropriate:*

100 g $CaCO_3$ gives 24 dm$^3$ $CO_2$ at rtp

*Do the simple proportion sum:*

If       100 g $CaCO_3$ gives 24 dm$^3$ $CO_2$ at rtp

then    1 g $CaCO_3$ gives $\frac{1}{100} \times 24$ dm$^3$

        = 0.24 dm$^3$

So 1 g of calcium carbonate gives 0.24 dm$^3$ of carbon dioxide.

**Important:** The commonest mistake in a sum of this kind is to work out the **mass** of 1 mole of $CO_2$. Once you have that figure of 44 g, you feel you have to do something with it, and will probably work out the mass of $CO_2$ produced instead of the volume.

4. What does Avogadro's Law tell us?
5. Convert a volume of 10 litres into: **a)** cubic centimetres, **b)** millilitres.
6. What is the molar volume of a gas at room temperature and normal pressure?

## Calculation of molar volume from experimental data

When magnesium is added to sulphuric acid, hydrogen gas is produced.

$$Mg + H_2SO_4 \rightarrow MgSO_4 + H_2$$

0.6 g of magnesium produced 599 cm$^3$ of hydrogen at room temperature and pressure. What is the molar volume of hydrogen at rtp?

Number of moles of magnesium    $= \frac{0.6}{24} = 0.025$ mole

Hence 599 cm$^3$ hydrogen         $= 0.025$ mole at rtp

So l mole                  $= \frac{1}{0.025} \times 599$

                     = 23 960 cm$^3$ of hydrogen

This is very close to the accepted value of 24 000 cm$^3$ of gas at rtp.

### A problem involving making hydrogen

What is the maximum mass of aluminium which you could add to an excess of dilute hydrochloric acid so that you produced no more than 100 cm$^3$ of hydrogen at room temperature and pressure?

($A_r$: Al = 27. Molar volume = 24 000 cm$^3$ at rtp)

What you are being asked is what mass of aluminium will give 100 cm$^3$ of hydrogen at rtp.

*First write the equation:*

$$2Al(s) + 6HCl(aq) \rightarrow 2AlCl_3(aq) + 3H_2(g)$$

*Interpret the equation in terms of moles:*

2 mol Al gives 3 mol H$_2$

*Substitute masses and volumes where appropriate:*

2 × 27 g Al gives 3 × 24 000 cm$^3$ H$_2$
54 g Al gives 72 000 cm$^3$ H$_2$

*Do the simple proportion sum:*

If        72 000 cm$^3$ H$_2$ comes from 54 g Al

then      1 cm$^3$ H$_2$ comes from $\dfrac{54}{72\,000}$ g Al          = 0.000 75 g

and       100 cm$^3$ H$_2$ comes from 100 × 0.000 75 g Al     = 0.075 g Al

To get 100 cm$^3$ of hydrogen, you would need 0.075 g of aluminium.

### An industrial example

Crude oil contains sulphur compounds as impurities. Low sulphur crude oils are more expensive to buy. When oil burns, sulphur dioxide is produced. To remove it from the waste gases from power stations, the sulphur dioxide is reacted with limestone (calcium carbonate) and air.

$$2CaCO_3(s) + 2SO_2(g) + O_2(g) \rightarrow 2CaSO_4(s) + 2CO_2(g)$$

The calcium sulphate produced can be used to make plaster board for building.

One large oil-fired power station uses 10 000 tonnes of crushed limestone every week. We are going to calculate what mass of calcium sulphate is produced and what volume of sulphur dioxide the limestone removes from the flue gases (assuming the sulphur dioxide is at rtp).

($A_r$s: C = 12; O = 16; S = 32; Ca = 40. Molar volume = 24 dm$^3$ at rtp)

*First write the equation:*

$$2CaCO_3(s) + 2SO_2(g) + O_2(g) \rightarrow 2CaSO_4(s) + 2CO_2(g)$$

*Interpret the equation in terms of moles:*

> 2 mol $CaCO_3$ reacts with 2 mol $SO_2$ and produces 2 mol $CaSO_4$

That's exactly the same as saying that:

> 1 mol $CaCO_3$ reacts with 1 mol $SO_2$ and produces 1 mol $CaSO_4$

*Substitute masses and volumes where appropriate:*

> 100 g $CaCO_3$ reacts with 24 dm$^3$ $SO_2$ and produces 136 g $CaSO_4$

*Do the simple proportion sum for the calcium sulphate:*

If      100 g $CaCO_3$ produces 136 g $CaSO_4$

then    100 tonnes $CaCO_3$ produces 136 tonnes $CaSO_4$

and     1 tonne $CaCO_3$ produces 1.36 tonnes $CaSO_4$

So      10 000 tonnes $CaCO_3$ produces 10 000 × 1.36 tonnes $CaSO_4$
        = 13 600 tonnes

10 000 tonnes of calcium carbonate produces 13 600 tonnes of $CaSO_4$.

*Do the simple proportion sum for the $SO_2$:*

If      100 g $CaCO_3$ reacts with 24 dm$^3$ $SO_2$

then    1 g $CaCO_3$ reacts with 0.24 dm$^3$ $SO_2$

and     10 000 000 000 g $CaCO_3$ reacts with 10 000 000 000 × 0.24 dm$^3$ $SO_2$
        = 2 400 000 000 dm$^3$ $SO_2$

10 000 tonnes of calcium carbonate reacts with 2 400 000 000 dm$^3$ $SO_2$.

10 000 tonnes of calcium carbonate reacts with $2.4 \times 10^9$ dm$^3$ $SO_2$.

---

Unfortunately, this time you can't work directly in tonnes – you have to work in grams. 1 tonne is a million grams. 10 000 tonnes is 10 000 000 000 grams.

It would be much more sensible to do this in scientific notation! 10 000 tonnes is $1 \times 10^{10}$ g.

The volume of $SO_2$ then turns out to be $2.4 \times 10^9$ dm$^3$.

---

> **7.** Why does burning coal release sulphur dioxide gas?
> **8.** How can you remove sulphur dioxide from waste gases?
> **9.** What is the volume of 10 moles of sulphur dioxide gas at rtp?

# End of Chapter Checklist

In this chapter you have learnt that:

● you can work out the numbers of moles of reactants and products from a balanced equation

● you can convert moles into masses and back again

● gas volumes can be measured in cubic decimetres or cubic centimetres or in millilitres or litres

● Avogadro's Law tells us that equal volumes of all gases contain equal numbers of molecules (at the same temperature and pressure)

● 1 mole of any gas occupies 24 000 cm³ at rtp or 22 400 cm³ at stp. We call this the molar volume of a gas.

# Questions

1. 1 mole of sulphur dioxide has a mass of:

   A   32 g
   B   96 g
   C   16 g
   D   64 g                                              (1)

2. Which of the following statements is/are true?

   I     molar volume of a gas is 24 l at rtp
   II    molar volume of a gas is 24 000 ml at rtp
   III   molar volume of a gas is 100 ml at rtp
   IV    molar volume of all gases is the same at rtp

   A   I, II, III and IV
   B   I and II
   C   I, II and IV
   D   III and IV                                         (1)

3. a)  Write the formula of calcium carbonate.           (1)

   b)  Calculate the $M_r$.                               (2)

   c)  Calculate the mass of 2 moles.                     (1)

   d)  Calculate the mass of carbon dioxide released by heating 2 moles of calcium carbonate.   (4)

   e)  Calculate the mass of calcium oxide produced in part d).                                  (2)

4. a)  Write a balanced symbol equation for burning graphite in excess air.                      (2)

   b)  Add state symbols.                                 (2)

   c)  Calculate the numbers of moles of each material in the equation.                          (2)

   d)  Calculate the mass of carbon dioxide produced by burning 12 g graphite.                   (4)

5. a)  A certain volume of oxygen contains X molecules. How many molecules are there under the same conditions in: *i)* the same volume of argon *ii)* double the volume of helium *iii)* ten times the volume of nitrogen?   (3)

   b)  State and explain Avogadro's Law.                 (2)

   c)  Candles release water vapour which can be cooled to water before measuring the volume. Does Avogadro's Law apply to this example? Explain your answer.   (5)

6. a)  Calculate the volumes of each of the following masses of carbon dioxide at rtp: *i)* 44 g *ii)* 11 g *iii)* 88 g *iv)* 4.4 g.    (4)

   b)  Calculate both the volume and the number of moles in 4 g hydrogen at rtp.                 (4)

   c)  Why do we bother to calculate volumes of gases at rtp?                                    (2)

7. a)  Plan an experiment to measure the change in mass on burning magnesium in air.            (6)

   b)  Draw up a blank results table.                    (2)

   c)  How could you work out the number of moles of magnesium oxide produced?                  (2)

8. 2.67 g of aluminium chloride was dissolved in water and an excess of silver nitrate solution was added to give a precipitate of silver chloride.

   $AlCl_3(aq) + 3AgNO_3(aq) \rightarrow Al(NO_3)_3(aq) + 3AgCl(s)$

   What mass of silver chloride precipitate would be formed?
   ($A_r$s: Al = 27; Cl = 35.5; Ag = 108)                 (5)

9. Calcium hydroxide is manufactured by heating calcium carbonate strongly to produce calcium oxide, and then adding a controlled amount of water to produce calcium hydroxide.

$$CaCO_3(s) \rightarrow CaO(s) + CO_2(g)$$

$$CaO(s) + H_2O(l) \rightarrow Ca(OH)_2(s)$$

a) What mass of calcium oxide would you produce from 1 tonne of calcium carbonate? (2)

b) What mass of water would you need to add to that calcium oxide? (2)

c) What mass of calcium hydroxide would you eventually produce? ($A_r$s: H = 1; C = 12; O = 16; Ca = 40) (2)

10. Copper(II) sulphate crystals, $CuSO_4 \cdot 5H_2O$, can be made by heating copper(II) oxide with dilute sulphuric acid and then crystallising the solution formed. Calculate the maximum mass of crystals that could be made from 4.00 g of copper(II) oxide using an excess of sulphuric acid.

$$CuO(s) + H_2SO_4(aq) \rightarrow CuSO_4(aq) + H_2O(l)$$

$$CuSO_4(aq) + 5H_2O(l) \rightarrow CuSO_4 \cdot 5H_2O(s)$$

($A_r$s: H = 1; O = 16; S = 32; Cu = 64) (6)

11. Chromium is manufactured by heating a mixture of chromium(III) oxide with aluminium powder.

$$Cr_2O_3(s) + 2Al(s) \rightarrow 2Cr(s) + Al_2O_3(s)$$

a) Calculate the mass of aluminium needed to react with 1 tonne of chromium(III) oxide. ($A_r$s: O = 16; Al = 27; Cr = 52) (2)

b) Calculate the mass of chromium produced from 1 tonne of chromium(III) oxide. (2)

12. If the mineral pyrite, $FeS_2$, is heated strongly in air, iron(III) oxide and sulphur dioxide are produced. What mass of a) iron(III) oxide, and b) sulphur dioxide could be made by heating 1 tonne of an ore which contained 50% by mass of pyrite? ($A_r$s: O = 16; S = 32; Fe = 56)

$$4FeS_2(s) + 11O_2(g) \rightarrow 2Fe_2O_3(s) + 8SO_2(g) \qquad (4)$$

13. Carbon monoxide burns according to the equation:

$$2CO(g) + O_2(g) \rightarrow 2CO_2(g)$$

a) Calculate the volume of oxygen needed for the complete combustion of 100 cm³ of carbon monoxide. (2)

b) What volume of carbon dioxide will be formed? (2)

14. What volume of air is needed for the complete combustion of 1 dm³ of propane?

$$C_3H_8(g) + 5O_2(g) \rightarrow 3CO_2(g) + 4H_2O(l) \qquad (4)$$

15. Take the molar volume to be 24 dm³ (24 000 cm³) at rtp.

a) Calculate the mass of 200 cm³ of chlorine gas ($Cl_2$) at rtp. ($A_r$: Cl = 35.5) (2)

b) Calculate the volume occupied by 0.16 g of oxygen ($O_2$) at rtp. ($A_r$: O = 16) (2)

c) If 1 dm³ of a gas at rtp weighs 1.42 g, calculate the mass of 1 mole of the gas. (2)

16. Calculate the volume of hydrogen (measured at room temperature and pressure) obtainable by reacting 0.240 g of magnesium with an excess of dilute sulphuric acid. ($A_r$: Mg = 24. Molar volume = 24 000 cm³ at rtp)

$$Mg(s) + H_2SO_4(aq) \rightarrow MgSO_4(aq) + H_2(g) \qquad (6)$$

# End of Section Questions

**1.** **a)** What do you understand by the term *relative atomic mass* of an element? *(2 marks)*

**b)** Show that the relative atomic mass of chlorine is 35.5, given that an average sample of chlorine contains 75% $^{35}Cl$ and 25% $^{37}Cl$. *(2 marks)*

**c)** Chlorine gas was bubbled through a solution containing 4.15 g of potassium iodide until no further reaction occurred. Calculate the mass of iodine produced by the reaction:

$$Cl_2(g) + 2KI(aq) \rightarrow 2KCl(aq) + I_2(s)$$

(Relative atomic masses: K = 39; I = 127)
*(4 marks)*

**d)** Calculate the density of chlorine gas at room temperature and pressure in $g\,dm^{-3}$. (Volume of 1 mole of a gas at room temperature and pressure = 24.0 $dm^3$) *(2 marks)*

***Total 10 marks***

**2.** In an experiment to find the empirical formula of some lead oxide, a small porcelain dish was weighed, filled with lead oxide and weighed again. The dish was placed in a tube, and was heated in a stream of hydrogen. The hydrogen reduced the lead oxide to a bead of metallic lead. When the apparatus was cool, the dish with its bead of lead was weighed.

| | |
|---|---|
| Mass of porcelain dish | = 17.95 g |
| Mass of porcelain dish + lead oxide | = 24.80 g |
| Mass of porcelain dish + lead | = 24.16 g |

(Relative atomic masses: O = 16; Pb = 207)

**a)** Calculate the mass of lead in the lead oxide. *(2 marks)*

**b)** Calculate the mass of oxygen in the lead oxide. *(2 marks)*

**c)** There are three different oxides of lead: PbO, $PbO_2$ and $Pb_3O_4$. Use your results from *a)* and *b)* to find the empirical formula of the oxide used in the experiment. *(3 marks)*

**d)** Calculate the percentage by mass of lead in the oxide $PbO_2$. *(3 marks)*

***Total 10 marks***

**3.** In an experiment to find the percentage of calcium carbonate in sand from a beach, 1.86 g of sand reacted with an excess of dilute hydrochloric acid to give 0.55 g of carbon dioxide.

$$CaCO_3(s) + 2HCl(aq) \rightarrow$$
$$CaCl_2(aq) + CO_2(g) + H_2O(l)$$

**a)** How many moles of each material are there in the equation? *(2 marks)*

**b)** Calculate the number of moles of carbon dioxide present in 0.55 g of $CO_2$. (Relative atomic masses: C = 12; O = 16) *(2 marks)*

**c)** How many moles of calcium carbonate must have been present in the sand to produce this amount of carbon dioxide? *(2 marks)*

**d)** Calculate the mass of calcium carbonate present in the sand. (Relative atomic masses: C = 12; O = 16; Ca = 40) *(2 marks)*

**e)** Calculate the percentage of calcium carbonate in the sand. *(2 marks)*

***Total 10 marks***

**4.** **a)** Chalcopyrite is a copper-containing mineral with a formula $CuFeS_2$.

**i)** Calculate the percentage by mass of copper in pure chalcopyrite. *(2 marks)*

(Relative atomic masses: S = 32; Fe = 56; Cu = 64)

**ii)** Analysis of a copper ore showed that it contained 50% chalcopyrite by mass. Assuming that all the copper can be extracted, what mass of copper could be obtained from 1 tonne (1000 kg) of the copper ore? *(3 marks)*

**b)** Copper reacts with concentrated nitric acid to give copper(II) nitrate solution and nitrogen dioxide gas.

$$Cu(s) + 4HNO_3(aq) \rightarrow$$
$$Cu(NO_3)_2(aq) + 2NO_2(g) + 2H_2O(l)$$

**i)** Calculate the maximum mass of copper(II) nitrate, $Cu(NO_3)_2$, which could be obtained from 8.00 g of copper. (Relative atomic masses: N = 14; O = 16; Cu = 64) *(3 marks)*

**ii)** Calculate the volume of nitrogen dioxide produced at room temperature and pressure using 8.00 g of copper. (Volume of 1 mole of a gas at room temperature and pressure = 24.0 $dm^3$) *(2 marks)*

***Total 10 marks***

**5.** If pyrite, $FeS_2$, is heated strongly in air it reacts according to the equation:

$$4FeS_2(s) + 11O_2(g) \rightarrow 2Fe_2O_3(s) + 8SO_2(g)$$

Iron can be extracted from the iron(III) oxide produced, and the sulphur dioxide can be converted into sulphuric acid.

**a)** Calculate the mass of iron(III) oxide which can be obtained from 480 kg of pure pyrite. (Relative atomic masses: O = 16; S = 32; Fe = 56) *(2 marks)*

**b)** What mass of iron could be obtained by the reduction of the iron(III) oxide formed from 480 kg of pyrite? *(2 marks)*

**c)** Calculate the volume of sulphur dioxide (measured at room temperature and pressure) produced from 480 kg of pyrite. (Volume of 1 mole of a gas at room temperature and pressure = 24.0 dm$^3$) *(4 marks)*

**d)** The next stage of the manufacture of sulphuric acid is to convert the sulphur dioxide into sulphur trioxide.

$$2SO_2(g) + O_2(g) \rightarrow 2SO_3(g)$$

Calculate the volume of oxygen (measured at room temperature and pressure) needed for the complete conversion of the sulphur dioxide produced in part *c)* into sulphur trioxide. *(2 marks)*

***Total 10 marks***

**6.** Calcium carbonate decomposes when heated strongly to give calcium oxide and carbon dioxide gas.

**a)** Write word and balanced symbol equations for this reaction. *(2 marks)*

**b)** Calculate the mass of 1 mole of each material in the equation.

(Relative atomic masses: Ca = 40, C = 12, O = 16.) *(3 marks)*

**c)** Calculate the volume of gas liberated at room temperature and pressure by heating 20 g of calcium carbonate. (Volume of 1 mole of a gas at room temperature and pressure = 24.0 dm$^3$) *(3 marks)*

***Total 8 marks***

# Section C: Salts, Reactivity and Titrations

## Chapter 9: The Reactivity Series

**When you have completed this chapter, you will be able to:**

- use the Reactivity Series of metals
- carry out experiments to investigate the Reactivity Series
- understand oxidation and reduction
- write electron half-equations
- understand displacement reactions
- describe how metals react with water
- describe how hydrogen can be a reducing agent
- make predictions based on the Reactivity Series.

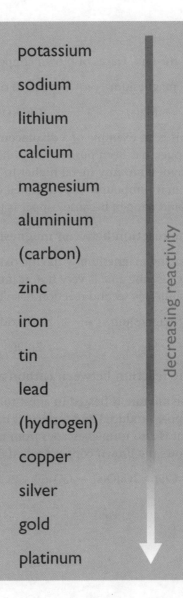

potassium

sodium

lithium

calcium

magnesium

aluminium

(carbon)

zinc

iron

tin

lead

(hydrogen)

copper

silver

gold

platinum

decreasing reactivity

The Reactivity Series lists elements (mainly metals) in order of decreasing reactivity.

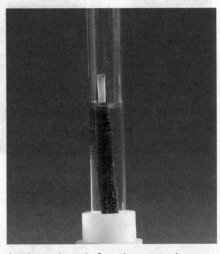

Lead crystals made from the reaction between lead(II) nitrate solution and zinc.

Copper is used in rum stills because it doesn't react with water or alcohol.

# Displacement reactions involving metal oxides

### The reaction between magnesium and copper(II) oxide

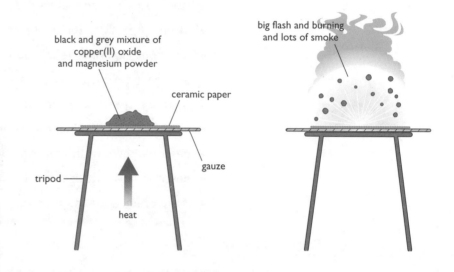

black and grey mixture of copper(II) oxide and magnesium powder

ceramic paper

gauze

tripod

heat

big flash and burning and lots of smoke

*The reaction between magnesium and zinc oxide.*

At the end, traces of brown copper are left on the ceramic paper.

magnesium + copper(II) oxide → magnesium oxide + copper

$$Mg(s) + CuO(s) \rightarrow MgO(s) + Cu(s)$$

This is an example of a **displacement reaction**. The less reactive metal, copper, has been pushed out of its compound by the more reactive magnesium. Any metal higher in the series will displace one lower down from a compound. If you heated copper with magnesium oxide, nothing would happen because copper is less reactive than magnesium.

### The reaction between magnesium and zinc oxide

Heating magnesium with zinc oxide produces zinc metal. This time, though, because the zinc is very hot, it immediately burns in air to form zinc oxide again! This second reaction hasn't been included in the equations below.

magnesium + zinc oxide → magnesium oxide + zinc

$$Mg(s) + ZnO(s) \rightarrow MgO(s) + Zn(s)$$

### The reaction between carbon and copper(II) oxide

The mixture is heated in a test tube to avoid the air getting at the hot copper produced and turning it back to copper(II) oxide. The carbon dioxide that is also formed escapes from the tube as a gas. The photograph clearly shows the brown copper formed from the black mixture.

$$C(s) + 2CuO(s) \rightarrow CO_2(g) + 2Cu(s)$$

*The reaction between carbon and copper(II) oxide. Notice the brown copper formed at the end of the reaction.*

## Activity 1: Extracting copper from its oxide

**Investigative skills**

| O | A |
|---|---|

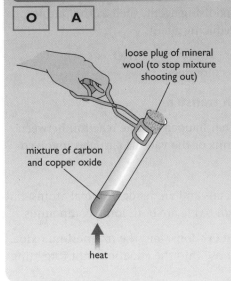

loose plug of mineral wool (to stop mixture shooting out)

mixture of carbon and copper oxide

heat

**You will need:**

- test-tube
- beaker
- copper oxide
- test-tube holder
- mineral wool
- carbon powder
- spatula

**Carry out the following:**

1. Mix one spatula full of each solid in a dry test-tube.
2. Plug the tube with mineral wool to stop the contents shooting out.
3. Heat gently, gradually increasing the temperature.
4. Look for signs of a red-brown powder (copper).
5. Leave to cool.
6. Pour the mixture into a beaker of water. Any copper will sink to the bottom.
7. Is carbon more or less reactive than copper? Explain your answer.

---

1. Name two metals that are more reactive than carbon.
2. What are the products when copper oxide is heated with magnesium metal?
3. Can you use copper to extract zinc from zinc oxide?

---

## Oxidation and reduction

### Oxidation and reduction – oxygen transfer

A substance has been **oxidised** if it gains oxygen. **Oxidation** is gain of oxygen.

A substance has been **reduced** if it loses oxygen. **Reduction** is loss of oxygen.

In the case of magnesium reacting with copper(II) oxide:

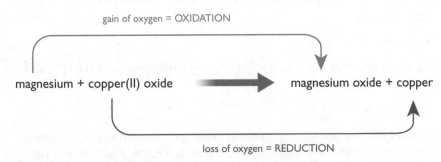

gain of oxygen = OXIDATION

magnesium + copper(II) oxide ⟶ magnesium oxide + copper

loss of oxygen = REDUCTION

A **redox** reaction is one in which both **red**uction and **ox**idation are occurring. Oxidation and reduction always go hand-in-hand.

A **reducing agent** is a substance which reduces something else. In this case, the magnesium is the reducing agent.

An **oxidising agent** is a substance which oxidises something else. The copper(II) oxide is the oxidising agent in this reaction.

Hydrogen peroxide is unusual. It can act as both an oxidising agent *and* as a reducing agent. For example, it can oxidise lead sulphide to lead sulphate.

$$PbS + 4H_2O_2 \rightarrow PbSO_4 + 4H_2O$$

But if it reacts with an even stronger oxidising agent, such as the purple manganate(VII) ion, then it acts as a reducing agent.

$$2MnO_4^- + 5H_2O_2 + 6H^+ \rightarrow 2Mn_2^+ + 8H_2O + 5O_2$$
$$\text{purple} \qquad\qquad \text{acidic} \qquad \text{colourless}$$

## Oxidation and reduction – electron transfer

We are going to look very closely at what happens in the reaction between magnesium and copper(II) oxide in terms of the various particles involved.

$$Mg(s) + CuO(s) \rightarrow MgO(s) + Cu(s)$$

The magnesium and the copper are metals and are made of metal atoms, but the copper(II) oxide and the magnesium oxide are both ionic compounds.

The copper(II) oxide contains $Cu^{2+}$ and $O^{2-}$ ions, and the magnesium oxide contains $Mg^{2+}$ and $O^{2-}$ ions. Writing those into the equation (not forgetting the state symbols) gives:

$$Mg(s) + Cu^{2+}(s) + O^{2-}(s) \rightarrow Mg^{2+}(s) + O^{2-}(s) + Cu(s)$$

Look very carefully at this equation to see what is being changed. Notice that something odd is going on – the oxide ion ($O^{2-}$) is completely unaffected by the reaction. It ends up with a different partner but is totally unchanged itself. An ion like this is described as a **spectator ion**.

You don't write the spectator ions into the equation because they aren't changed in the reaction. The equation showing just those things being changed looks like this:

$$Mg(s) + Cu^{2+}(s) \rightarrow Mg^{2+}(s) + Cu(s)$$

This is known as an **ionic equation** and shows the reaction in a quite different light. It shows that the reaction has nothing to do with the oxygen.

What is actually happening is that magnesium *atoms* are turning into magnesium *ions*. The magnesium atoms lose electrons to form magnesium ions.

$$Mg(s) \rightarrow Mg^{2+}(s) + 2e^-$$

Those electrons have been gained by the copper(II) ions to make the atoms present in metallic copper.

$$Cu^{2+}(s) + 2e^- \rightarrow Cu(s)$$

Remember that we are talking about the reaction between copper(II) oxide and magnesium. We've already described this as a redox reaction, but the equations no longer have any oxygen in them! We now need a wider definition of oxidation and reduction.

*How do you know whether a substance contains ions or not? As a rough guide, if it is a metal (or ammonium) compound, or an acid in solution, it will be ionic – otherwise it's not.*

These equations are called **half equations** or **electron half equations**. They show just part of a reaction from the point of view of one of the substances present.

**O**xidation **I**s **L**oss of electrons; **R**eduction **I**s **G**ain of electrons

OILRIG

In this case:

$$Mg(s) \rightarrow Mg^{2+}(s) + 2e^-$$ Mg is oxidised.

$$Cu^{2+}(s) + 2e^- \rightarrow Cu(s)$$ $Cu^{2+}$ is reduced.

> 4. Define reduction in terms of electrons.
> 5. If copper oxide is turned into copper metal, is this oxidation or reduction?
> 6. If zinc ions (charge 2+) turn into zinc metal, is this oxidation or reduction?

## Using hydrogen as a reducing agent

### The reduction of copper(II) oxide to copper

Copper won't react with water because copper is below hydrogen in the Reactivity Series, but that means that you *can* get a reaction between hydrogen and copper(II) oxide.

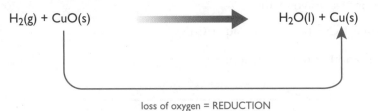

$$H_2(g) + CuO(s) \xrightarrow{\hspace{3cm}} H_2O(l) + Cu(s)$$

loss of oxygen = REDUCTION

The hydrogen removes the oxygen from the copper(II) oxide, so the hydrogen is a reducing agent.

In the experiment in the diagram, hydrogen is passed over hot copper(II) oxide. The oxide glows red hot and continues glowing, even if you remove the Bunsen burner. Lots of heat is released during the reaction.

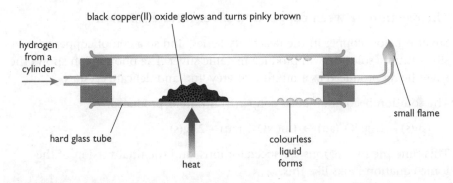

black copper(II) oxide glows and turns pinky brown

hydrogen from a cylinder

small flame

hard glass tube

colourless liquid forms

heat

> **Warning!** This is potentially a very dangerous experiment unless it is carried out by someone knowing exactly what they are doing. Things have to be done in a particular order to avoid explosion, and the hydrogen must be taken from a cylinder and not chemically produced.

The colourless liquid is, of course, water, and the pinky brown solid is copper. The small flame at the end of the apparatus is excess hydrogen being burnt off.

*Zinc displaces copper from copper(II) sulphate solution.*

It wouldn't matter which copper(II) salt you started with, as long as it was soluble in water. Copper(II) chloride or copper(II) nitrate would react in exactly the same way with zinc, because the chloride ions or the nitrate ions would once again be spectator ions, taking no part in the reaction.

*Displacing silver from silver nitrate solution.*

# Displacement reactions involving solutions of salts

Salts are compounds like copper(II) sulphate, silver nitrate, or sodium chloride. You will find a proper definition of what a salt is on page 93. This section explores some reactions between metals and solutions of salts in water.

## The reaction between zinc and copper(II) sulphate solution

The more reactive zinc displaces the less reactive copper. The blue colour of the copper(II) sulphate solution fades as colourless zinc sulphate solution is formed.

$$Zn(s) + CuSO_4(aq) \rightarrow ZnSO_4(aq) + Cu(s)$$

The zinc and the copper are simply metals consisting of atoms, but the copper(II) sulphate and the zinc sulphate are metal compounds and so are ionic.

If you rewrite the equation showing the ions, you will find that the sulphate ions are spectator ions.

$$Zn(s) + Cu^{2+}(aq) + SO_4^{2-}(aq) \rightarrow Zn^{2+}(aq) + SO_4^{2-}(aq) + Cu(s)$$

Removing the spectator ions (because they aren't changed during the reaction) leaves you with:

$$Zn(s) + Cu^{2+}(aq) \rightarrow Zn^{2+}(aq) + Cu(s)$$

This is another redox reaction:

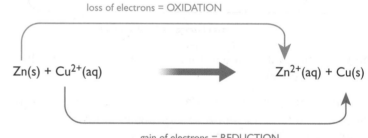

loss of electrons = OXIDATION

$$Zn(s) + Cu^{2+}(aq) \longrightarrow Zn^{2+}(aq) + Cu(s)$$

gain of electrons = REDUCTION

## The reaction between copper and silver nitrate solution

Silver is below copper in the Reactivity Series, and so a coil of copper wire in silver nitrate solution will produce metallic silver. The photograph shows the silver being produced as a mixture of grey 'fur' and delicate crystals.

The solution becomes blue as copper(II) nitrate is produced.

$$Cu(s) + 2AgNO_3(aq) \rightarrow Cu(NO_3)_2(aq) + 2Ag(s)$$

This time the nitrate ions are spectator ions, and the final version of the ionic equation looks like this:

$$Cu(s) + 2Ag^+(aq) \rightarrow Cu^{2+}(aq) + 2Ag(s)$$

This is another redox reaction:

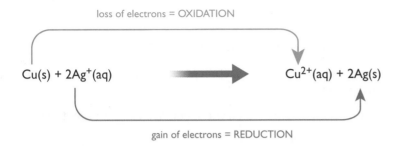

loss of electrons = OXIDATION

$$Cu(s) + 2Ag^+(aq) \longrightarrow Cu^{2+}(aq) + 2Ag(s)$$

gain of electrons = REDUCTION

## Activity 2: Using displacement to establish a reactivity series

**Investigative skills**

| A | E |

The rule is that a more reactive metal can displace a less reactive metal from a solution of its salts.

**You will need:**

- spotting tile (or small test-tubes)
- solutions as shown in the table below
- metal powders or granules: copper, lead, iron, zinc, magnesium

**Carry out the following:**

1. Set out the solutions on the tile or in separate tubes, following this pattern:

|  | Copper | Iron | Lead | Magnesium | Zinc |
|---|---|---|---|---|---|
| **Copper sulphate** | Do not try X |  |  |  |  |
| **Iron sulphate** |  | X |  |  |  |
| **Lead nitrate** |  |  | X |  |  |
| **Magnesium sulphate** |  |  |  | X |  |
| **Zinc sulphate** |  |  |  |  | X |

2. Add one metal to solutions of all the others, except where marked X.

3. Look for evidence of displacement such as colour changes.

4. Use your results to put these metals in order of their reactivity, most reactive metal first.

5. Write word equations for each successful reaction.

6. Explain why some combinations are marked X.

Metals at the top of the Reactivity Series cannot be tested like this since they react with water itself.

## The Reactivity Series and displacement

The rule for the displacement of metals from solution is very simple. Any metal can displace another that is lower down (less reactive). For example, this one will work:

magnesium + copper sulphate →
magnesium sulphate + copper

But this one won't work:

copper + magnesium sulphate →
copper sulphate + magnesium

# Reactions of metals with water

## A general summary

### Metals above hydrogen in the Reactivity Series

Metals above hydrogen in the Reactivity Series react with water (or steam) to produce hydrogen.

If the metal reacts with cold water, the metal *hydroxide* and hydrogen are formed.

> **metal + cold water → metal hydroxide + hydrogen**

If the metal reacts with steam, the metal *oxide* and hydrogen are formed.

> **metal + steam → metal oxide + hydrogen**

As you go down the Reactivity Series, the reactions become less and less vigorous.

### Metals below hydrogen in the Reactivity Series

Metals below hydrogen in the Reactivity Series (such as copper) don't react with water or steam. That is why copper can be used for both hot and cold water pipes.

## Potassium or sodium and cold water (demonstration only)

These reactions are very vigorous.

$$2K(s) + 2H_2O(l) \rightarrow 2KOH(aq) + H_2(g)$$

$$2Na(s) + 2H_2O(l) \rightarrow 2NaOH(aq) + H_2(g)$$

Lithium also reacts with cold water. It floats, fizzes and leaves an alkaline solution of lithium hydroxide. It is slower to react than sodium. See if you can place lithium, potassium and sodium in order of their reactivity.

# Activity 3: Reacting metals with cold water

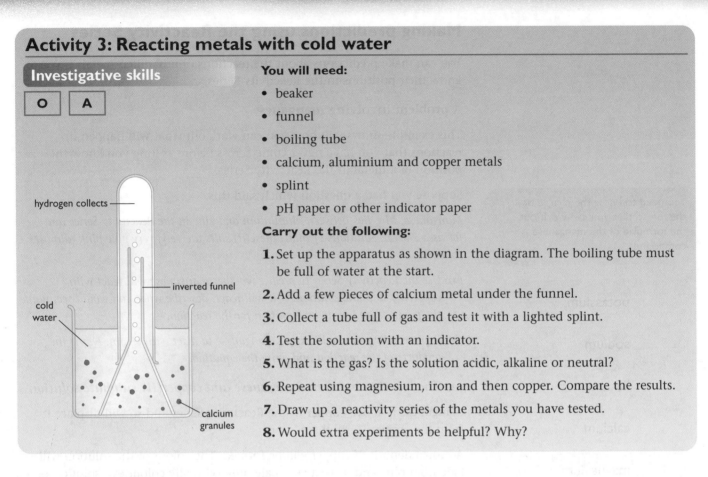

**You will need:**

- beaker
- funnel
- boiling tube
- calcium, aluminium and copper metals
- splint
- pH paper or other indicator paper

**Carry out the following:**

1. Set up the apparatus as shown in the diagram. The boiling tube must be full of water at the start.
2. Add a few pieces of calcium metal under the funnel.
3. Collect a tube full of gas and test it with a lighted splint.
4. Test the solution with an indicator.
5. What is the gas? Is the solution acidic, alkaline or neutral?
6. Repeat using magnesium, iron and then copper. Compare the results.
7. Draw up a reactivity series of the metals you have tested.
8. Would extra experiments be helpful? Why?

## Magnesium and cold water

There is almost no reaction. If the magnesium is very clean, a few bubbles of hydrogen form on it, but the reaction soon stops again. This is because the magnesium gets coated with insoluble magnesium hydroxide which prevents water coming into contact with the magnesium.

## Magnesium and steam

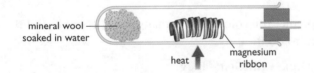

The mineral wool isn't heated directly. Enough heat spreads back along the test tube to vaporise the water.

The magnesium burns with a bright white flame in the steam, producing hydrogen which can be ignited at the end of the delivery tube. White magnesium oxide is formed.

$$Mg(s) + H_2O(g) \rightarrow MgO(s) + H_2(g)$$

In fact you also get a lot of black product in the tube. This is where the magnesium has reacted with the glass.

# Making predictions using the Reactivity Series

You can make predictions about the reactions of unfamiliar metals if you know their positions in the Reactivity Series.

### A problem involving manganese

This example shows you how you can work out what will happen in reactions that you have never come across before as long you know the position of a metal in the Reactivity Series.

Suppose you had a question which said this:

*Manganese, Mn, lies between aluminium and zinc in the Reactivity Series and forms a 2+ ion. Solutions of manganese(II) salts are very, very pale pink (almost colourless).*

*(a) Use the Reactivity Series to predict whether manganese will react with copper(II) sulphate solution. If it will react, describe what you would see, name the products and write an equation for the reaction.*

*(b) Explain why you would expect manganese to react with steam. Name the products of the reaction and write the equation.*

### (a) The reaction between manganese and copper(II) sulphate solution

Manganese is above copper in the Reactivity Series and so will displace it from the copper(II) sulphate.

A brown deposit of copper will be formed. The colour of the solution will fade from blue and leave a very pale pink (virtually colourless) solution of manganese(II) sulphate.

$$Mn(s) + CuSO_4(aq) \rightarrow MnSO_4(aq) + Cu(s)$$

### (b) The reaction between manganese and steam

Manganese is above hydrogen in the Reactivity Series and so reacts with steam to give hydrogen and the metal oxide – in this case, manganese(II) oxide.

You couldn't predict the colour of the manganese(II) oxide and the question doesn't ask you to do it.

$$Mn(s) + H_2O(g) \rightarrow MnO(s) + H_2(g)$$

> You need to know the charge on the ion so that you can work out the formulae of the manganese compounds.

potassium

sodium

lithium

calcium

magnesium

aluminium

manganese

zinc

iron

lead

(hydrogen)

copper

**7.** The table shows some data on the reactivity of metals.

| Number | Metal code | Cold water | Steam | Acid |
|---|---|---|---|---|
| I | Q | no change | no change | few bubbles |
| 2 | R | floats and fizzes | violent reaction | very violent reaction |
| 3 | T | no change | slow, forms oxide + hydrogen | bubbles |
| 4 | Z | no change | no change | no change |
| 5 | A | slow, few bubbles | burns to oxide + hydrogen | rapid, becomes hot, bubbles |

Use the data to put the metals in order of their reactivity.

**8.** Which metal might be suitable for jewellery? Why?

## Activity 4: Testing unknown metals

**Investigative skills**

Plan and design an experiment to investigate the reactions of an unknown metal with water and acids. What safety precautions would you take? How would you record your observations?

## Summary of the Reactivity Series

| Metal | Cold air | Cold water | Dilute acid (HCl) | Dilute acid (nitric) |
|---|---|---|---|---|
| K | burns easily | fast | violent | violent |
| Na | burns easily | fast | violent | violent |
| Li | burns easily | slower than sodium, still fast | violent | violent |
| Ca | burns easily | fast | violent | violent |
| Mg | burns easily | steam needed | fast | fast |
| Al | slow, heat needed | steam needed | reacts well | reacts well |
| Zn | slow, heat needed | steam needed | reacts well | reacts well |
| Fe | slow, heat needed | steam needed/reversible | reacts well | reacts well |
| Pb | slow, heat needed | no reaction | no reaction | slow reaction |
| Cu | slow, heat needed | no reaction | no reaction | slow reaction |
| Au | no reaction | no reaction | no reaction | no reaction |

# End of Chapter Checklist

**In this chapter you have learnt that:**

- metals can be compared and put into a Reactivity Series
- metals can be reacted with oxygen, water and acids to determine their reactivity
- oxidation occurs when metals combine with oxygen
- oxidation is loss of electrons
- reduction is the opposite of oxidation, i.e. gain of electrons
- more reactive metals can displace less reactive metals from their compounds
- some metals react with water or with steam, giving hydrogen
- hydrogen is a reducing agent and can reduce metal oxides to the metal
- we can use a Reactivity Series or data table to predict how metals will react.

# Questions

1. An example of a very reactive metal is:

   A  lead
   B  potassium
   C  iron
   D  copper (1)

2. When one metal takes the place of another it is called:

   A  precipitation
   B  displacement
   C  hydration
   D  dissolving (1)

3. Which of the following statements is/are true?

   I    oxidation is combining with oxygen
   II   reduction is removing oxygen
   III  oxidation is loss of electrons
   IV   oxidation is gain of electrons

   A  I and III
   B  I and II
   C  I, II, III and IV
   D  I, II and III (1)

4. An oxidising agent is itself:

   A  oxidised
   B  unchanged
   C  reduced
   D  exothermic (1)

5. Adding zinc to copper sulphate gives:

   A  no reaction
   B  copper and zinc
   C  copper and zinc sulphate
   D  no colour change (1)

6. Use this table of data to answer the questions which follow:

| Metal | Reaction with magnesium sulphate solution | Reaction with copper sulphate solution | Reaction with Q sulphate solution |
|-------|-------------------------------------------|----------------------------------------|-----------------------------------|
| Q | no change | brown solid, colourless solution | not tried |
| J | not tried | not tried | grey powder, colourless solution |

   a) What was the brown powder? Write a word equation for the reaction. (3)

   b) What can you conclude from the reaction of Q with the other two solutions? (3)

   c) What is the grey powder? (1)

   d) Which is the more reactive metal, Q or J? Explain your answer. (3)

7. a) Draw a diagram to show how hydrogen may be used as a reducing agent. (4)

   b) Write word and balanced symbol equations for the reaction. (2)

   c) What changes would you observe? (3)

   d) How could you use up any spare hydrogen safely? (1)

8.  a)  When a zinc strip is placed in lead nitrate solution, crystals grow. Explain why. (4)

    b)  What happens when a lead strip is placed in silver nitrate solution? (2)

    c)  Write word and balanced symbol equations for the reaction in part a). (2)

    d)  Why is the reactivity series of metals useful to us? (2)

9.  a)  List the following metals in order of decreasing reactivity: aluminium, copper, iron, sodium. (2)

    b)  Some magnesium powder was mixed with some copper(II) oxide and heated strongly. There was a vigorous reaction, producing lots of sparks and a bright flash of light.

        i)   Name the products of the reaction. (1)

        ii)  Write a balanced symbol equation for the reaction. (2)

        iii) Which substance in the reaction has been reduced? (1)

        iv)  Which substance is the oxidising agent? (1)

    c)  If a mixture of zinc powder and cobalt(II) oxide is heated, the following reaction occurs:

        $$Zn(s) + CoO(s) \rightarrow ZnO(s) + Co(s)$$

        i)   Which metal is higher in the Reactivity Series? (1)

        ii)  The zinc can be described as a reducing agent. Using this example, explain what is meant by the term *reducing agent*. (4)

        iii) Which substance in this reaction has been oxidised? (1)

    d)  Aluminium, chromium and manganese are all moderately reactive metals. (Care! We are talking about manganese, *not magnesium*.) Use the following information to arrange them in the correct Reactivity Series order, starting with the most reactive one.

        •  Chromium is manufactured by heating chromium(III) oxide with aluminium.

        •  If manganese is heated with aluminium oxide there is no reaction.

        •  If manganese is heated with chromium(III) oxide, chromium is produced. (6)

10. The equation for the reaction when solid magnesium and solid lead(II) oxide are heated together is:

    $$Mg(s) + PbO(s) \rightarrow MgO(s) + Pb(s)$$

    a)  Write down any *two* things that you might expect to see during this reaction. (2)

    b)  Rewrite the equation as an ionic equation. (2)

11. Some iron filings were shaken with some copper(II) sulphate solution. The ionic equation for the reaction is:

    $$Fe(s) + Cu^{2+}(aq) \rightarrow Fe^{2+}(aq) + Cu(s)$$

    a)  Write down any *one* change that you would observe during this reaction. (2)

    b)  Which substance has been oxidised in this reaction? (2)

    c)  Write down the full (not ionic) equation for this reaction. (2)

# Chapter 10: Acids, Bases and pH

Concentrated sulphuric acid can dissolve flesh and bones, so it's no wonder protective clothing has to be worn to clean up spills of the acid!

**When you have completed this chapter, you will be able to:**

- understand the pH scale and how pH meters work
- describe the colours of different indicators
- understand how acids react with metals, metal oxides, hydroxides and carbonates
- name salts
- write equations for the reactions of acids
- use the Bronsted–Lowry theory to describe acids and bases
- understand the difference between strong and weak acids and alkalis
- describe the uses of hydrogen carbonates
- describe how to make ammonia gas and test it.

But not all acids are dangerous – oranges contain citric acid.

## pH and indicators

### The pH scale

The pH scale ranges from about 0 to about 14 and tells you how acidic or how alkaline a solution is.

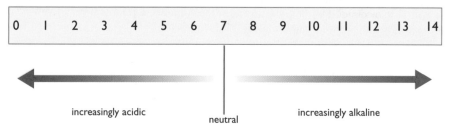

| 0 | 1 | 2 | 3 | 4 | 5 | 6 | 7 | 8 | 9 | 10 | 11 | 12 | 13 | 14 |

increasingly acidic                    neutral                    increasingly alkaline

### Measuring pH

#### Using universal indicator

Universal indicator is made from a mixture of dyes which change colour in a gradual way over a range of pHs. It can be used as a solution or as a paper. The commonest form is known as *wide range* universal indicator. It changes through a variety of colours from pH 1 right up to pH 14, but isn't very accurate.

#### Using a pH meter

You can measure pH much more accurately using a pH meter. Before you can use a pH meter, you have to adjust it to make sure that it is reading accurately. To do this you put it into a solution whose pH is known and adjust the reading so that it gives exactly that value.

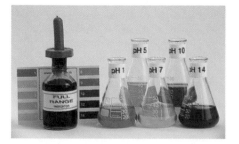

Using universal indicator solution to measure the pH of various solutions.

1. What is the pH scale?
2. What is the pH of the strongest acid?
3. What is meant by pH 7?

## Simple indicators

Any substance which has more than one colour form, depending on the pH, can be used as an indicator. One of the commonest ones is **litmus**.

| acidic | neutral | alkaline |
|---|---|---|

Litmus is red in acidic solutions and blue in alkaline ones. The neutral purple colour is an equal mixture of the red and blue forms.

*Measuring the pH of dilute sulphuric acid.*

| Indicator | Colour in acid | Colour in alkali |
|---|---|---|
| litmus | red | blue |
| methyl orange | red | yellow |
| phenolphthalein | colourless | red |
| universal | red | purple |

## Reacting acids with metals

Simple dilute acids react with metals depending on their positions in the Reactivity Series.

- Metals below hydrogen in the series don't react with dilute acids.

- Metals above hydrogen in the series react to produce hydrogen gas.

- The higher the metal is in the reactivity series, the more vigorous the reaction. You would never mix metals like sodium or potassium with acids, because their reactions are too violent.

**A summary equation for metals above hydrogen in the Reactivity Series**

> **metal + acid → salt + hydrogen**

## Salts

All simple acids contain hydrogen. When that hydrogen is replaced by a metal, the compound formed is called a **salt**. Magnesium sulphate is a salt, so is zinc chloride, and so is lead(II) nitrate.

One of the common acids in the lab, nitric acid, has much more complex reactions with metals.

Common salt is sodium chloride. This is produced if the hydrogen in hydrochloric acid is replaced by sodium.

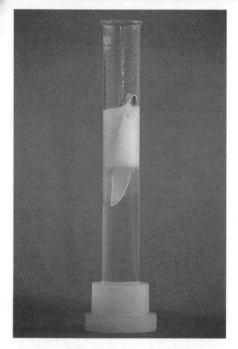

*Magnesium reacting with dilute sulphuric acid.*

Sulphuric acid can be thought of as the **parent acid** of all the sulphates.

| Parent acid | Salts |
|---|---|
| sulphuric acid | sulphates |
| hydrochloric acid | chlorides |
| nitric acid | nitrates |
| phosphoric acid | phosphates |
| ethanoic (acetic) acid | ethanoates (acetates) |
| carbonic acid | carbonates |

It doesn't matter if the replacement can't be done directly. For example, you can't make copper(II) sulphate from copper and dilute sulphuric acid because they don't react. There are, however, other ways of making it from sulphuric acid. Copper(II) sulphate is still a salt.

### Reactions involving magnesium and acids

#### With dilute sulphuric acid

There is rapid fizzing and a colourless gas is evolved which pops with a lighted splint (the test for hydrogen). The reaction mixture gets very warm as heat is produced. The magnesium gradually disappears to leave a colourless solution of magnesium sulphate.

$$Mg(s) + H_2SO_4(aq) \rightarrow MgSO_4(aq) + H_2(g)$$

This is a displacement reaction. The more reactive magnesium has displaced the less reactive hydrogen.

#### With dilute hydrochloric acid

The reaction looks exactly the same. The only difference is that this time a solution of magnesium chloride is formed.

$$Mg(s) + 2HCl(aq) \rightarrow MgCl_2(aq) + H_2(g)$$

#### Why are the reactions so similar?

Acids in solution are ionic. Dilute sulphuric acid contains hydrogen ions and sulphate ions. Dilute hydrochloric acid contains hydrogen ions and chloride ions.

You can rewrite the equations as ionic equations. In the sulphuric acid case:

$$Mg(s) + 2H^+(aq) + SO_4^{2-}(aq) \rightarrow Mg^{2+}(aq) + SO_4^{2-}(aq) + H_2(g)$$

You can see that the sulphate ion hasn't been changed by the reaction. It is a spectator ion, and so we leave it out of the ionic equation:

$$Mg(s) + 2H^+(aq) \rightarrow Mg^{2+}(aq) + H_2(g)$$

Repeating this with hydrochloric acid, you find that the chloride ions are also spectator ions.

$$Mg(s) + 2H^+(aq) + 2Cl^-(aq) \rightarrow Mg^{2+}(aq) + 2Cl^-(aq) + H_2(g)$$

Leaving the spectator ions out produces the ionic equation:

$$Mg(s) + 2H^+(aq) \rightarrow Mg^{2+}(aq) + H_2(g)$$

The reactions look the same because they are the same. All acids in solution contain hydrogen ions. That means that magnesium will react with any simple dilute acid in the same way.

## Reactions involving zinc and acids

Again the reactions between zinc and the two acids look exactly the same. The reactions are slower because zinc is lower down the Reactivity Series than magnesium. The reaction can be speeded up if it is heated or if the zinc is impure. A little copper(II) sulphate solution is often added to these reactions to make the zinc impure.

The full equations are:

$$Zn(s) + H_2SO_4(aq) \rightarrow ZnSO_4(aq) + H_2(g)$$

$$Zn(s) + 2HCl(aq) \rightarrow ZnCl_2(aq) + H_2(g)$$

The ionic equations turn out to be identical because the sulphate ions and chloride ions are spectator ions:

$$Zn(s) + 2H^+(aq) \rightarrow Zn^{2+}(aq) + H_2(g)$$

*Making hydrogen from the reaction between zinc and dilute sulphuric acid.*

> 4. From which acid do we obtain sulphates?
> 5. What do you see when magnesium reacts with an acid?
> 6. How can you speed up the reaction between zinc and an acid?

# Reacting acids with metal oxides

The metal magnesium reacts with dilute sulphuric acid; the metal copper doesn't. However, both magnesium oxide and copper(II) oxide react similarly with acids. All the metal oxide and acid combinations that you will meet behave in exactly the same way.

### A summary equation for acids and metal oxides

**metal oxide + acid → salt + water**

### Reacting dilute sulphuric acid with copper(II) oxide

The black powder reacts with hot dilute sulphuric acid to produce a blue solution of copper(II) sulphate.

$$CuO(s) + H_2SO_4(aq) \rightarrow CuSO_4(aq) + H_2O(l)$$

Most metal oxide and acid combinations need to be heated to get the reaction started.

### The ionic equation for an acid/metal oxide reaction

In the reaction between copper(II) oxide and dilute sulphuric acid, everything in the equation is ionic apart from the water.

$$Cu^{2+}(s) + O^{2-}(s) + 2H^+(aq) + SO_4^{2-}(aq) \rightarrow Cu^{2+}(aq) + SO_4^{2-}(aq) + H_2O(l)$$

Look carefully to find the spectator ions. The sulphate ion isn't changed at all, and the $Cu^{2+}$ ion has only changed to the extent that it started as solid

*Copper(II) oxide reacting with hot dilute sulphuric acid.*

> Remember that acids in solution and metal (and ammonium) compounds are ionic.

and ends up in solution. In this particular reaction we count that as unchanged. Leaving the spectator ions out gives:

$$O^{2-}(s) + 2H^+(aq) \rightarrow H_2O(l)$$

This would be equally true of any simple metal oxide reacting with any acid. Oxide ions combine with hydrogen ions to make water. This is a good example of a **neutralisation reaction**. The presence of the hydrogen ions is what makes the sulphuric acid acidic. If something combines with these and removes them from solution, then obviously the acid has been neutralised.

### Bases

Bases are defined as substances which combine with hydrogen ions. In the ionic equation above, an oxide ion is acting as a base because it combines with hydrogen ions to make water. The simple metal oxides you will meet are described as **basic oxides**.

Some bases can dissolve in water, for example sodium hydroxide. We give these solutions a special name – **alkalis**. They have a pH greater than 7. Sodium hydroxide is a base and an alkali since it is soluble in water. All alkalis are bases, but only some bases are alkalis.

> The bases you will meet include: metal oxides (because they contain oxide ions), metal hydroxides (containing hydroxide ions), metal carbonates (containing carbonate ions) and ammonia. All of these have the ability to combine with hydrogen ions.

## Reacting acids with metal hydroxides

All metal hydroxides react with acids, but the ones most commonly used in the lab are the soluble hydroxides – usually sodium, potassium or calcium hydroxide solutions.

### A summary equation for acids and metal hydroxides

**metal hydroxide + acid → salt + water**

### Reacting dilute hydrochloric acid with sodium hydroxide solution

Mixing sodium hydroxide solution and dilute hydrochloric acid produces a colourless solution – not much seems to have happened. But if you repeat the reaction with a thermometer in the tube, the temperature rises several degrees, showing that there has been a chemical change. Sodium chloride solution has been formed.

$$NaOH(aq) + HCl(aq) \rightarrow NaCl(aq) + H_2O(l)$$

The ionic equation for this shows that the underlying reaction is between hydroxide ions and hydrogen ions in solution to produce water.

$$OH^-(aq) + H^+(aq) \rightarrow H_2O(l)$$

This is another good example of a neutralisation reaction. The hydroxide ion is a base because it combines with hydrogen ions. Sodium hydroxide is a soluble base.

### Following the course of a neutralisation reaction

If everything involved in a neutralisation reaction is a colourless solution, how can you tell when exactly enough acid has been added to an alkali to produce a neutral solution?

burette containing dilute hydrochloric acid

conical flask containing sodium hydroxide solution + indicator

### Using an indicator

Some indicators will change colour when you have added even one drop too much acid. You normally avoid litmus because its colour change isn't very sharp and distinct. A common alternative is **methyl orange**.

Methyl orange is yellow in alkaline solutions and red in acids. You run acid in from the burette, swirling the flask all the time. The alkali is neutralised when the solution shows the first trace of orange. If it turns red, you have added too much acid.

### Using a pH meter

Instead of using an indicator, you could use a pH meter.

The sodium hydroxide solution would be put in a beaker rather than a flask so that there is room to fit a pH meter in as well. The acid would be added a little at a time and the pH recorded after each addition. The mixture would have to be well stirred. The results can be drawn on a graph.

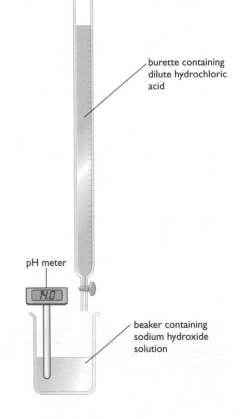

burette containing dilute hydrochloric acid

pH meter

beaker containing sodium hydroxide solution

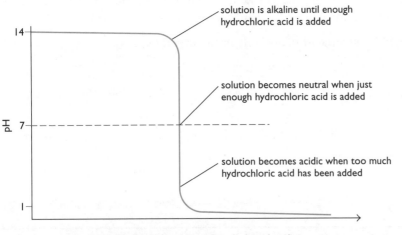

solution is alkaline until enough hydrochloric acid is added

solution becomes neutral when just enough hydrochloric acid is added

solution becomes acidic when too much hydrochloric acid has been added

pH

volume of acid added to the sodium hydroxide solution

Notice how fast the pH changes around the neutral point.

## Reacting acids with carbonates

Carbonates react with cold acids to produce carbon dioxide gas.

### A summary equation for acids and carbonates

**carbonate + acid → salt + carbon dioxide + water**

### Reacting dilute sulphuric acid with copper(II) carbonate

A colourless gas is given off which turns lime water milky. This is the test for carbon dioxide. The green copper(II) carbonate reacts to give a blue solution of copper(II) sulphate.

$$CuCO_3(s) + H_2SO_4(aq) \rightarrow CuSO_4(aq) + CO_2(g) + H_2O(l)$$

Occasionally, you might come across an acid reacting with a carbonate solution. In that case, replace (s) by (aq).

**The ionic equation for carbonate/acid reactions**

$$CO_3^{2-}(s) + 2H^+(aq) \rightarrow CO_2(g) + H_2O(l)$$

Because carbonate ions are combining with hydrogen ions, carbonate ions are bases.

## Activity 1: Reactions of hydrogencarbonates

### Investigative skills

O  A

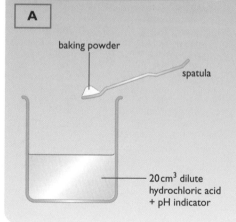

stirring rod

dilute hydrochloric acid

beaker

bubbles

antacid tablet

**You will need:**

- a beaker and stirring rod
- dilute hydrochloric acid
- universal (pH) indicator
- antacid tablets

**Carry out the following:**

1. Pour 20 cm³ dilute hydrochloric acid into a beaker and add three drops of indicator. This is the acid that we have in our stomachs.

2. Add one antacid tablet (seltzer tablet) and stir.

3. Observe the colour changes.

4. Add a second tablet or more until the colour shows the solution to be neutral.

5. Explain why antacid tablets are useful if your stomach is upset by being too acidic.

## Activity 2: Baking powder

### Investigative skills

A

baking powder

spatula

20 cm³ dilute hydrochloric acid + pH indicator

**You will need:**

- a beaker and spatula
- baking powder
- dilute hydrochloric acid
- universal (pH) indicator

**Carry out the following:**

1. Pour 20 cm³ acid into a beaker and add three drops of indicator.

2. Add one spatula full of baking powder.

3. What do you notice? Is there any evidence of a gas being given off? Does the indicator change colour?

4. Can you identify the gas?

5. Why do you think baking powder is used in cooking? (*Hint*: think about the difference in texture between biscuits and bread.)

## Ammonium salts

Ammonia gas, $NH_3$, dissolves easily in water to give ammonia solution. It is a weak base. The solution reacts with acids to give salts such as ammonium chloride or ammonium sulphate. There is a general reaction for ammonium salts:

**any ammonium salt + base → ammonia gas**

Ammonia gas is colourless and has a strong, unpleasant smell. It is the only common alkaline gas. It changes the colour of indicators. For example, moist red litmus paper turns blue in the presence of ammonia.

## Activity 3: Reaction of ammonium salts with bases

**Investigative skills**

O    A

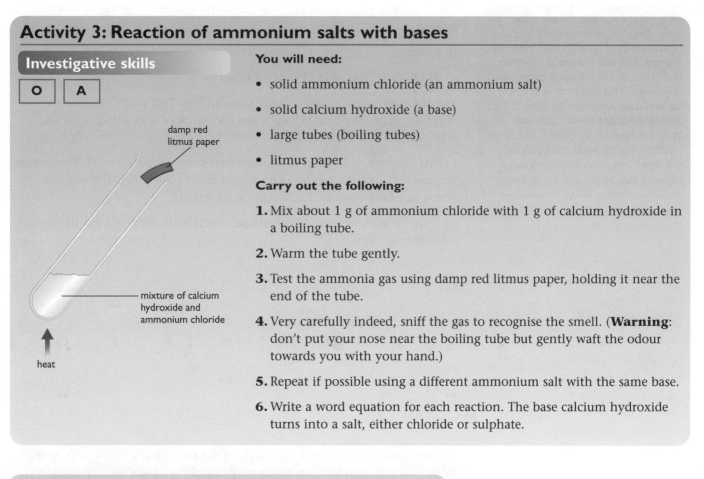

damp red litmus paper

mixture of calcium hydroxide and ammonium chloride

heat

**You will need:**

- solid ammonium chloride (an ammonium salt)
- solid calcium hydroxide (a base)
- large tubes (boiling tubes)
- litmus paper

**Carry out the following:**

1. Mix about 1 g of ammonium chloride with 1 g of calcium hydroxide in a boiling tube.

2. Warm the tube gently.

3. Test the ammonia gas using damp red litmus paper, holding it near the end of the tube.

4. Very carefully indeed, sniff the gas to recognise the smell. (**Warning**: don't put your nose near the boiling tube but gently waft the odour towards you with your hand.)

5. Repeat if possible using a different ammonium salt with the same base.

6. Write a word equation for each reaction. The base calcium hydroxide turns into a salt, either chloride or sulphate.

7. Complete the reaction: carbonate + acid → ?
8. Give two uses of bicarbonates.
9. What is the test for ammonia gas?

## The Bronsted–Lowry theory

Bronsted was a Danish chemist; Lowry an English one. They defined acids and bases as follows:

- An **acid** is a proton (hydrogen ion) donor.

- A **base** is a proton (hydrogen ion) acceptor.

A hydrogen ion is just a proton – the hydrogen nucleus minus its electron.

In this theory, when hydrogen chloride dissolves in water to give hydrochloric acid, there is a transfer of a proton from the HCl to the water.

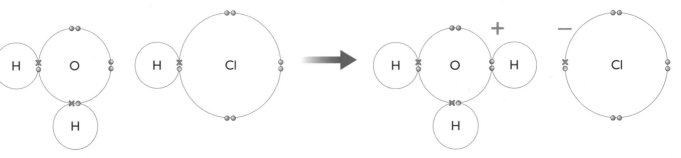

Only the outer electrons are shown in these diagrams to avoid confusion. Notice the new bond formed between the hydrogen and the water molecule. Both electrons in the bond come from the oxygen. This is described as a **co-ordinate covalent bond** or a **dative covalent bond**. Once the bond has been made, there is absolutely no difference between this bond and the other two normal covalent bonds between the oxygen and the hydrogens.

The hydrogen nucleus breaks away from the chlorine, leaving its electron behind.

In symbols:

$$H_2O(l) + HCl(g) \rightarrow H_3O^+(aq) + Cl^-(aq)$$

The $H_3O^+(aq)$ ion is called a **hydroxonium ion**. This is the ion that we normally write simply as $H^+(aq)$ – you can think of it as a hydrogen ion riding on a water molecule.

In this example, according to the Bronsted–Lowry theory, the HCl is an acid because it is giving a proton (a hydrogen ion) to the water. The water is acting as a base because it is accepting the proton.

In a similar way, hydrogen chloride gas reacts with ammonia gas to produce ammonium chloride:

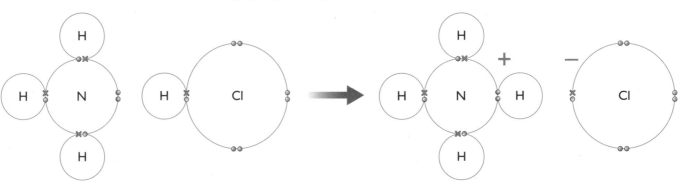

The ammonia acts as a base by accepting the proton; the HCl acts as an acid by donating it. This time an ammonium ion, $NH_4^+$, is formed. Notice the co-ordinate bond that is formed between the nitrogen and the new hydrogen.

$$NH_3(g) + HCl(g) \rightarrow NH_4^+(s) + Cl^-(s)$$

### Acids in solution

Acids in solution are acidic because of the presence of the hydroxonium ion. We would normally write a neutralisation reaction between an acid and a hydroxide, for example, as:

$$H^+(aq) + OH^-(aq) \rightarrow H_2O(l)$$

What actually happens is that the hydroxonium ion donates a proton to the base, $OH^-$.

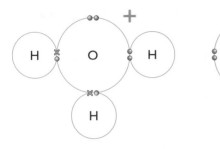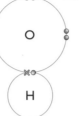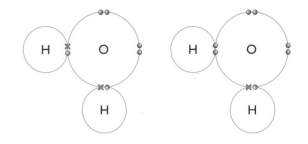

$$H_3O^+(aq) + OH^-(aq) \rightarrow 2H_2O(l)$$

We normally use the simplified version:

$$H^+(aq) + OH^-(aq) \rightarrow H_2O(l)$$

## Strong and weak acids and alkalis

### Strong acids

Hydrochloric acid is a strong acid. This means that when hydrogen chloride gas comes into contact with water, it reacts completely to give hydroxonium ions and chloride ions.

$$H_2O(l) + HCl(g) \rightarrow H_3O^+(aq) + Cl^-(aq)$$

Sulphuric acid and nitric acid are also strong acids. There are no un-ionised acid molecules left in their solutions. An acid which is 100% ionised in solution is called a **strong acid**.

### Weak acids

An acid which is only partially ionised in solution is described as a **weak acid**. Ethanoic acid (in vinegar), citric acid (in oranges and lemons) or carbonic acid (formed when carbon dioxide dissolves is water) are all examples of weak acids.

In dilute ethanoic acid only about 1% of the ethanoic acid molecules have actually formed ions at any one time. As fast as the molecules react with water to produce ions, the ions react back again to give the original molecules. You have a reversible reaction.

$$CH_3COOH(aq) + H_2O(l) \rightleftharpoons CH_3COO^-(aq) + H_3O^+(aq)$$

### Recognising strong and weak acids

If you had a strong and a weak acid of *equal concentration*, the weak acid will have a lower concentration of hydrogen ions in solution. That means:

- Its pH won't be so low. The greater the concentration of hydrogen ions in solution, the lower the pH. Strong acids have lower pHs than weak acids of the same concentration.

- The reactions will be slower. For example, weak acids react with metals or carbonates more slowly than strong acids of the same concentration do.

Be careful to distinguish between the words *strong* and *weak* as opposed to the words *concentrated* and *dilute*.

> If you are meeting this for the first time, don't worry too much about the formula for ethanoic acid for now.

> For simplicity, we normally talk about 'hydrogen ions in solution', even if we should really be calling them hydroxonium ions.

- *Concentrated* and *dilute* tell you about the amount of acid which has gone into the solution.

- *Strong* and *weak* tell you about the proportion of the acid which has reacted with the water to form ions.

lots of acid molecules – but hardly any have formed ions

small amounts of acid – but all of it has formed ions

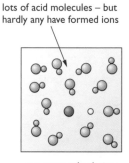

concentrated solution
of a weak acid

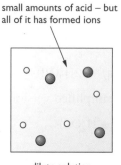

dilute solution
of a strong acid

### Strong alkalis

> Remember that an alkali is a soluble base – a soluble substance which will combine with hydrogen ions.

A strong alkali is one which is fully ionised in solution. Examples of strong alkalis include sodium and potassium hydroxide. These are ionic compounds containing $Na^+$ or $K^+$ ions along with $OH^-$ ions. These ions are present in the solid, but dissolving them in water makes them free to move around. The presence of lots of hydroxide ions makes the solution strongly alkaline with a very high pH.

### Weak alkalis

The simplest example of a weak alkali is ammonia, $NH_3$ in solution. When ammonia dissolves in water, a very small proportion of it (typically about 1%) reacts.

$$NH_3(g) + H_2O(l) \rightleftharpoons NH_4^+(aq) + OH^-(aq)$$

The hydroxide ions tend to recapture the $H^+$ from the ammonium ions, $NH_4^+$, as fast as they are formed.

The relatively small numbers of hydroxide ions in solution (compared with sodium hydroxide solution of the same concentration) gives the ammonia solution a pH of around 11.

# End of Chapter Checklist

**In this chapter you have learnt that:**

- the pH scale measures the strengths of acids and alkalis

- different indicators give different colours with acids, alkalis and neutral solutions

- acids react with metals to give hydrogen, with metal oxides or hydroxides to give salts and with carbonates to give salts and carbon dioxide gas

- one way to understand acids and alkalis is described by the Bronsted–Lowry theory

- strong acids and strong alkalis are completely ionised, but weak ones are not

- hydrogencarbonates are used as antacids in baking

- ammonia is the only common alkaline gas.

# Questions

1. Which of the following statements is/are true?

   I    metal oxides are bases
   II   metals hydroxides are bases
   III  soluble bases are alkalis
   IV   alkalis have a pH less than 7

   A    I
   B    I, II and III
   C    IV
   D    III and IV                                          (1)

2. The only common alkaline gas is called:

   A    ammonia
   B    hydrogen
   C    oxygen
   D    carbon dioxide                                      (1)

3. a) Write the word equation for the general reaction for metals reacting with acids.          (2)

   b) Write word and balanced symbol equations for the reaction of magnesium with sulphuric acid.   (4)

   c) What is the test for hydrogen?                        (2)

   d) How could you obtain crystals from the reaction in part b)?                                  (2)

4. a) How could you collect a sample of the gas from the reaction of zinc with sulphuric acid?      (3)

   b) Draw a labelled diagram of the apparatus you would use.                                    (4)

   c) Is hydrogen soluble in water? Explain your answer.   (2)

   d) What is produced when hydrogen burns?                (1)

5. a) What do you see when carbonates react with acids?                                          (2)

   b) How could you identify the gas produced? Give the result of the test you would use.          (1)

   c) Write a word equation for the reaction of calcium carbonate with hydrochloric acid.          (2)

   d) Calcium sulphate is insoluble. What would you expect to happen if you reacted calcium carbonate with sulphuric acid? Is this a good method of producing calcium sulphate?                            (4)

   e) Name a hydrogencarbonate used in cooking.            (1)

6. Which of the following equations represent reactions between acids and bases? For each acid–base reaction, state which substance is the acid and which the base.

   i)    $MgO(s) + H_2SO_4(aq) \rightarrow MgSO_4(aq) + H_2O(l)$    (2)

   ii)   $CO_3^{2-}(s) + 2H^+(aq) \rightarrow CO_2(g) + H_2O(l)$    (2)

   iii)  $2Al(s) + 6HCl(aq) \rightarrow 2AlCl_3(aq) + 3H_2(g)$     (2)

   iv)   $H_2O(l) + HCl(g) \rightarrow H_3O^+(aq) + Cl^-(aq)$      (2)

   v)    $Zn(s) + Cu^{2+}(aq) \rightarrow Zn^{2+}(aq) + Cu(s)$     (2)

   vi)   $NH_3(g) + HCl(g) \rightarrow NH_4^+(s) + Cl^-(s)$        (2)

   vii)  $NH_3(g) + H_2O(l) \rightleftharpoons NH_4^+(aq) + OH^-(aq)$   (2)

   viii) $NaOH(aq) + HCl(aq) \rightarrow NaCl(aq) + H_2O(l)$       (2)

# Chapter 11: Making Salts

> **When you have completed this chapter, you will be able to:**
> - name different salts
> - work out whether a salt is soluble or not
> - make a range of salts
> - explain the meaning of the words 'anhydrous' and 'hydrated'
> - use titration to make a salt
> - use precipitation to prepare insoluble salts
> - make salts by direct combination of elements.

## Soluble and insoluble salts

### The importance of knowing whether a salt is soluble or insoluble in water

Remember that acids react with carbonates to give a salt, carbon dioxide and water. In the case of calcium carbonate (for example, marble chips) reacting with dilute hydrochloric acid, calcium chloride solution is produced.

$$CaCO_3(s) + 2HCl(aq) \rightarrow CaCl_2(aq) + CO_2(g) + H_2O(l)$$

If you try the reaction between calcium carbonate and dilute sulphuric acid, nothing much seems to happen if you use large marble chips. You will get a few bubbles when you first add the acid, but the reaction soon stops.

Calcium carbonate reacting with dilute hydrochloric acid.

Calcium carbonate not reacting with dilute sulphuric acid.

The problem is that the calcium sulphate produced in the reaction is almost insoluble in water. As soon as the reaction starts, a layer of calcium sulphate is formed around the calcium carbonate, stopping any further reaction.

layer of insoluble calcium sulphate

acid particles can't get at the calcium carbonate

original calcium carbonate

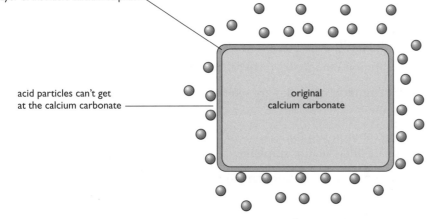

How quickly the reaction stops may well depend on the size of the marble chips (because that affects the surface area), the concentration of the acid, the volume of acid added, and the amount the flask is shaken. This would make an interesting investigation — particularly if you could then explain any pattern in your results.

Any attempt to produce an insoluble salt from the reaction between a solid and a liquid will fail for this reason.

## Solubility patterns

| | nitrate | chloride | sulphate | ethanoate | carbonate | hydroxide |
|---|---|---|---|---|---|---|
| ammonium | soluble | soluble | soluble | soluble | soluble | soluble |
| potassium | soluble | soluble | soluble | soluble | soluble | soluble |
| sodium | soluble | soluble | soluble | soluble | soluble | soluble |
| barium | soluble | soluble | insoluble | soluble | insoluble | almost insoluble |
| calcium | soluble | soluble | almost insoluble | soluble | insoluble | almost insoluble |
| magnesium | soluble | soluble | soluble | soluble | insoluble | insoluble |
| aluminium | soluble | soluble | soluble | soluble | insoluble | insoluble |
| zinc | soluble | soluble | soluble | soluble | insoluble | insoluble |
| iron | soluble | soluble | soluble | soluble | insoluble | insoluble |
| lead | soluble | insoluble | insoluble | soluble | insoluble | insoluble |
| copper | soluble | soluble | soluble | soluble | insoluble | insoluble |
| silver | soluble | insoluble | soluble | soluble | insoluble | insoluble |

key:

soluble    insoluble    almost insoluble (slightly soluble)

To keep the table simple, it includes one or two compounds (like aluminium carbonate, for example) which don't actually exist. Don't worry about these.

Hydroxides have been included for the sake of completeness although they are bases and not salts.

The list is in Reactivity Series order, apart from the ammonium group. Ammonium compounds often have similarities with sodium and potassium compounds, and so are included near them.

Ethanoates are salts made from ethanoic acid.

There is no clear cut-off between 'insoluble' and 'almost insoluble' compounds. The ones picked out as 'almost insoluble' include the more common ones that you will need to know about elsewhere in the course.

Note that:

- All sodium, potassium and ammonium compounds are *soluble*.

- All nitrates are *soluble*.

- All common ethanoates (also called acetates) are *soluble*.

- Most common chlorides are *soluble*, except lead(II) chloride and silver chloride.

- Most common sulphates are *soluble*, except lead(II) sulphate, barium sulphate and calcium sulphate.

- Most common carbonates are *insoluble*, except sodium, potassium and ammonium carbonates.

- Most metal hydroxides are *insoluble* (or *almost insoluble*), except sodium, potassium and ammonium hydroxides.

1. Which of these salts are insoluble?
     lead nitrate    barium sulphate    silver chloride
2. Are all sodium and potassium salts soluble or insoluble?
3. Which part of an acid is replaced when making a salt?

## Making soluble salts (except sodium, potassium and ammonium salts)

These all involve reacting a solid with an acid. You can use any of the following mixtures:

- **acid + metal** (but only for the moderately reactive metals from magnesium to iron in the reactivity series)

- **acid + metal oxide** or **hydroxide**

- **acid + carbonate**

Whatever mixture you use, the method is essentially the same.

*Dilute sulphuric acid with excess magnesium.*

### Why not just evaporate the solution to dryness? Water of crystallisation

It would seem much easier to just boil off all the water rather than crystallising the solution slowly, but evaporating to dryness wouldn't give you magnesium sulphate crystals. Instead, you would produce a white powder of **anhydrous** magnesium sulphate.

'Anhydrous' means 'without water'. When many salts form their crystals, water from the solution becomes chemically bound up with the salt. This is called **water of crystallisation**. A salt which contains water of crystallisation is said to be **hydrated**.

$$MgSO_4(aq) + 7H_2O(l) \rightarrow MgSO_4 \cdot 7H_2O(s)$$

The extra water in the equation comes from the water in the solution.

# Activity 1: Making magnesium sulphate crystals

Magnesium reacts with dilute sulphuric acid to form the salt magnesium sulphate.

$$Mg(s) + H_2SO_4(aq) \rightarrow MgSO_4(aq) + H_2(g)$$

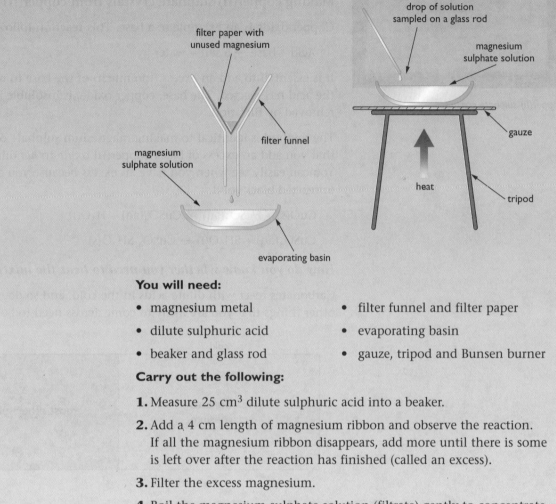

## You will need:

- magnesium metal
- dilute sulphuric acid
- beaker and glass rod
- filter funnel and filter paper
- evaporating basin
- gauze, tripod and Bunsen burner

## Carry out the following:

1. Measure 25 cm$^3$ dilute sulphuric acid into a beaker.

2. Add a 4 cm length of magnesium ribbon and observe the reaction. If all the magnesium ribbon disappears, add more until there is some is left over after the reaction has finished (called an excess).

3. Filter the excess magnesium.

4. Boil the magnesium sulphate solution (filtrate) gently to concentrate it, then leave it to cool and form crystals. You can test to see if it is concentrated enough to form crystals by cooling a drop on a glass rod first.

5. Pour off any solution that remains and dry the crystals with tissue.

6. Why is it necessary to use excess magnesium?

## Finding the percentage water of crystallisation in a hydrated salt

When a hydrated salt is heated, the water of crystallisation is lost. The mass of the substance becomes less. Here are some sample results:

| | |
|---|---|
| Mass of hydrated magnesium sulphate crystals | = 8.50 g |
| Mass of anhydrous magnesium sulphate after heating | = 4.15 g |
| Loss in mass on heating | = 4.35 g |

*Copper(II) sulphate crystals.*

The loss in mass is equal to the mass of the water of crystallisation.

So, percentage water of crystallisation $= \dfrac{4.35}{8.50} \times 100\%$

$= 51.2\%$

### Making copper(II) sulphate crystals from copper(II) oxide

Copper oxide is an example of a base. This reaction follows the pattern:

Acid + base → a salt + water

It is essential to add an excess (too much) of the base to make sure that all of the acid has reacted. The base, copper oxide, is insoluble in water and can be removed by filtration.

The method is identical to making magnesium sulphate on page 107 except that you add an excess of black copper(II) oxide to *hot* dilute sulphuric acid. You can easily see when you have an excess because you are left with some unreacted black solid.

$CuO(s) + H_2SO_4(aq) \rightarrow CuSO_4(aq) + H_2O(l)$

$CuSO_4(aq) + 5H_2O(l) \rightarrow CuSO_4 \cdot 5H_2O(s)$

### *How do you know whether you need to heat the mixture?*

Carbonates react with dilute acids in the cold, and so does magnesium. Most other things that you are likely to come across need to be heated.

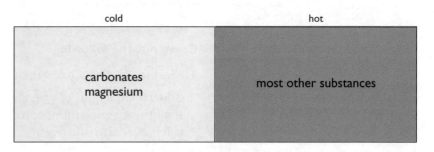

4. What is an anhydrous salt?
5. What is water of crystallisation?
6. Write out the formula for hydrated copper(II) sulphate crystals.

## Making sodium, potassium and ammonium salts

### The need for a different method

In the method we've just been looking at, you add an excess of a solid to an acid, and then filter off the unreacted solid. You do this to make sure that all the acid is used up.

The problem is that all sodium, potassium and ammonium compounds are soluble in water. The solid you added to the acid would not only react with the acid, but any excess would just dissolve in the water present. You wouldn't have any visible excess to filter off. There's no simple way of seeing when you have added just enough of the solid to neutralise the acid.

## Solving the problem by doing a titration

You normally make these salts from sodium or potassium hydroxide or ammonia solution, but you can also use the carbonates. Fortunately, solutions of all these are alkaline. This means that you can find out when you have a neutral solution by using an indicator.

The method of finding out exactly how much of two solutions you need to neutralise each other is called a titration. The point at which an indicator changes colour during the **titration** is called the **end point** of the titration.

Having found out how much acid and alkali are needed, you can make a pure solution of the salt by mixing these same volumes again but without the indicator.

*Apparatus for carrying out a titration.*

---

## Activity 2: Making sodium sulphate crystals

### Investigative skills

| O | A |
|---|---|

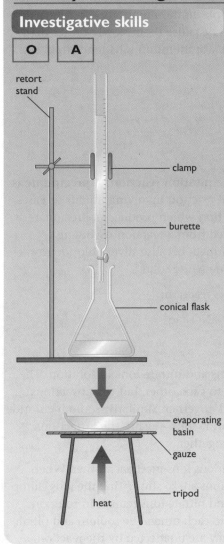

retort stand — clamp — burette — conical flask — evaporating basin — gauze — tripod — heat

Sodium hydroxide solution reacts with sulphuric acid to form the salt sodium sulphate and water. The water can be evaporated to give sodium sulphate crystals.

$$2NaOH(aq) + H_2SO_4(aq) \rightarrow Na_2SO_4(aq) + 2H_2O(l)$$

$$Na_2SO_4(aq) + 10H_2O(l) \rightarrow Na_2SO_4 \cdot 10H_2O(s)$$

**You will need:**

- burette
- retort stand and clamp
- funnel
- conical flask
- pipette
- dilute solution of sodium hydroxide
- dilute sulphuric acid
- methyl orange indicator
- evaporating basin
- tripod and gauze
- Bunsen burner

**Carry out the following:**

1. Using a funnel, fill the burette with the dilute sulphuric acid and read the level carefully on the scale.

2. Using a pipette, transfer exactly 25.0 cm³ of sodium hydroxide (an alkali) to the conical flask.

3. Add three drops of methyl orange indicator and note the colour.

4. Add acid slowly from the burette until the solution is neutral (see indicator colour guide). Note the new burette reading to see how much acid was added.

5. Repeat steps 2 to 4 using the same volumes of acid and alkali, but without any indicator.

6. Evaporate the solution carefully to give crystals.

7. Why do you leave out the indicator the second time?

| not enough acid | just right! | too much acid |
|:---:|:---:|:---:|

*Colour changes for methyl orange.*

### Making sodium chloride crystals

$$NaOH(aq) + HCl(aq) \rightarrow NaCl(aq) + H_2O(l)$$

You would need to do the titration using dilute hydrochloric acid rather than dilute sulphuric acid. However, once you have re-mixed the acid and the alkali without the indicator, you can then evaporate the sodium chloride solution to dryness rather than crystallising it slowly. Sodium chloride crystals don't contain any water of crystallisation, so you can save time by evaporating all the water in one go. The disadvantage is that you end up with either a powder or very tiny crystals.

### Making ammonium sulphate crystals

$$2NH_3(aq) + H_2SO_4(aq) \rightarrow (NH_4)_2SO_4(aq)$$

Using ammonia solution rather than sodium hydroxide solution makes no difference to the method. Although simple ammonium salts don't have water of crystallisation, you would still crystallise them slowly rather than evaporating them to dryness. Heating dry ammonium salts tends to break them up.

## Making insoluble salts

### Precipitation reactions

To make an insoluble salt, you do a **precipitation reaction**. A **precipitate** is a fine solid that is formed by a chemical reaction involving liquids or gases. A precipitation reaction is simply a reaction which produces a precipitate. For example, if silver chloride is produced from a reaction involving solutions, you get a white precipitate formed, because silver chloride won't dissolve in water – and so is seen as a fine white solid.

The photograph shows the results of this reaction:

$$AgNO_3(aq) + NaCl(aq) \rightarrow AgCl(s) + NaNO_3(aq)$$

### *Explaining what's happening*

Silver nitrate solution contains silver ions and nitrate ions in solution. The positive and negative ions are attracted to each other, but the attractions aren't strong enough to make them stick together. Similarly, sodium chloride solution contains sodium ions and chloride ions – again, the attractions aren't strong enough for them to stick together.

When you mix the two solutions, the various ions meet each other. When silver ions meet chloride ions, the attractions are so strong that the ions clump together and form a solid. The sodium and nitrate ions remain in solution because they aren't sufficiently attracted to each other. The sodium and nitrate ions are called **spectator ions** because they are unaffected by the reaction.

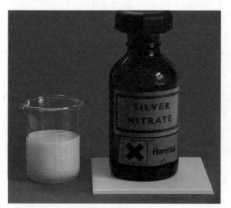

*A precipitate of silver chloride.*

Remember that all nitrates are soluble. So are all the salts of Group 1 metals such as sodium and potassium.

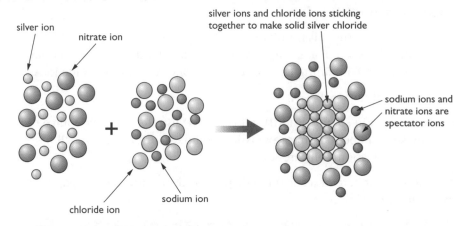

silver ion

nitrate ion

silver ions and chloride ions sticking together to make solid silver chloride

sodium ions and nitrate ions are spectator ions

chloride ion

sodium ion

The water molecules in the solutions have been left out to avoid cluttering the diagram.

### Writing ionic equations for precipitation reactions

The ionic equation for a precipitation reaction is much easier to write than the full equation. All that is happening in one of these reactions is that the ions of an insoluble salt are clumping together to form the solid. The ionic equation simply shows that happening. You don't need to worry at all about the spectator ions – they aren't doing anything.

$$Ag^+(aq) + Cl^-(aq) \rightarrow AgCl(s)$$

This means that if you mix any solution containing silver ions with any solution containing chloride ions you will get the same white precipitate of silver chloride.

Ionic equations for precipitation reactions are simple to write. Write down the formula for the precipitate on the right-hand side of the equation. Write down the formulae for the ions that have clumped together to produce it on the left-hand side. Don't forget the state symbols.

## Activity 3: Making barium sulphate

**Investigative skills**

O

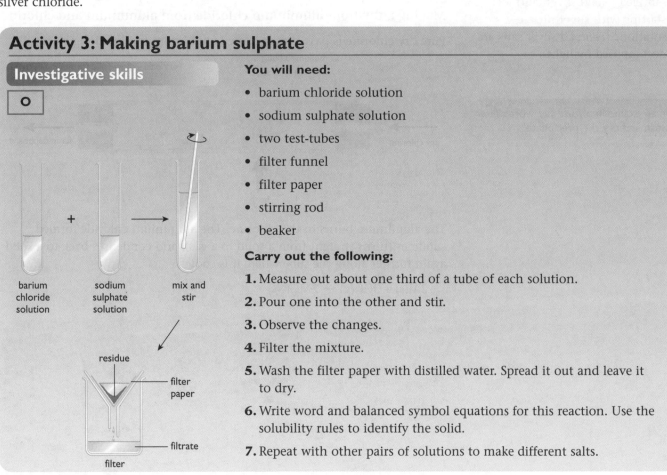

barium chloride solution

sodium sulphate solution

mix and stir

residue

filter paper

filtrate

filter

**You will need:**

- barium chloride solution
- sodium sulphate solution
- two test-tubes
- filter funnel
- filter paper
- stirring rod
- beaker

**Carry out the following:**

1. Measure out about one third of a tube of each solution.
2. Pour one into the other and stir.
3. Observe the changes.
4. Filter the mixture.
5. Wash the filter paper with distilled water. Spread it out and leave it to dry.
6. Write word and balanced symbol equations for this reaction. Use the solubility rules to identify the solid.
7. Repeat with other pairs of solutions to make different salts.

*A precipitate of lead(II) iodide.*

## Making pure lead(II) iodide

It doesn't matter if the salt is unfamiliar to you, as long as you are told that it is insoluble in water. For example, to make lead(II) iodide you would have to mix a solution containing lead(II) ions with one containing iodide ions.

$$Pb^{2+}(aq) + 2I^-(aq) \rightarrow PbI_2(s)$$

The most common soluble lead(II) salt is lead(II) nitrate. A simple source of iodide ions would be sodium or potassium iodide solution, because all sodium and potassium salts are soluble.

The yellow precipitate of lead(II) iodide produced can now be filtered, washed and dried.

> 7. What is a spectator ion?
> 8. Why should you wash a precipitate with distilled water?
> 9. Which solutions would you mix to prepare lead(II) iodide?

## Activity 4: Making pure silver chloride

### Investigative skills

P

How could you make a pure sample of solid silver chloride, starting with silver nitrate solution? (*Hint*: all silver salts are decomposed by light.)

In an exam, simply use the words 'filter, wash and dry the precipitate'.

## Making salts by direct combination

A few salts can be made directly from their elements. Aluminium chloride and iron(III) chloride are good examples. Both can be made by passing dry chlorine over the heated metal. Because everything is done in the absence of water, the anhydrous salts are formed.

### Making anhydrous aluminium chloride from aluminium and chlorine

Pure dry chlorine is passed over heated aluminium foil in a long glass tube.

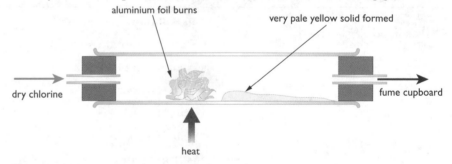

The aluminium burns in the chlorine. The aluminium chloride formed sublimes (turns straight from a solid to a gas), and condenses back to a solid again further along the tube where it is cooler.

$$2Al(s) + 3Cl_2(g) \rightarrow 2AlCl_3(s)$$

# Summarising the methods of making salts

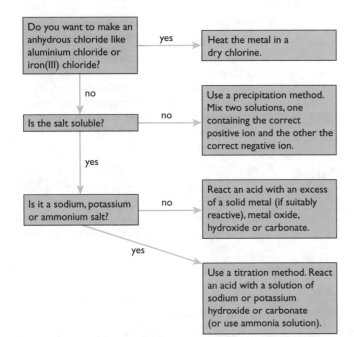

# End of Chapter Checklist

**In this chapter you have learnt that:**

- salts are named after the acid that produces them, e.g. sulphuric acid gives sulphate salts
- soluble salts dissolve in water but insoluble salts form suspensions
- there are rules to help identify soluble and insoluble salts
- anhydrous salts contain no water of crystallisation, e.g. sodium chloride
- hydrated salts contain water of crystallisation within the crystal, e.g. hydrated copper(II) sulphate
- in a titration we use a burette and a pipette to measure volumes of liquids accurately
- insoluble salts can be prepared by mixing solutions of soluble salts
- some salts can be prepared directly from elements, e.g. aluminium chloride from aluminium and chlorine.

# Questions

1. Hydrochloric acid gives salts called:

   A   hydrates
   B   nitrates
   C   sulphates
   D   chlorides                                         (1)

2. A metal oxide is an example of:

   A   an acid
   B   a salt
   C   a base
   D   an anhydrous salt                                 (1)

3. Which of the following statements is/are true?

   I     silver nitrate is soluble in water
   II    barium chloride is soluble in water
   III   silver chloride is insoluble
   IV    silver chloride is changed by light

   A   I and II
   B   II and III
   C   III and IV
   D   I, II, III and IV                                 (1)

4. *a)* Give an example of a hydrated salt, including the formula.                                    (2)

   *b)* Draw a labelled diagram of the apparatus you would use to see whether water is lost on heating this salt.                                      (4)

   *c)* Is the change in b) exothermic or endothermic? Explain your answer.                             (2)

   *d)* What happens when water is added to an anhydrous salt?                                   (2)

5. *a)* Draw and label the apparatus you would use for a titration.                                     (6)

   *b)* How accurately can you measure volumes with this apparatus?                                     (2)

   *c)* Describe how *one* indicator is used in titrations.   (2)

6. *a)* Describe how you would prepare the insoluble salt barium sulphate from soluble salts.          (5)

   *b)* How could you purify the product?                (2)

   *c)* Write word and balanced symbol equations for the reaction.                                       (2)

   *d)* Identify the spectator ions.                     (1)

7. *a)* Draw particle diagrams to explain the making of insoluble salts by precipitation.              (4)

   *b)* How could you recover the spectator ions from a precipitation reaction?                         (3)

   *c)* Give *two* examples of salts that can be prepared by this method and write word equations for each reaction.                                        (2)

   *d)* Add state symbols to the equations.              (1)

8. Sort the following compounds into two lists – those which are soluble in water, and those which are insoluble.

   sodium chloride, lead(II) sulphate, zinc nitrate, calcium carbonate, iron(III) sulphate, lead(II) chloride, potassium sulphate, copper(II) carbonate, silver chloride, aluminium nitrate, barium sulphate, ammonium chloride, magnesium nitrate, calcium sulphate, sodium phosphate, nickel(II) carbonate, chromium(III) hydroxide, potassium dichromate(VI)                      (8)

9. *a)* Describe in detail the preparation of a pure, dry sample of copper(II) sulphate crystals, $CuSO_4 \cdot 5H_2O$, starting from copper(II) oxide.                          (6)

   *b)* Write full equations for i) the reaction producing copper(II) sulphate solution, ii) the crystallisation reaction.                                       (4)

10. *a)* Read the following description of a method for making sodium sulphate crystals, $Na_2SO_4 \cdot 10H_2O$, and then explain the reasons for each of the underlined phrases or sentences.

    25.0 cm³ of sodium carbonate solution was transferred to a conical flask <u>using a pipette</u>, and a <u>few drops of methyl orange were added</u>. Dilute sulphuric acid was run in from a burette <u>until the solution became orange</u>. The volume of acid added was noted. That same volume of dilute sulphuric acid was added to a fresh 25.0 cm³ sample of sodium carbonate solution in a clean flask, but <u>without the methyl orange</u>. The mixture was <u>evaporated until a sample taken on the end of a glass rod crystallised on cooling in the air</u>. <u>The solution was left to cool</u>. The crystals formed were separated from the remaining solution and dried.

    *b)* Write equations for i) the reaction producing sodium sulphate solution, ii) the crystallisation reaction.        (6)

11. There are four main methods of making salts:

    A   reacting an acid with an excess of a suitable solid

    B   using a titration

    C   using a precipitation reaction

    D   by direct combination

    For each of the following salts, write down the letter of the appropriate method, and name the substances you would react together. You should state whether they are used as solids, solutions or gases. Write an equation (full or ionic as appropriate) for each reaction.

    *a)* zinc sulphate                                   (2)

    *b)* barium sulphate                                 (2)

    *c)* potassium nitrate (nitric acid is $HNO_3$)      (2)

    *d)* anhydrous aluminium chloride                    (2)

    *e)* copper(II) nitrate                              (2)

    *f)* lead(II) chromate(VI) (a bright yellow insoluble solid; chromate(VI) ions have the formula $CrO_4^{2-}$).    (2)

## Chapter 12: Titration Calculations

**When you have completed this chapter, you will be able to:**

- calculate the concentrations of solutions
- convert concentration units
- calculate concentrations using experimental data
- use the titration rules to calculate molarities
- titrate acids with bases.

## Working with solution concentrations

### Concentrations of solutions

Concentrations can be measured in either

- $g\,dm^{-3}$
- $mol\,dm^{-3}$

These are exactly the same as writing $g/dm^3$ and $mol/dm^3$. You read them as 'grams per cubic decimetre' and 'moles per cubic decimetre'. 1 cubic decimetre is the same as 1 litre.

You have to be able to convert between $g\,dm^{-3}$ and $mol\,dm^{-3}$. This is no different from converting moles into grams and vice versa. When you are doing the conversions in concentration sums, the amount of substance you are talking about happens to be dissolved in $1\,dm^3$ of solution. That doesn't affect the sum in any way.

Remember:

$$\text{number of moles} = \frac{\text{mass (g)}}{\text{mass of 1 mole (g)}}$$

### Converting from $g\,dm^{-3}$ to $mol\,dm^{-3}$

A sample of sea water had a concentration of sodium chloride of $35.1\,g\,dm^{-3}$. Find its concentration in $mol\,dm^{-3}$. ($A_rS$: Na = 23; Cl = 35.5)

> 1 mol NaCl weighs 58.5 g
>
> 35.1 g is $\frac{35.1}{58.5}$ mol = 0.6 mol

The concentration of the NaCl is $0.6\,mol\,dm^{-3}$.

*Sea water contains about 0.6 moles of NaCl per cubic decimetre.*

### Converting from $mol\,dm^{-3}$ to $g\,dm^{-3}$

What is the concentration of a $0.050\,mol\,dm^{-3}$ solution of sodium carbonate, $Na_2CO_3$, in $g\,dm^{-3}$? ($A_rS$: C = 12; O = 16; Na = 23)

> 1 mol $Na_2CO_3$ weighs 106 g
> 0.050 mol weighs 0.050 × 106 g = 5.3 g
> $0.050\,mol\,dm^{-3}$ is therefore $5.3\,g\,dm^{-3}$

## Making it as tricky as possible!

What is the concentration in mol dm$^{-3}$ of a solution containing 2.1 g of sodium hydrogencarbonate, NaHCO$_3$, in 250 cm$^3$ of solution?
(A$_r$S: H = 1; C = 12; O = 16; Na = 23)

The problem here is that the volume is wrong. The solid is dissolved in 250 cm$^3$ instead of 1000 cm$^3$ (1 dm$^3$).

250 cm$^3$ is $\frac{1}{4}$ of 1000 cm$^3$ (1 dm$^3$)

Therefore a solution containing 2.1 g in 250 cm$^3$ has the same concentration as one containing 4 × 2.1 g in 1000 cm$^3$.

4 × 2.1 g is 8.4 g
1 mol NaHCO$_3$ weighs 84 g
8.4 g is $\frac{8.4}{84}$ mol = 0.10 mol

The concentration is therefore 0.10 mol dm$^{-3}$.

## Calculations from titrations

### A reminder about acid–alkali titrations

A solution of the alkali is measured into a conical flask using a pipette. The acid is run in from the burette – swirling the flask constantly. Towards the end, the acid is run in a drop at a time until the indicator just changes colour. If the indicator changes to red (its acidic colour), you have added too much acid.

### The standard calculation

A simple titration problem will look like this:

25.0 cm$^3$ of 0.100 mol dm$^{-3}$ sodium hydroxide solution required 23.5 cm$^3$ of dilute hydrochloric acid for neutralisation. Calculate the concentration of the hydrochloric acid.

$$NaOH(aq) + HCl(aq) \rightarrow NaCl(aq) + H_2O(l)$$

You do a titration to find the concentration of one solution, knowing the concentration of the other one.

### Planning a route through the calculation

- Start with what you know most about. In this case, you know both the volume and the concentration of the sodium hydroxide solution. Work out how many moles of this you have got.

- Look at the equation to work out how many moles of hydrochloric acid that amount of sodium hydroxide reacts with.

- Work out the concentration of the hydrochloric acid.

## Doing the calculation

The experiment used 25.0 cm$^3$ of 0.100 mol dm$^{-3}$ NaOH solution.

Number of moles of NaOH $= \dfrac{25.0}{1000} \times 0.100$ mol $= 0.00250$ mol

The equation says that 1 mol NaOH reacts with 1 mol HCl

Therefore 0.002 50 mol NaOH reacts with 0.002 50 mol HCl

That 0.002 50 mol HCl must have been in the 23.5 cm$^3$ of hydrochloric acid that was added during the titration – otherwise neutralisation wouldn't have occurred.

All you need to do now is to find out how many moles there would be in 1000 cm$^3$ (1 dm$^3$) of this solution.

If 23.5 cm$^3$ contain 0.002 50 mol HCl

1000 cm$^3$ contain $\dfrac{1000}{23.5} \times 0.00250$ mol HCl $= 0.106$ mol

The concentration is therefore 0.106 mol dm$^{-3}$.

> Put in an extra step if you need to. Work out how many moles there are in 1 cm$^3$ by dividing 0.100 by 1000. (The concentration is 0.100 mol in 1000 cm$^3$.) Then multiply by 25 to find out how many there are in 25 cm$^3$.

> Again, insert an extra step if you need to. Work out the number of moles in 1 cm$^3$ by dividing by 23.5, and then multiply by 1000 to find out how many moles there are in 1000 cm$^3$.

## Activity 1: Acid–base titration

**Investigative skills**

| O | A |
|---|---|

**You will need:**

- burette
- retort stand
- funnel
- pipette
- white tile
- bottle of distilled water
- standard 0.1 mol dm$^{-3}$ alkali, potassium hydroxide
- approximately 0.1 mol dm$^{-3}$ hydrochloric acid
- phenolphthalein indicator
- beaker
- conical flask

**Carry out the following:**

1. Fill the burette with acid and note the reading.
2. Pipette 25.0 cm$^3$ alkali into the conical flask.
3. Add 3 drops of indicator.
4. Add the acid to the alkali, swirling the mixture constantly.
5. Stop when the indicator changes colour, the end-point.
6. Record the new reading on the burette (see sample data table on next page).
7. Repeat the titration three times.

## Activity 1: Acid–base titration (continued)

*Sample data.*

| Trial | | 1 | 2 | 3 |
|---|---|---|---|---|
| final reading/cm$^3$ | 25.50 | 25.10 | 25.40 | 25.10 |
| initial reading/cm$^3$ | 0.10 | 0.00 | 0.20 | 0.00 |
| volume used/cm$^3$ | 25.40 | 25.10 | 25.20 | 25.10 |

Average (ignoring trial) = (25.10 + 25.20 + 25.10) / 3 = 25.10 cm$^3$

*Rules you can use in calculations:*

$$\frac{\text{(concentration in mol dm}^{-3} \times \text{volume) acid}}{\text{(concentration in mol dm}^{-3} \times \text{volume) alkali}}$$

= mole ratio from equation **Rule A**

number of moles in solution = $\dfrac{\text{volume given (cm}^3)}{1000 \text{ (cm}^3)}$

$\times$ concentration in mol dm$^{-3}$ **Rule B**

concentration in g dm$^{-3}$ = molar mass $\times$ concentration in mol dm$^{-3}$ **Rule C**

*Calculation of the concentration in mol dm$^{-3}$ of the acid using the sample data:*

*Equation:* KOH + HCl → KCl + H$_2$O
*Mole ratio:* 1 1

Volume of acid used = 25.10 cm$^3$

Concentration of alkali = 0.1 mol dm$^{-3}$

*Use Rule A:*

$25.0 \text{ cm}^3 = \dfrac{0.1}{1000} \times 25$

= 0.0025 moles

Ratio = 1 : 1

therefore 25.10 cm$^3$ acid = 0.0025

$1000 \text{ cm}^3 \text{ acid} = \dfrac{0.0025}{25.1} \times 1000$

= 0.100 mol dm$^{-3}$

*Use Rule C:*

Molar mass of alkali = 56

Concentration = 56 $\times$ 0.1 = 5.6 g dm$^{-3}$

**Note:** Burette readings should be given to the nearest 0.05 cm$^3$. Pipette readings should be given to .0 cm$^3$ accuracy.

This chapter shows you more than one way to calculate concentrations of solutions in titrations. Try each of the methods yourself.

## A very slightly harder calculation

25.0 cm$^3$ of sodium hydroxide solution of unknown concentration was titrated with dilute sulphuric acid of concentration 0.050 mol dm$^{-3}$. 20.0 cm$^3$ of the acid was required to neutralise the alkali. Find the concentration of the sodium hydroxide solution in mol dm$^{-3}$.

$$2NaOH(aq) + H_2SO_4(aq) \rightarrow Na_2SO_4(aq) + 2H_2O(l)$$

This time you know everything about the sulphuric acid.

The experiment used 20.0 cm$^3$ of 0.050 mol dm$^{-3}$ H$_2$SO$_4$.

$$\text{Number of moles of sulphuric acid used} = \frac{20.0}{1000} \times 0.050 \text{ mol}$$
$$= 0.0010 \text{ mol}$$

The equation proportions aren't 1:1 this time. That's what makes the calculation slightly different from the last one. The equation says that each mole of sulphuric acid reacts with 2 moles of sodium hydroxide.

$$\text{Number of moles of sodium hydroxide} = 2 \times 0.0010 \text{ mol}$$
$$= 0.0020 \text{ mol}$$

That 0.0020 mol must have been in the 25 cm$^3$ of sodium hydroxide solution.

$$\text{Concentration} = \frac{1000}{25} \times 0.0020 \text{ mol dm}^{-3} = 0.080 \text{ mol dm}^{-3}$$

> Use more steps for this, and similar future problems, if you need to.

> 1. Define the term 'concentration'. How is this definition related to the different units of concentration?
> 2. What is meant by the end-point of a titration? How can you determine when you have reached the end-point?
> 3. Why are indicators necessary for titrations?

## A straightforward titration sum with a sting in the tail

Washing soda crystals have the formula Na$_2$CO$_3$·nH$_2$O. The object of this calculation is to find the number of molecules of water of crystallisation, 'n'.

28.6 g of washing soda crystals were dissolved in pure water. More pure water was added to make the total volume of the solution up to 1000 cm$^3$.

A 25.0 cm$^3$ sample of this solution was neutralised by 40.0 cm$^3$ of 0.125 mol dm$^{-3}$ hydrochloric acid using methyl orange as indicator.

$$Na_2CO_3(aq) + 2HCl(aq) \rightarrow 2NaCl(aq) + CO_2(g) + H_2O(l)$$

**(a)** Calculate the concentration of the sodium carbonate solution in moles of sodium carbonate (Na$_2$CO$_3$) per cubic decimetre.

**(b)** Calculate the mass of Na$_2$CO$_3$ and mass of water in the washing soda crystals, and use these results to find a value for 'n' in the formula Na$_2$CO$_3$·nH$_2$O. (A$_r$S: H = 1; C = 12; O = 16; Na = 23)

Part (a) is the straightforward titration calculation. Part (b) is an extra bit.

*Washing soda crystals.*

### Part (a) – the titration calculation

You know the volume and concentration of the hydrochloric acid.

$$\text{Number of moles of HCl} = \frac{40.0}{1000} \times 0.125 \text{ mol} = 0.00500 \text{ mol}$$

From the equation you can see that you only need half that number of moles of sodium carbonate.

$$\text{Number of moles of Na}_2\text{CO}_3 = \frac{0.005\,00}{2} \text{ mol} = 0.002\,50 \text{ mol}$$

The sodium carbonate solution contained 0.002 50 mol in 25.0 cm³.

$$\text{Concentration of Na}_2\text{CO}_3 = \frac{1000}{25.0} \times 0.002\,50 \text{ mol dm}^{-3}$$
$$= 0.100 \text{ mol dm}^{-3}$$

### Part (b)

First calculate the mass of $Na_2CO_3$ in the total 1000 cm³ (1 dm³) of solution. Remember that you have just worked out that the solution is 0.100 mol dm⁻³.

1 mol $Na_2CO_3$ weighs 106 g

0.100 mol $Na_2CO_3$ weighs 0.100 × 106 g = 10.6 g

Now for the mass of water in the crystals:

The original mass of the crystals dissolved in the water was 28.6 g. Of this, we have worked out that 10.6 g is $Na_2CO_3$.

Mass of water = 28.6 – 10.6 g = 18.0 g
But 1 mol $H_2O$ weighs 18 g

There is therefore 1 mol of $H_2O$ in the crystals together with 0.100 mol of $Na_2CO_3$.

Since there are ten times as many moles of $H_2O$ as of $Na_2CO_3$, the formula is $Na_2CO_3 \cdot 10H_2O$.

## Calculations from equations involving solutions

### A calculation involving dissolved salts

A sample of river water contained 0.002 mol dm⁻³ of calcium hydrogencarbonate, $Ca(HCO_3)_2$. When this is heated, it decomposes to make calcium carbonate. This forms as a white precipitate known as 'limescale'. Calculate the mass of calcium carbonate which could be formed when 100 dm³ (100 litres) of the hard water is heated. ($A_r$S: C = 12; O = 16; Ca = 40)

$$Ca(HCO_3)_2(aq) \rightarrow CaCO_3(s) + H_2O(l) + CO_2(g)$$

1 dm³ of river water contains 0.002 mol of $Ca(HCO_3)_2$
100 dm³ contains 100 × 0.002 mol = 0.2 mol

The equation says that 1 mol of $Ca(HCO_3)_2$ gives 1 mol of $CaCO_3$
That means that 0.2 mol of $Ca(HCO_3)_2$ gives 0.2 mol of $CaCO_3$
1 mol $CaCO_3$ weighs 100 g
0.2 mol weighs 0.2 × 100 g = 20 g

20 g of calcium carbonate would be formed.

Notice that we've only just got around to using the figure of 28.6 g, despite the fact that it was the very first number in the question. That is quite common in titration sums. You always start from the volume and concentration of the substance you know most about.

When carbon dioxide in rain water reacts with limestone, some calcium salts dissolve.

*Another calculation involving calcium carbonate*

Limescale can be removed from, for example, electric kettles by reacting it with a dilute acid such as the ethanoic acid present in vinegar.

$$CaCO_3(s) + 2CH_3COOH(aq) \rightarrow (CH_3COO)_2Ca(aq) + CO_2(g) + H_2O(l)$$

What mass of calcium carbonate can be removed by 50 cm$^3$ of a solution containing 2 mol dm$^{-3}$ of ethanoic acid? ($A_rS$: C = 12; O = 16; Ca = 40)

In any question of this sort, it is always a good policy to start by working out the number of moles of any substance where you know both the volume and the concentration. In this case, we know both of these for the ethanoic acid.

Number of moles of ethanoic acid $= \dfrac{50}{1000} \times 2 = 0.1$ mol

Now look at the equation. (Don't be scared by this equation! As long as you know that CH$_3$COOH is ethanoic acid, that's all you need worry about for this calculation.)

The equation says that 1 mol CaCO$_3$ reacts with 2 mol ethanoic acid.

That means that however many moles of ethanoic acid there are in the reaction, there will only be half as many moles of calcium carbonate.

Number of moles of CaCO$_3 = \dfrac{0.1}{2} = 0.05$ mol
1 mol CaCO$_3$ weighs 100 g
0.05 mol weighs 0.05 × 100 g = 5 g

The ethanoic acid would react with 5 g of calcium carbonate.

## Percentage purity

We can use titrations to find out how pure a material is. Here is some sample data for sodium hydroxide, an alkali.

Concentration of solution = 4.0 g dm$^{-3}$ by weighing out 4.0 g solid

Concentration in mol dm$^{-3}$ using Rule C
$= 4.0/40 = 0.1$ mol dm$^{-3}$ since $M_r$ of NaOH is 40.

When this solution was titrated with a standard acid, its concentration was found to be only 3.5g dm$^{-3}$.

Percentage purity $= \dfrac{\text{concentration found by titration}}{\text{concentration by weighing}} = \dfrac{3.5}{4.0} \times 100$

$= 87.5\%$

Some of the material weighed out in the 4.0 g sample must have been an impurity and not sodium hydroxide.

> 4. What is a pure substance?
> 5. What is meant by percentage purity?
> 6. Give one reason why we need to know if a substance is pure.

If you aren't happy with this, put in an extra step. If there are 2 moles in 1000 cm$^3$ (because it is 2 mol dm$^{-3}$), work out how many there are in 1 cm$^3$ by dividing by 1000. That gives 0.002 mol. Multiply that by 50 to find out how many moles there are in 50 cm$^3$.

# End of Chapter Checklist

**In this chapter you have learnt that:**

- concentrations of solutions may be expressed as $g/dm^3$ ($gm\ dm^{-3}$) or as $mol\ dm^{-3}$
- there are three rules that help with titration calculations
- the end-point can be detected using an indicator
- we repeat titrations to improve the reliability of the results
- the equation gives us the mole ratio needed for the calculation
- we can carry out acid–base titrations.

# Questions

1. The volume of a burette is usually:

   A  $5\ cm^3$
   B  $50\ cm^3$
   C  $500\ cm^3$
   D  $5\ dm^3$ (1)

2. The first experiment in a titration is called the:

   A  standard value
   B  accurate value
   C  mean value
   D  trial value (1)

3. **a)** Draw and label a burette. (3)

   **b)** How accurately can you read the scale? (1)

   **c)** Why should a burette be fixed vertically? (3)

   **d)** Why should you first rinse the burette with the solution you will be using? (3)

4. **a)** Describe how you could measure the percentage purity of a sample of solid sodium hydroxide. (6)

   **b)** What apparatus would you use? (2)

   **c)** Show how you would calculate the percentage purity. (2)

5. **a)** Draw a blank results table for an acid–base titration, including the units needed. (3)

   **b)** Why should you repeat a titration? (2)

   **c)** How would you find the end-point? Explain. (2)

   **d)** How can you convert a concentration from $g\ dm^{-3}$ into $mol\ dm^{-3}$? (3)

6. Some dilute sulphuric acid, $H_2SO_4$, had a concentration of $4.90\ g\ dm^{-3}$. What is its concentration in $mol\ dm^{-3}$? ($A_rS$: H = 1; O = 16; S = 32) (3)

7. What is the concentration in $g\ dm^{-3}$ of some potassium hydroxide, KOH, solution with a concentration of $0.200\ mol\ dm^{-3}$? ($A_rS$: H = 1; O = 16; K = 39) (4)

8. What mass of sodium carbonate, $Na_2CO_3$, would be dissolved in $100\ cm^3$ of solution in order to get a concentration of $0.100\ mol\ dm^{-3}$? ($A_rS$: C = 12; O = 16; Na = 23) (4)

9. What mass of barium sulphate would be produced by adding excess barium chloride solution to $20.0\ cm^3$ of copper(II) sulphate solution of concentration $0.100\ mol\ dm^{-3}$? ($A_rS$: O = 16; S = 32; Ba = 137)

   $$BaCl_2(aq) + CuSO_4(aq) \rightarrow BaSO_4(s) + CuCl_2(aq)$$ (6)

10. What is the maximum mass of calcium carbonate which will react with $25.0\ cm^3$ of $2.00\ mol\ dm^{-3}$ hydrochloric acid? ($A_rS$: C = 12; O = 16; Ca = 40)

    $$CaCO_3(s) + 2HCl(aq) \rightarrow CaCl_2(aq) + H_2O(l) + CO_2(g)$$ (4)

11. In each of these questions concerning simple titrations, calculate the unknown concentration in $mol\ dm^{-3}$.

    **a)** $25.0\ cm^3$ of $0.100\ mol\ dm^{-3}$ sodium hydroxide solution was neutralised by $20.0\ cm^3$ of dilute nitric acid of unknown concentration.

    $$NaOH(aq) + HNO_3(aq) \rightarrow NaNO_3(aq) + H_2O(l)$$ (2)

    **b)** $25.0\ cm^3$ of sodium carbonate solution of unknown concentration was neutralised by $30.0\ cm^3$ of $0.100\ mol\ dm^{-3}$ nitric acid.

    $$Na_2CO_3(aq) + 2HNO_3(aq) \rightarrow$$
    $$2NaNO_3(aq) + CO_2(g) + H_2O(l)$$ (2)

    **c)** $25.0\ cm^3$ of $0.250\ mol\ dm^{-3}$ potassium carbonate solution was neutralised by $12.5\ cm^3$ of ethanoic acid of unknown concentration.

    $$2CH_3COOH(aq) + K_2CO_3(aq) \rightarrow$$
    $$2CH_3COOK(aq) + CO_2(g) + H_2O(l)$$ (2)

# End of Section Questions

**1.** **a)** Complete the general equation for the reaction of an acid with a base:

acid + base → *(1 mark)*

**b)** Write word and balanced symbol equations for the reactions of copper oxide with i) hydrochloric acid, ii) nitric acid. *(4 marks)*

**c)** Draw and label the apparatus needed to produce crystals of a hydrated salt from its aqueous solution. *(3 marks)*

**d)** When blue copper sulphate crystals are heated, they turn white and a colourless vapour escapes from the tube. Explain these observations. *(2 marks)*

**Total 10 marks**

**2.** **a)** Explain how to use a titration to produce crystals of common salt from solutions of sodium hydroxide and hydrochloric acid. *(4 marks)*

**b)** What would be the approximate pH of each of these solutions at the start of the titration? *(2 marks)*

**c)** How would the pH change during the titration? *(2 marks)*

**d)** How could you monitor the pH changes during the titration? *(2 marks)*

**Total 10 marks**

**3.** **a)** How could you prepare pure dry samples of the following:

i) barium sulphate from barium chloride solution

ii) copper sulphate from solid copper carbonate? *(4 marks)*

**b)** How could you measure the solubility of potassium chloride in water at different temperatures? *(2 marks)*

**c)** Sketch the solubility curve you would expect to find in b). *(2 marks)*

**d)** Name *two* types of salt that are always soluble in water. *(2 marks)*

**Total 10 marks**

**4.** This question is about salts.

**a)** What would you add to dilute hydrochloric acid to make each of the following salts? In each case, say whether you would add it as a solid or in solution.

i) copper(II) chloride

ii) sodium chloride

iii) silver chloride. *(3 marks)*

**b)** Silver chloride is a white solid which turns greyish on exposure to light. State *one* use for silver chloride based on this property. *(1 mark)*

**c)** Potassium sulphate is produced when dilute sulphuric acid reacts with potassium carbonate solution.

$$K_2CO_3(aq) + H_2SO_4(aq) \rightarrow$$
$$K_2SO_4(aq) + CO_2(g) + H_2O(l)$$

Given solutions of potassium carbonate and dilute sulphuric acid, an indicator, and suitable titration apparatus, describe how you would make a pure, neutral solution of potassium sulphate by a titration method. You should name the indicator you would choose to use, and state any important colour change(s). *(6 marks)*

**Total 10 marks**

# Chapter 13: Rates of Reaction

*Some reactions are very fast.*

**When you have completed this chapter, you will be able to:**

* explain the meaning of 'rate of reaction'
* plan and carry out experiments to measure rates of reaction
* understand collision theory
* describe the ways we can change rates of reaction
* understand how catalysts work
* plot graphs that show rates of reaction
* explain the meaning of activation energy.

## An investigation of the reaction between marble chips and dilute hydrochloric acid

Marble chips are made of calcium carbonate and react with hydrochloric acid to produce carbon dioxide gas. Calcium chloride solution is also formed.

$$CaCO_3(s) + 2HCl(aq) \rightarrow CaCl_2(aq) + H_2O(l) + CO_2(g)$$

The first diagram shows some apparatus which can be used to measure how the mass of carbon dioxide produced changes with time. The apparatus is drawn as it would look before the reaction starts.

The flask is stoppered with cotton wool and contains marble chips. The cotton wool is to allow the carbon dioxide to escape, but to stop any acid spraying out. The measuring cylinder contains dilute hydrochloric acid. The marble is in large excess – most of it will be left over when the acid is all used up. Everything is placed on a top pan balance which is reset to zero.

The second diagram shows what happens during the reaction. The acid has been poured into the flask and everything has been replaced on the balance.

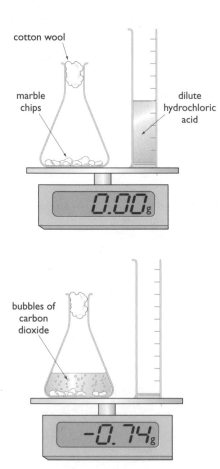

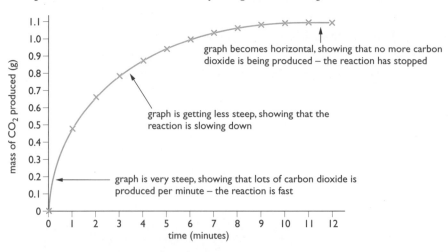

graph becomes horizontal, showing that no more carbon dioxide is being produced – the reaction has stopped

graph is getting less steep, showing that the reaction is slowing down

graph is very steep, showing that lots of carbon dioxide is produced per minute – the reaction is fast

## Activity 1: Rate of reaction and loss in mass

**Investigative skills**

| O | A |
|---|---|

**You will need:**

- conical flask with cotton wool plug
- accurate balance
- measuring cylinder
- stopclock
- dilute hydrochloric acid
- marble chips

**Carry out the following:**

1. Measure out about 10 g marble chips and 30 cm$^3$ acid.

2. Note the exact mass.

3. Mix the marble and acid in the conical flask on the balance and insert the cotton wool plug.

4. Note the mass and start the clock.

5. Note the mass every 30 seconds for at least 15 minutes.

6. Plot a line graph of your results (see example illustrated on page 127).

7. Explain the shape of the graph.

Notice that once the reaction starts, the balance shows a negative mass. This measures the carbon dioxide escaping through the cotton wool.

The reaction is fastest at the beginning. It then slows down until it eventually stops because all the hydrochloric acid has been used up. There will still be unreacted marble chips in the flask.

You can measure how fast the reaction is going at any point by finding the slope of the line at that point. This is called the **rate of the reaction** at that point. You might find that at a particular time, the carbon dioxide was being lost at the rate of 0.12 g per minute.

> You can use a graph to find actual values for the rate of reaction at any particular time by drawing a tangent to the line at the time you are interested in and finding its slope.

### Explaining what is happening

We can explain the shape of the curve by thinking about the particles present and how they interact. This is called the **collision theory**. Everything is made of particles. Particles must bump into each other (collide) before a reaction can happen. However, not every collision between particles is successful. Sometimes particles don't have enough energy to react together. They just bounce off each other. If we want a fast reaction, we need three things:

- lots of collisions
- enough energy for the collisions to be successful
- correct orientation (steric factor).

The minimum energy needed for a particular reaction is called the activation energy. If no particles have this much energy, no reaction will occur. If just a few particles have energy equal to or greater than the **activation energy**, reaction will be slow.

In a concentrated solution there will be lots of particles to collide. Reaction will still be slow if these colliding particles don't have enough energy. As a reaction proceeds, some particles react and are used up. There will now be fewer collisions and so the reaction will start to slow down.

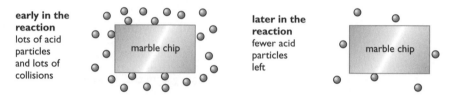

**early in the reaction**
lots of acid particles and lots of collisions

marble chip

**later in the reaction**
fewer acid particles left

marble chip

The marble is in such large excess that its shape doesn't change much during the reaction.

### A different form of graph

You normally plot graphs showing the mass or volume of product formed during a reaction. It is possible that you may come across graphs showing the fall in the concentration of one of the reactants – in this case, the concentration of the dilute hydrochloric acid.

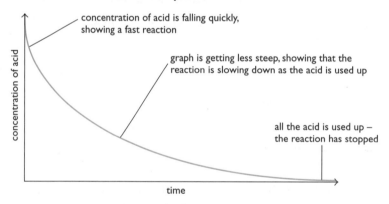

concentration of acid

concentration of acid is falling quickly, showing a fast reaction

graph is getting less steep, showing that the reaction is slowing down as the acid is used up

all the acid is used up – the reaction has stopped

time

Where the graph is falling most quickly, it shows that the reaction is fastest.

Eventually, the graph becomes horizontal because the reaction has stopped.

You could measure the rate of this reaction in a different way, by collecting the gas produced. By measuring the volume of gas produced in a given time, you can work out the rate of the reaction. Look at the examples of apparatus used for the oxygen reaction on pages 131 and 132.

### Changing the conditions in the experiment

#### Using smaller marble chips

You can easily repeat the experiment using exactly the same quantities of everything, but using much smaller marble chips. The reaction with the small chips happens faster.

Both sets of results are plotted on the same graph. Notice that the same mass of carbon dioxide is produced because you are using the same quantities of

*Some reactions happen over several minutes.*

If you are going to investigate the effect of changing the size of the marble chips, it is important that everything else stays exactly the same.

everything in both experiments. However, the reaction with the smaller chips starts off much more quickly and finishes sooner.

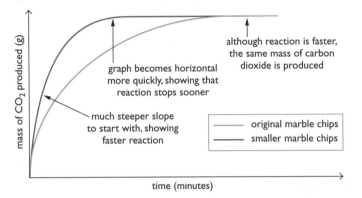

mass of CO₂ produced (g)

although reaction is faster, the same mass of carbon dioxide is produced

graph becomes horizontal more quickly, showing that reaction stops sooner

much steeper slope to start with, showing faster reaction

— original marble chips
— smaller marble chips

time (minutes)

Reactions between solids and liquids (or solids and gases) are faster if the solids are present as lots of small bits rather than a few big ones. The more finely divided the solid, the faster the reaction, because the surface area in contact with the gas or liquid is much greater.

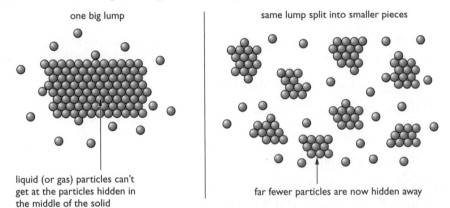

one big lump

same lump split into smaller pieces

liquid (or gas) particles can't get at the particles hidden in the middle of the solid

far fewer particles are now hidden away

High surface areas are frequently used to speed up reactions outside the lab. For example, a **catalytic converter** for a car uses expensive metals like platinum, palladium and rhodium coated onto a honeycomb structure in a very thin layer to keep costs down.

In the presence of these metals harmful substances like carbon monoxide and nitrogen oxides are converted into relatively harmless carbon dioxide and nitrogen. The high surface area means the reaction is very rapid. This is important because the gases in the exhaust system are only in contact with the catalytic converter for a very short time.

1. Why does the reaction between marble chips and acid slow down?
2. Write a balanced symbol equation for the reaction of marble with hydrochloric acid.
3. Why does the size of the marble chips change during the reaction?

**The problem with powders**

Very fine powders have a large surface area. If the powder is in paint, this is very useful. The paint will cover a large area. If the powder is made of a flammable material, such as coal dust or flour, then there is a problem. Dusty atmospheres in coal mines or flour mills can be disastrous. A single spark or lighted match can set off a violent explosion. Many miners have been killed by dust explosions.

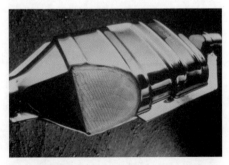

*In a catalytic converter, the honeycomb structure gives a very large surface area.*

### Changing the concentration of the acid

You could repeat the experiment using the original marble chips, but using hydrochloric acid which is of a different concentration. Keep everything else the same – the mass of marble, and the volume of the acid.

# Activity 2: Rate of reaction and concentration

## Investigative skills

O    A

## You will need:

- conical flask with cotton wool plug
- accurate balance
- measuring cylinder
- stopclock
- dilute hydrochloric acid
- marble chips

## Carry out the following:

1. Make up at least three different concentrations of acid by dilution with water, for example 1 part acid: 1 part water; 1 part acid: 2 parts water; and so on.

2. Measure out 30 cm³ of one acid concentration and about 10 g marble chips.

3. Mix the marble and acid in the conical flask on the balance and insert the cotton wool plug.

4. Note the mass and start the stopclock.

5. Note the mass every 30 seconds for at least 15 minutes.

6. Repeat for at least two more acid concentrations.

7. Plot your results on one graph.

8. Explain the shapes of the curves. To help you, look at the example graph below where the concentration of one acid was half that of the other.

The reason you get half the mass of carbon dioxide is because you only have half the amount of acid present. (You have the same volume of acid, but it is only half as concentrated.) The amount of carbon dioxide is controlled by the amount of acid because that's what runs out first.

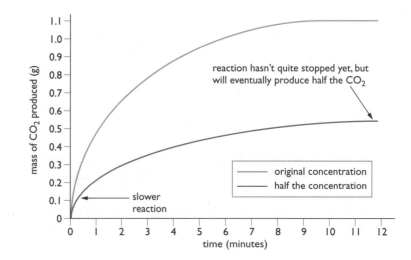

If you increase the concentration of the reactants, the reaction becomes faster. Increasing the concentration increases the chances of particles hitting each other.

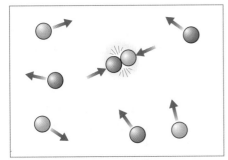
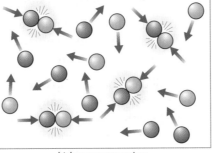

| lower concentration | higher concentration |

*Changing the temperature of the reaction*

## Activity 3: Rate of reaction and temperature

### Investigative skills

| O | A |

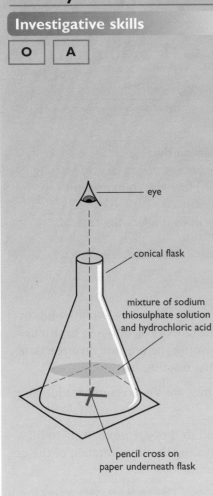

eye

conical flask

mixture of sodium
thiosulphate solution
and hydrochloric acid

pencil cross on
paper underneath flask

When acid is added to sodium thiosulphate solution, the mixture gradually turns cloudy. If the flask is placed over a pencil cross, the cross gradually disappears if viewed from above. You can measure the time it takes for the cross to disappear at different temperatures.

**You will need:**

- conical flask
- measuring cylinder
- stopclock
- thermometer
- Bunsen burner
- tripod
- gauze
- dilute hydrochloric acid
- sodium thiosulphate solution (about 40 g dm$^{-3}$)

**Carry out the following:**

1. Draw a pencil cross on a piece of white paper.

2. Place the conical flask over the cross.

3. Add 50 cm$^3$ sodium thiosulphate solution and measure the temperature.

4. Add 5 cm$^3$ dilute hydochloric acid to start the reaction and start timing.

5. Stop the clock when you can no longer see the pencil mark.

6. Repeat at least five times at a range of temperatures by warming the thiosulphate before adding the acid.

7. Draw a line graph of temperature against time.

8. Explain the shape of your graph.

As a rough approximation, a 10°C increase in temperature doubles the rate of a reaction.

There are two factors at work here.

- Increasing the temperature means that the particles are moving faster and so hit each other more often. That will make the reaction go faster, but it only accounts for a small part of the increase in rate.

- Not all collisions end up causing a reaction. Many particles just bounce off each other. In order for anything interesting to happen, the particles have to collide with a minimum amount of energy called **activation energy**. A relatively small increase in temperature produces a very large increase in the number of collisions which have enough energy for a reaction to occur.

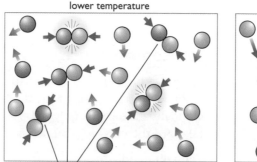
lower temperature

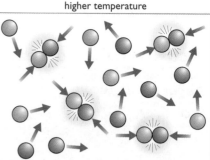
higher temperature

lots of collisions don't produce a reaction

higher number of useful collisions

4. What are the main ideas in the collision theory?
5. How could you produce different concentrations of acid starting with 2 mol dm$^{-3}$ hydrochloric acid? How will you know the concentrations?
6. What is activation energy and how does it affect the rate of a reaction?

### Changing the pressure on the reaction

Changing the pressure on a reaction where the reactants are only solids or liquids makes virtually no difference to the rate of reaction – so in this case, the graphs would be unchanged. But increasing the pressure on a reaction where the reactants are gases does speed the reaction up.

If you have a fixed amount of a gas, you increase the pressure by squeezing it into a smaller volume.

This forces the particles closer together and so they are more likely to hit each other. This is exactly the same as increasing the concentration of the gas.

**3-D effects**

The three-dimensional shape of a molecule can influence its rate of reaction. If a large, bulky group of atoms gets in the way, a molecule may react more slowly. We call this effect **steric hindrance**.

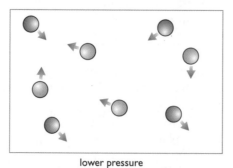
lower pressure

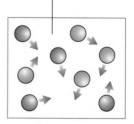
same number of particles squeezed into smaller volume

higher pressure

# Catalysts

## What are catalysts?

Catalysts are substances which speed up chemical reactions, but aren't used up in the process. They are still there chemically unchanged at the end of the reaction. Because they don't get used up, small amounts of catalyst can be used to process lots and lots of reactant particles – whether atoms or molecules or ions. Different reactions need different catalysts.

There are different kinds of catalyst. In **heterogeneous catalysis**, the catalyst is in a different physical state to the reactants. For example, a solid catalyst may be used with gases or liquids. The industrial production of ammonia is an example.

In **homogeneous catalysis**, the catalyst and reactants are in the same physical state. Enzyme catalysts reacting with solutions are examples of homogeneous catalysis.

Think of a catalyst as being rather like a machine tool in a factory. Because the tool doesn't get used up, one tool can process huge amounts of stainless steel into teaspoons. A different tool could turn virtually endless quantities of plastic into yoghurt pots.

## The catalytic decomposition of hydrogen peroxide

Bombardier beetles defend themselves by spraying a hot, unpleasant liquid at their attackers. Part of the reaction involves splitting hydrogen peroxide into water and oxygen using the enzyme catalase. This reaction happens almost explosively, and produces a lot of heat.

Enzymes are biological catalysts. There are lots of other things which also catalyse this reaction. One of these is manganese(IV) oxide, $MnO_2$ – also called manganese dioxide. This is what is normally used in the lab to speed up the decomposition of hydrogen peroxide.

*Bombardier beetles use hydrogen peroxide as part of their defence mechanism.*

The reaction happening with the hydrogen peroxide is:

hydrogen peroxide $\rightarrow$ water + oxygen

$$2H_2O_2(aq) \rightarrow 2H_2O(l) + O_2(g)$$

Notice that you don't write catalysts into the equation because they are chemically unchanged at the end of the reaction. If you like, you can write their name over the top of the arrow.

### *Measuring the volume of oxygen evolved*

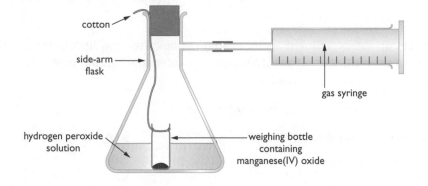

cotton

side-arm flask

gas syringe

hydrogen peroxide solution

weighing bottle containing manganese(IV) oxide

Using a weighing bottle like this is a simple way of mixing the chemicals together without losing any oxygen before you can get the bung in. It is also impossible to get the bung in quickly without forcing a bit of air into the gas syringe. That would give a misleading reading on the syringe.

When you are ready to start the reaction, shake the flask so that the weighing bottle falls over and the manganese(IV) oxide comes into contact with the hydrogen peroxide. You need to keep shaking so that an even mixture is formed.

You could use this apparatus to find out what happens to the rate of reaction if you:

- change the mass of the catalyst
- change how 'lumpy' the catalyst is
- use a different catalyst
- change the concentration of the hydrogen peroxide solution
- change the temperature of the solution.

In each case you could measure the volume of oxygen produced at regular intervals and produce graphs just like the ones earlier in the chapter. However, if you wanted to look in a more detailed way at how a change in concentration or temperature, for example, affects the rate of the reaction, there is a much easier way of doing it.

### Exploring the very beginning of the reaction

An easy way of comparing rates under different conditions is to time how long it takes to produce a small, but constant, volume of gas – say, 5 cm³ – as you vary the conditions. You take measurements of the rate at the beginning of the reaction – the so-called **initial rate**.

This experiment uses an upturned measuring cylinder to measure the volume of the gas. The cylinder is initially full of water.

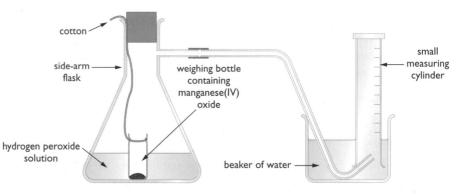

You would shake the flask exactly as before to mix the hydrogen peroxide and manganese(IV) oxide.

This time you record how long it takes for 5 cm³ of oxygen to be collected in the measuring cylinder.

Then you set up the experiment again, changing one of the conditions (for example, the concentration of the hydrogen peroxide) and find out how long it takes to produce the same 5 cm³.

### Varying the concentration – some sample results

Concentrations are measured in mol/dm³ (moles per cubic decimetre – also written as mol dm⁻³).

| concentration /mol dm⁻³ | 2.00 | 1.00 | 0.50 | 0.25 |
|---|---|---|---|---|
| time to collect 5 cm³ of oxygen /s | 10 | 20 | 40 | 80 |

You will see that every time you halve the concentration, it takes twice as long to produce the 5 cm³ of gas. That means that the rate of reaction has also been halved. You can see this more easily if you work out the initial rate for each reaction.

The rate of the reaction would be worked out in terms of the volume of oxygen produced per second. If it takes 10 seconds to produce 5 cm³ at the beginning of the reaction, then:

initial rate = $\frac{5}{10}$ = 0.5 cm³/sec

If you do this for all the experiments, and then redraw the table, you get these figures:

| concentration /mol dm⁻³ | 2.00 | 1.00 | 0.50 | 0.25 |
|---|---|---|---|---|
| initial rate /cm³ s⁻¹ | 0.5 | 0.25 | 0.125 | 0.0625 |

The graph shows that the rate of the reaction is proportional to the concentration – whatever you do to the concentration also happens to the rate. If you double the concentration, the rate doubles – and so on.

### Varying the temperature – a sample graph

You could repeat the experiment, starting with the hydrogen peroxide solution at a range of different temperatures from room temperature up to about 50°C.

This time, the graph isn't a straight line.

As a rough approximation, the rate of a reaction doubles for every 10°C temperature rise.

### Showing that a substance is a catalyst

It isn't difficult to show that manganese(IV) oxide speeds up the decomposition of hydrogen peroxide to produce oxygen. The photograph shows two flasks, both of which contain hydrogen peroxide solution. Without the catalyst, there is only a trace of bubbles in the solution. With it, oxygen is given off quickly.

How can you show that the manganese(IV) oxide is chemically unchanged by the reaction? It still looks the same, but has any been used up? You can only find out by weighing it before you add it to the hydrogen peroxide solution, and then reweighing it at the end.

You can separate it from the liquid by filtering through a weighed filter paper, allowing the paper and residue to dry and then reweighing it to work out the mass of the remaining manganese(IV) oxide. You should find that the mass hasn't changed.

7. What is a catalyst?
8. Give an example of a catalyst and how it is used.
9. What do we mean by heterogeneous catalysis?

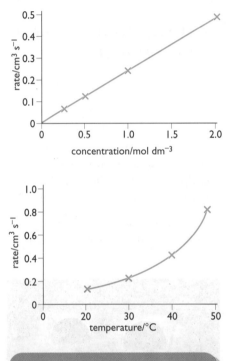

The approximation that reaction rates double for every 10°C rise in temperature works for quite a lot of reactions as long as you are within a few tens of degrees of room temperature.

*Manganese(IV) oxide speeds up the production of oxygen from hydrogen peroxide solution.*

## How does a catalyst work?

You will remember that not all collisions result in a reaction happening. Collisions have to involve at least a certain minimum amount of energy, called **activation energy**.

You can show this on an energy profile. In order for anything interesting to happen, the reactants have to gain enough energy to overcome the activation energy barrier.

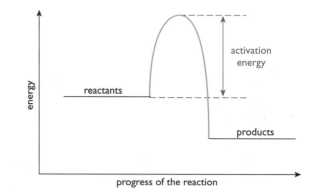

If a reaction is slow, it means that very few collisions have this amount of energy.

Adding a catalyst gives the reaction an alternative and easier way for it to happen – involving a lower activation energy.

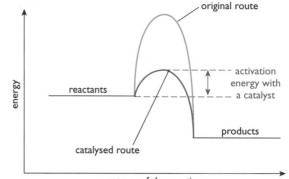

If the activation energy is lower, many more collisions are likely to be successful. The reaction happens faster because the alternative route is easier.

Catalysts work by providing an alternative route for the reaction, involving a lower activation energy.

You can illustrate this with a simple everyday example. Suppose you have a mountain between two valleys. Only a few very energetic people will climb over the mountain from one valley to the next.

Now imagine building a road tunnel through the mountain. Lots of people will be able to travel easily from one valley to the next.

### Catalysts in industry

Catalysts are especially important in industrial reactions because they help substances to react quickly at lower temperatures and pressures than would otherwise be needed. This saves money.

*Traffic passes easily through a road tunnel under a mountain.*

**WARNING!** Be careful how you phrase the statement explaining how a catalyst works. You should say 'Catalysts provide an alternative route with a lower activation energy.'

They do *not* 'lower the activation energy' – any more than building a tunnel lowers the mountain. The original route is still there, and if particles collide with enough energy they will still use it, just as very energetic people will still choose to climb over the top of the mountain.

# End of Chapter Checklist

**In this chapter you have learnt that:**

- not all chemical reactions go at the same speed or rate

- the collision theory explains the changes in rates of reaction by considering how particles collide

- rates of reaction can be changed by altering concentration, temperature, particle size or the use of a catalyst

- catalysts speed up reactions but are not part of the products

- activation energy is the minimum energy needed by particles before a reaction can occur.

# Questions

1. The fastest rate of reaction is shown by:

   A fruit going bad
   B explosions
   C iron rusting
   D metals dissolving in acids (1)

2. You can speed up a reaction by:

   A using less concentrated solutions
   B using a lower temperature
   C using more concentrated solutions
   D stopping stirring (1)

3. When the catalyst and reactants are in different physical states, we call the catalysis:

   A rate
   B heterogeneous
   C homogeneous
   D concentrated (1)

4. When marble chips dissolve in acid, the combined mass:

   A goes down
   B stays the same
   C goes up
   D cannot be measured (1)

5. Which of the following statements is/are true?

   I catalysts speed up reactions
   II catalysts are not part of the products
   III catalysts can be used again
   IV catalysts take no part in a reaction

   A I and IV
   B I and II
   C I, II and III
   D III and IV (1)

6. a) Draw the apparatus you would use to investigate the reaction between marble (calcium carbonate) chips and hydrochloric acid. (5)

   b) Describe *i)* the measurements and *ii)* the safety precautions you would take when carrying out this investigation. (4)

   c) Which gas is produced as a result of the reaction? (1)

7. a) When a mixture of solid manganese(IV) oxide and solid potassium chlorate is heated, oxygen is given off. Design and draw the apparatus you could use to collect the gas. (5)

   b) One of the materials is a catalyst. What further experiments could you carry out to identify it? (4)

   c) What is the test for oxygen gas? (1)

8. a) Sodium thiosulphate produces sulphur when mixed with an acid, which makes the mixture go cloudy. How could you measure the rate of the reaction? (4)

   b) Draw a blank results table to show what measurements you would need to take. (2)

   c) Describe one way to speed up this reaction. (3)

   d) What safety precautions would you take when carrying out this activity? (1)

9. a) Hydrogen peroxide decomposes in sunlight. Explain why it is sold in sealed, dark glass bottles. (4)

   b) Hydrogen peroxide decomposes to give oxygen gas and water. Write a word equation for the reaction. (2)

   c) How could you measure the rate of decomposition when powdered metal is added to hydrogen peroxide? (4)

10. a) When an antacid tablet dissolves in water, the combined mass of water and tablet goes down. Why is this? (3)

b) How could you measure the rate of this reaction? (3)

c) Sketch a graph to show the results you would expect. (3)

d) The reaction rate slows down. Explain why. (1)

11. A student carried out an experiment to investigate the rate of reaction between an excess of dolomite (magnesium carbonate) and 50 cm³ of dilute hydrochloric acid. The dolomite was in small pieces. The reaction is:

$$MgCO_3(s) + 2HCl(aq) \rightarrow MgCl_2(aq) + H_2O(l) + CO_2(g)$$

He measured the volume of carbon dioxide given off at regular intervals, with these results:

| Time (sec) | 0 | 30 | 60 | 90 | 120 | 150 | 180 | 210 | 240 | 270 | 300 | 330 | 360 |
|---|---|---|---|---|---|---|---|---|---|---|---|---|---|
| Vol (cm³) | 0 | 27 | 45 | 59 | 70 | 78 | 85 | 90 | 94 | 97 | 99 | 100 | 100 |

a) Draw a diagram of the apparatus you would use for this experiment, and explain briefly what you would do. (2)

b) Plot these results on graph paper, with time on the x-axis and volume of gas on the y-axis. (2)

c) At what time is the gas being given off most quickly? Explain why the reaction is fastest at that time. (2)

d) Use your graph to find out how long it took to produce 50 cm³ of gas. (2)

e) In each of the following questions, decide what would happen to the initial rate of the reaction and to the total volume of gas given off if various changes were made to the experiment.

i) The mass of dolomite and the volume and concentration of acid were kept constant, but the dolomite was in one big lump instead of small bits. (2)

ii) The mass of dolomite was unchanged and it was still in small pieces. 50 cm³ of hydrochloric acid was used which had half the original concentration. (2)

iii) The dolomite was unchanged again. This time 25 cm³ of the original acid was used instead of 50 cm³. (2)

iv) The acid was heated to 40°C before the dolomite was added to it. (2)

12. The effect of concentration and temperature on the rate of a reaction can be explored using the reaction between magnesium ribbon and dilute sulphuric acid.

$$Mg(s) + H_2SO_4(aq) \rightarrow MgSO_4(aq) + H_2(g)$$

A student dropped a 2 cm length of magnesium ribbon into 25 cm³ of dilute sulphuric acid in a boiling tube (a large excess of acid). She stirred the contents of the tube continuously and timed how long it took for the magnesium to disappear.

a) What would you expect to happen to the time taken for the reaction if she repeated the experiment using the same length of magnesium with a mixture of 20 cm³ of acid and 5 cm³ of water? Explain your answer in terms of the collision theory. (3)

b) What would you expect to happen to the time taken for the reaction if she repeated the experiment using the original quantities of magnesium and acid, but first heated the acid to 50°C? Explain your answer in terms of the collision theory. (3)

c) Why is it important to keep the reaction mixture stirred continuously? (1)

13. Catalysts speed up reactions but can be recovered chemically unchanged at the end of the reaction. (3)

a) Explain briefly how a catalyst has this effect on a reaction.

b) Describe how you would find out whether copper(II) oxide was a catalyst for the decomposition of hydrogen peroxide solution. You need to show not only that it speeds the reaction up, but that it is chemically unchanged at the end. (4)

14. By doing an internet (or other) search, find an industrial process which uses a catalyst. Write a short article on the process, using a computer. You should use your own words – pages copied directly from the internet are not acceptable. Your article must fit on a single side of A4 paper, and must not exceed 300 words. It should include at least two pictures or diagrams. You can produce your artwork yourself on the computer, scan it in, or take pictures directly from the internet. (10)

## Chapter 14: Energy Changes in Reactions

> **When you have completed this chapter, you will be able to:**
>
> - understand the difference between exothermic and endothermic reactions
> - explain heat changes using energy diagrams
> - understand the sign convention for energy changes
> - calculate the heat change in a reaction
> - carry out experiments to measure heat changes.

## Exothermic and endothermic changes

### Exothermic reactions

It is common experience that lots of chemical reactions give out energy – often in the form of heat. A reaction which gives out energy is said to be **exothermic**.

### *Combustion reactions*

Any reaction which produces a flame must be exothermic.

*Burning fuel produces enough energy to launch a rocket.*

*The flare stack at an oil refinery is a safety device. If a process goes wrong (for example, if pressures get too high), reactants and products can be vented to the flare stack where they burn off safely.*

You will be familiar with the test for hydrogen by lighting it and getting a squeaky pop. That is an obvious sign of energy being released – an exothermic change. This can be harnessed in oxy-hydrogen cutting equipment which can be used underwater.

$$2H_2(g) + O_2(g) \rightarrow 2H_2O(l)$$

Apart from burning, other simple exothermic changes include:

- The reactions of metals with acids.
- Neutralisation reactions.
- Adding water to calcium oxide.

### The reactions of metals with acids

For example, when magnesium reacts with dilute sulphuric acid, the mixture gets very warm.

$$Mg(s) + H_2SO_4(aq) \rightarrow MgSO_4(aq) + H_2(g)$$

### Neutralisation reactions

About the only interesting thing that you can observe happening when sodium hydroxide solution reacts with dilute hydrochloric acid is that the temperature rises!

$$NaOH(aq) + HCl(aq) \rightarrow NaCl(aq) + H_2O(l)$$

### Adding water to calcium oxide

If you add water to solid calcium oxide, the heat produced is enough to boil the water and produce steam.

$$CaO(s) + H_2O(l) \rightarrow Ca(OH)_2(s)$$

### Showing an exothermic change on an energy diagram

In an exothermic reaction the reactants have more energy than the products. As the reaction happens, energy is given out in the form of heat. That energy warms up both the reaction itself and the surroundings.

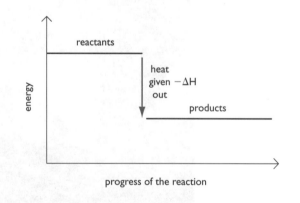

You can measure the amount of heat energy released. It is given the symbol $\Delta H$.

*The burning of hydrogen is used in oxy-hydrogen cutting equipment underwater.*

Remember that in a chemical reaction the reactants are the chemicals you start with.

ΔH is properly referred to as an enthalpy change, a heat change at constant pressure.

ΔH is given a minus or a plus sign to show whether heat is being given out or absorbed by the reaction. You always look at it from the point of view of the substances taking part. For an exothermic reaction, ΔH is given a *negative* number, because the reactants are *losing* energy as heat. That heat is transferred to the surroundings which then get warmer. ΔH is measured in units of $kJ\,mol^{-1}$ ('kilojoules per mole').

In an equation, this would be shown as, for example:

$$Mg(s) + H_2SO_4(aq) \rightarrow MgSO_4(aq) + H_2(g) \qquad \Delta H = -466.9 \text{ kJ mol}^{-1}$$

This means that 466.9 kJ of heat is given out when 1 mole of magnesium reacts in this way. You know it has been given out because ΔH has a negative sign.

> 1. What is meant by an exothermic reaction?
> 2. How could you show that a reaction is exothermic?
> 3. What happens to the energy released in an exothermic change?

ΔH is read as 'delta H'. The Greek letter Δ is used to mean 'change in'. ΔH means 'change in heat'.

The mole is a particular quantity of a substance. You can read about it on pages 264–267, but you probably don't need to worry about it at the moment. In the magnesium case, 1 mole weighs 24.3 g.

### Endothermic reactions

A reaction which absorbs energy is said to be **endothermic**. The energy absorbed may be in the form of light, heat taken from the surroundings, or electrical energy.

#### *Photosynthesis*

Photosynthesis uses energy in the form of light to convert carbon dioxide and water into carbohydrates (such as glucose) and oxygen.

$$6CO_2(g) + 6H_2O(l) \rightarrow C_6H_{12}O_6(aq) + 6O_2(g)$$

This reaction requires the green plant pigment, **chlorophyll**.

*Life on Earth depends on photosynthesis.*

#### *The effect of heat on calcium carbonate*

Calcium carbonate doesn't decompose (split up) unless it is heated at very high temperatures. The reaction is endothermic and needs a constant input of heat.

$$CaCO_3(s) \rightarrow CaO(s) + CO_2(g)$$

#### *Electrolysis*

The chemical changes during electrolysis need a constant input of electrical energy. If you stop the flow of electricity, the reaction stops at once. All electrolysis reactions must be endothermic.

> 4. What is meant by an endothermic change?
> 5. What sign convention do we use for endothermic changes?
> 6. Give one example of an endothermic change.

### The reaction between sodium carbonate and ethanoic acid

All acids react with carbonates to give off carbon dioxide, and ethanoic acid is no exception. Despite the vigorous reaction, the temperature of the mixture falls.

The reaction is endothermic, and takes heat from the water, the beaker and the surroundings. You can tell that the reaction is endothermic by putting a thermometer in the mixture, or because the beaker will feel cold as heat is taken from your skin.

$$Na_2CO_3(s) + 2CH_3COOH(aq) \rightarrow 2CH_3COONa(aq) + CO_2(g) + H_2O(l)$$

### Showing an endothermic change on an energy diagram

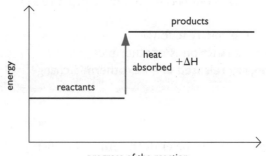

In an endothermic change, the products have more energy than the reactants. That extra energy has to come from somewhere, and it is absorbed from the surroundings – making everything cooler.

Because the reactants are *gaining* energy, $\Delta H$ is given a *positive* sign.

For example:

$$CaCO_3(s) \rightarrow CaO(s) + CO_2(g) \qquad \Delta H = +178 \text{ kJ mol}^{-1}$$

This means that it needs 178 kJ of heat energy to convert 1 mole of calcium carbonate (in this case 100 g) into calcium oxide and carbon dioxide.

### Conservation of energy

This principle states that when one form of energy is used up, it reappears in a different form. In other words, energy cannot be created or destroyed, it can only be changed in form. For example, petrol is a source of chemical energy. When the petrol runs out, the car engine stops. The chemical energy of the petrol has been turned into other forms, including:

- movement of the car on the road (kinetic energy)
- sound of the engine and tyres
- heat of the air and the cooling system
- friction in the engine and between tyres and the road
- electricity of the CD player and the car lights.

Again, don't worry if you don't understand the units of $\Delta H$. It isn't important for now.

# Calculations involving heat changes during reactions

### Bond energies (bond strengths)

It needs energy to break chemical bonds. The stronger the bond is, the more energy is needed to break it. Bond energy measures the amount of energy needed to break a particular bond.

| Bond | C–H | C–Cl | C–I | Cl–Cl | I–I | H–Cl | H–I |
|---|---|---|---|---|---|---|---|
| Bond energy (kJ mol$^{-1}$) | +413 | +346 | +234 | +243 | +151 | +432 | +298 |

You can see that some bonds are much stronger than others – for example, the bond between iodine and hydrogen is about twice as strong as the bond between two iodine atoms.

> Bond energies are measured in kilojoules per mole. For now, all you need to do is to realise that the 'mole' is related to the formula of a substance. For example, if you see '$Cl_2$' written in an equation, it will take 346 kJ to break all the bonds in it. If you see '2HCl', it will take 2×432 kJ to break all the bonds in that.

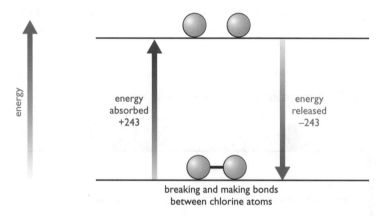

energy absorbed +243

energy released −243

breaking and making bonds between chlorine atoms

The diagram shows that it needs an input of 243 kJ per mole to break chlorine molecules into atoms. If the atoms recombine into their original molecules, then obviously exactly the same amount of energy will be released again. When bonds are made, energy is given out.

- Breaking bonds needs energy.

- Making bonds releases energy.

### Calculating the heat released or absorbed during a reaction

You can estimate the heat released or absorbed by working out how much energy would be needed to break the substances up into individual atoms, and then how much would be given out when those atoms recombine into new arrangements. For example:

| | | | |
|---|---|---|---|
| If | heat needed to break all the bonds | = | +1000 kJ |
| | heat released when new bonds are made | = | −1200 kJ |
| then | overall change | = | −200 kJ |

### *The reaction between methane and chlorine*

$$CH_4(g) + Cl_2(g) \rightarrow CH_3Cl(g) + HCl(g)$$

Methane reacts with chlorine in the presence of ultra-violet light to produce chloromethane and hydrogen chloride. You can picture all the bonds being broken in the methane and chlorine and then being reformed in new ways in the products.

You can work out the heat needed to break all the bonds, and the heat given out as new ones are made.

Bonds that need to be broken:

| | | | |
|---|---|---|---|
| 4 C–H bonds | = | 4 × (+413) | = +1652 kJ |
| 1 Cl–Cl bond | = | 1 × (+243) | = + <u>243</u> kJ |
| Total | | | = +1895 kJ |

New bonds made:

| | | | |
|---|---|---|---|
| 3 C–H bonds | = | 3 × (–413) | = –1239 kJ |
| 1 C–Cl bond | = | 1 × (–346) | = – 346 kJ |
| 1 H–Cl bond | = | 1 × (–432) | = – <u>432</u> kJ |
| Total | | | = –2017 kJ |

The overall energy change is +1895 + (–2017) kJ = –122 kJ

The negative sign of the answer shows that, overall, heat is given out as the bonds rearrange. More energy is released when the new bonds were made than is used to break the old ones.

The excess heat given out means that the reaction is exothermic.

You can show all this happening on an energy diagram:

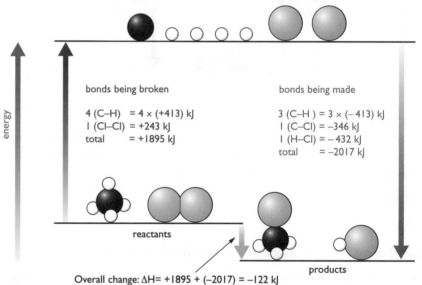

*The reaction between hydrogen and fluorine*

Calculate the heat released or absorbed when this reaction occurs:

$$H_2(g) + F_2(g) \rightarrow 2HF(g)$$

The bond energies (in kJ mol$^{-1}$) are: H–H 436, F–F 158, H–F 568.

Bonds that need to be broken:

| | | | |
|---|---|---|---|
| 1 H–H bond | = | 1 × (+436) | = +436 kJ |
| 1 F–F bond | = | 1 × (+158) | = <u>+158</u> kJ |
| Total | | | = +594 kJ |

New bonds made:

| | | | |
|---|---|---|---|
| 2 H–F bonds | = | 2 × (–568) | = –1136 kJ |

The overall energy change is +594 +(–1136) = –542 kJ

Because the answer is negative, the reaction is exothermic.

### Calculating heat given out during experimental work

If you do a reaction using a known mass of solution and measure the temperature rise, the amount of heat given out during the reaction is given by

> heat given out = mass × specific heat × temperature rise

**Important!** This isn't something you are likely to need in a Chemistry exam, but you might find it useful if you are asked to do an investigation involving energy changes.

The specific heat is the amount of heat which is needed to raise the temperature of 1 gram of a substance by 1°C. For water, the value is 4.18 $Jg^{-1}$ $°C^{-1}$ (joules per gram per degree Celsius).

You can normally assume that dilute solutions have the same specific heat (4.18 $Jg^{-1}$ $°C^{-1}$) and density (1 $gcm^{-3}$ – 1 gram per cubic centimetre) as water. You also assume that negligibly small amounts of heat are used to warm up the cup and the thermometer.

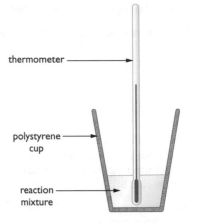

thermometer

polystyrene cup

reaction mixture

If you had, say, 50.0 $cm^3$ of solution, you could assume that it weighed 50.0 g. If the temperature rose by 11.2°C, the amount of heat given out would be as follows:

> heat given out  = 50.0 × 4.18 × 11.2 joules
> = 2340 joules

## Activity 1: Heat of reaction

**Investigative skills**

| O | A |
|---|---|

**You will need:**

- polystyrene drink cup, spatula and stirring rod
- thermometer
- 100 $cm^3$ measuring cylinder
- magnesium powder
- 0.5 mol $dm^{-3}$ copper sulphate solution
- access to a balance
- stopwatch

**Carry out the following:**

1. Measure 50 $cm^3$ copper sulphate into the cup.

2. Measure the initial temperature and then every minute.

3. Weigh out about 0.5 g powder.

4. Add the powder all at once.

5. Continue measuring the temperature until there is no further change.

6. Plot a line graph with temperature ($y$-axis) against time ($x$-axis)

7. Observe the colour changes to both solid and solution and try to work out what has happened.

8. Is this reaction exothermic or endothermic?

# End of Chapter Checklist

In this chapter you have learnt that:

● exothermic reactions give out heat to the surroundings so the sign used is negative ($-\Delta H$)

● endothermic reactions take in heat from the surroundings so the sign used is positive ($+\Delta H$)

● energy diagrams show how energy changes affect a reaction

● heat change = mass × specific heat × temperature change, in joules ($E = mC\Delta T$)

● you can measure energy changes and then calculate the energy change per mole of reactant.

# Questions

1. **a)** Explain what is meant by an exothermic reaction, and write balanced equations for any *two* exothermic changes (apart from the combustion of heptane given in part **b)**). (3)

   **b)** Heptane, $C_7H_{16}$, is a hydrocarbon found in petrol. The equation for the combustion of heptane is:

   $$C_7H_{16}(l) + 11O_2(g) \rightarrow 7CO_2(g) + 8H_2O(l)$$

   $$\Delta H = -4817 \text{ kJ mol}^{-1}$$

   Draw a simple energy diagram to show the combustion of heptane. Show clearly how the figure of $-4817$ fits onto your diagram. (4)

   **c)** Explain what is meant by an endothermic reaction, and write balanced equations for any *two* endothermic changes (apart from the photosynthesis reaction given in part **d)**). (3)

   **d)** Photosynthesis involves the conversion of carbon dioxide and water into carbohydrates such as glucose, $C_6H_{12}O_6$, and oxygen.

   $$6CO_2(g) + 6H_2O(l) \rightarrow C_6H_{12}O_6(aq) + 6O_2(g)$$

   $$\Delta H = +2820 \text{ kJ mol}^{-1}$$

   Draw a simple energy diagram to show the process of photosynthesis. Show clearly how the figure of $+2820$ fits onto your diagram. (4)

2. Use the bond energies in the table to estimate the amount of heat released or absorbed when the following reactions take place. In each case, say whether the change is exothermic or endothermic.

| Bond | C–H | C–Br | Br–Br | H–Br | H–H | Cl–Cl | H–Cl | O=O | O–H |
|------|-----|------|-------|------|-----|-------|------|-----|-----|
| Bond energy (kJ mol⁻¹) | +413 | +290 | +193 | +366 | +436 | +243 | +432 | +498 | +464 |

   **a)** $CH_4(g) + Br_2(g) \rightarrow CH_3Br(g) + HBr(g)$. (The structure of $CH_3Br$ is the same as that of $CH_3Cl$. See the diagram on page 142.) (2)

   **b)** $H_2(g) + Cl_2(g) \rightarrow 2HCl(g)$ (2)

   **c)** $2H_2(g) + O_2(g) \rightarrow 2H_2O(g)$ (2)

3. Self heating cans are used to provide warm food in situations where it is inconvenient to use a more conventional form of heat. By doing an internet (or other) search, find out how self heating cans work. Write a short explanation of your findings (not exceeding 200 words). You should include equation(s) for any reaction(s) involved, and a diagram or picture if it is useful. (10)

4. When nitrogen and hydrogen react together under suitable conditions, ammonia is formed.

   $$N_2(g) + 3H_2(g) \rightarrow 2NH_3(g)$$

   **a)** Use the following bond energies to calculate the energy change during the reaction.

| bond | N≡N | H–H | N–H |
|------|-----|-----|-----|
| bond energy (kJ mol⁻¹) | 945 | 436 | 391 |

   *Hint*: Be careful in adding up the number of N–H bonds in $2NH_3$.

   (3)

   **b)** Is the reaction exothermic or endothermic? Explain your answer. (2)

## Chapter 15: Industrial Chemistry and Reversible Reactions

> **When you have completed this chapter, you will be able to:**
>
> - understand how reactions can be reversible
> - explain the meaning of dynamic equilibrium
> - understand Le Chatelier's Principle
> - explain the effects of changing concentration, temperature or pressure
> - understand how ammonia and sulphuric acid are made and how we use them.

## Reversibility and dynamic equilibria

Two simple reversible reactions

### Heating copper(II) sulphate crystals

If you heat blue copper(II) sulphate crystals gently, the blue crystals turn to a white powder and water is driven off. Heating causes the crystals to lose their water of crystallisation, and white anhydrous copper(II) sulphate is formed. 'Anhydrous' simply means 'without water'.

$$CuSO_4 \cdot 5H_2O(s) \rightarrow CuSO_4(s) + 5H_2O(l)$$

Anhydrous copper(II) sulphate is used to test for the presence of water. If you add water to the white solid, it turns blue – and also gets very warm.

The original change has been exactly reversed. Even the heat that you put in originally has been given out again.

$$CuSO_4(s) + 5H_2O(l) \rightarrow CuSO_4 \cdot 5H_2O(s)$$

*Copper(II) sulphate crystals are split into anhydrous copper(II) sulphate and water on gentle heating.*

### Heating ammonium chloride

If you heat ammonium chloride, the white crystals disappear from the bottom of the tube and reappear further up. Heating ammonium chloride splits it into the colourless gases, ammonia and hydrogen chloride.

$$NH_4Cl(s) \rightarrow NH_3(g) + HCl(g)$$

These gases recombine further up the tube where it is cooler.

$$NH_3(g) + HCl(g) \rightarrow NH_4Cl(s)$$

The reaction reverses when the conditions are changed from hot to cool.

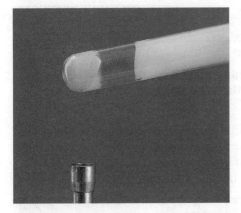

*Heating ammonium chloride.*

## Reversible reactions under 'closed' conditions

'Closed' conditions means that no substances are added to the reaction mixture and no substances escape from it. On the other hand, heat may be either given off or absorbed.

Imagine a substance which can exist in two forms – one of which we'll represent by a blue square and the other by a yellow one. Suppose you start of with a sample which is entirely blue.

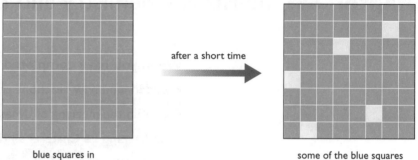

after a short time

blue squares in
a closed system

some of the blue squares
have turned yellow

Suppose you started with 64 blue squares, and in any second there was a 1 in 4 chance of each of them changing colour. In the first second, 16 would change colour, leaving 48 blue squares. In the next second, a quarter of these change colour – but that's only 12, leaving 36 blue ones. In the third second, 9 would change colour – and so on. The rate of change falls as the number of squares falls.

Because you are starting with a high concentration of blue squares, at the beginning of the reaction the rate at which they turn yellow will be relatively high in terms of the number of squares changing colour per second. The number changing colour per second (the rate of change) will fall as the blue gradually gets used up.

But the yellow squares can also change back to blue ones again – it is a reversible reaction. At the start, there aren't any yellow squares, so the rate of change from yellow into blue is zero. As their number increases, the rate at which yellow change to blue also increases.

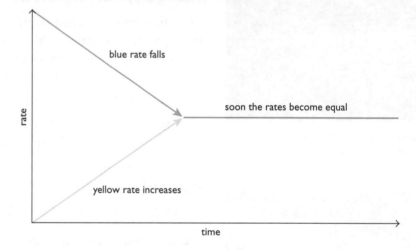

Soon the rates of both reactions become equal. At that point, blue ones are changing into yellow ones at exactly the same rate that yellow ones are turning blue.

What would you see in the reaction mixture when that happens? The total numbers of blue squares and of yellow squares would remain constant, but the reaction would still be going on. If you followed the fate of any one particular square, sometimes it would be blue and sometimes yellow.

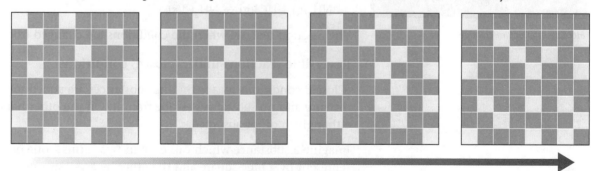

changes happening as time passes

This is an example of a **dynamic equilibrium**. It is *dynamic* in the sense that the reactions are still continuing. It is an *equilibrium* in the sense that the total amounts of the various things present are now constant.

Notice that you can only set up a dynamic equilibrium if the system is closed. If, for example, you removed the yellow ones as soon as they were formed, they would never get the chance to turn blue again. What was a reversible reaction will now go entirely in one direction as blue squares turn yellow without being replaced.

1. What is meant by a reversible reaction?
2. Explain why heating ammonium chloride is a reversible change.
3. What is meant by the term 'equilibrium'?

### Writing equations for reactions in dynamic equilibrium

Taking a general case where A and B react reversibly to give C and D:

$$A + 2B \rightleftharpoons C + D$$

The special two-way arrows show a reversible reaction in a state of dynamic equilibrium (see page 101, 'Strong and Weak Acids and Alkalis'). The reaction between A and B (the left to right reaction) is described as the **forward reaction**. The reaction between C and D (the right to left reaction) is called the **back reaction**.

> The reason for using '2B' will become obvious later on when we look at the effect of pressure on the reaction.

### Manipulating reversible reactions

If your aim in life was to produce substance C in the last equation as efficiently as possible, you might not be too pleased if it kept reacting back to produce A and B all the time.

This section looks at what can be done to alter the **position of equilibrium** so as to produce as much as possible of what you want in the equilibrium mixture. 'Position of equilibrium' is just a reference to the proportions of the various things in the equilibrium mixture.

If, for example, the equilibrium mixture contains a high proportion of C and D, we would say that the 'position of equilibrium lies towards C and D', or the 'position of equilibrium lies to the right'.

### *Le Chatelier's Principle*

> **If a dynamic equilibrium is disturbed by changing the conditions, the reaction moves to counteract the change.**

This is a useful guide to what happens if you change the conditions in a system in dynamic equilibrium. It is essentially a 'law of chemical cussedness'! The reaction sets about counteracting any changes you make.

The things that we might try to do to influence the reaction include:

- increasing or decreasing the concentrations of substances present

- changing the pressure

- changing the temperature.

> **Important!** Le Chatelier's Principle is no more than a useful 'rule-of-thumb' to help you to decide what happens if various conditions are changed. It is *not the reason* why the reaction responds in that way.

### Adding and removing substances

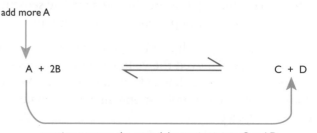

reaction removes the extra A by turning it into C and D

If you add more A, the system responds by removing it again. That produces more C and D – which is what you probably want. You might choose to increase the amount of A if it was essential to convert as much B as possible into products – because it was expensive, for example.

Alternatively, if you remove C as soon as it is formed, the reaction will respond by replacing it again by reacting more A and B. Removing a substance as soon as it is formed is a useful way of moving the position of equilibrium to generate more products.

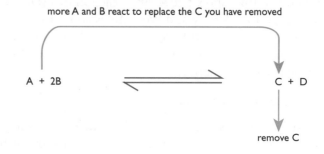

### Changing the pressure

This only really applies to gas reactions and where the total number of molecules is different on each side of the equation. In our example, there are three molecules on the left, but only two on the right.

$$A(g) + 2B(g) \rightleftharpoons C(g) + D(g)$$

Pressure is caused by molecules hitting the walls of their container. If you have fewer molecules at the same temperature, you will have a lower pressure.

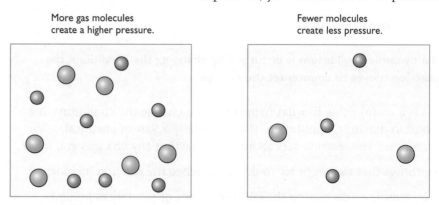

According to Le Chatelier's Principle, if you increase the pressure, the reaction will respond by reducing it again. It can reduce the pressure by producing fewer molecules to hit the walls of the container – in this case by creating more C and D. Increasing the pressure will always help the reaction go in the direction which produces the smaller number of molecules.

### Changing the temperature

Suppose the forward reaction is *exothermic*. This is shown in an equation by writing a negative sign in front of the quantity of heat energy. For example:

$$A + 2B \rightleftharpoons C + D \qquad \Delta H = -100 \text{ kJ mol}^{-1}$$

The back reaction would be *endothermic* by exactly the same amount.

Suppose you changed the conditions by decreasing the temperature of the equilibrium – for example, if the reaction was originally in equilibrium at 500°C, you lower the temperature to 100°C. The reaction will respond in such a way as to increase the temperature again. How can it do that?

If more C and D is produced, more heat is given out because of the exothemic change. That extra heat which is produced will warm the reaction mixture up again – as Le Chatelier suggests. In other words, decreasing the temperature will cause more C and D to be formed.

Increasing the temperature will have exactly the opposite effect. The reaction will move to get rid of the extra heat by absorbing it in an endothermic change. This time the back reaction is favoured.

> In a reversible reaction, the value of $\Delta H$ quoted always applies to the *forward* reaction as written in the equation. The value of $\Delta H$ is given as if the reaction was a one-way process.

4. What is Le Chatelier's Principle?
5. What is an exothermic reaction?
6. Explain what happens to a reversible reaction when you raise the temperature, using an exothermic reaction as your example.

## The manufacture and uses of ammonia

### The Haber Process

The Haber Process takes nitrogen from the air and hydrogen produced from natural gas, and combines them into ammonia, $NH_3$.

$$N_2(g) + 3H_2(g) \rightleftharpoons 2NH_3(g) \qquad \Delta H = -92 \text{ kJ mol}^{-1}$$

| | |
|---|---|
| *The raw materials:* | nitrogen (from the air) |
| | hydrogen (made from natural gas – methane) |
| *The proportions:* | 1 volume of nitrogen to 3 volumes of hydrogen |
| *The temperature:* | 450°C |
| *The pressure:* | 200 atmospheres |
| *The catalyst:* | iron |

> Remember that the negative sign for $\Delta H$ shows that the reaction is exothermic.

> The actual pressure varies in different manufacturing plants, but is always very high.

Each time the gases pass through the reaction vessel, only about 15% of the nitrogen and hydrogen combine to make ammonia. The reaction mixture is cooled and the ammonia condenses as a liquid. The unreacted nitrogen and hydrogen can simply be recycled through the reactor.

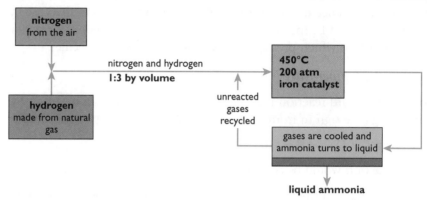

A flow scheme for the Haber Process.

### The reason for the proportions of nitrogen and hydrogen

Equation proportions are used – 1 of nitrogen to 3 of hydrogen. An excess of either would clutter the reaction vessel with molecules which wouldn't have anything to react with.

### The reason for the temperature

According to Le Chatelier's Principle, the forward reaction (an exothermic change) would be favoured by a low temperature, but the temperature used, 450°C, isn't a low temperature.

If the temperature was really low, the reaction would be so slow that it would take a very long time to produce much ammonia. 450°C is a **compromise temperature**, producing a reasonable yield of ammonia reasonably quickly.

### The reason for the pressure

There are four gas molecules on the left-hand side of the equation, but only two on the right-hand side. A reaction which produces fewer gaseous molecules is favoured by a high pressure. A high pressure would also produce a fast reaction rate because the molecules are brought closely together.

The 200 atmospheres actually used is high, but not *very high*. This is a compromise. Generating high pressures and building the vessels and pipes to contain them is very expensive. Pressures much higher than 200 atmospheres cost more to generate than you would get back in the value of the extra ammonia produced.

### The catalyst

The iron catalyst speeds the reaction up, but has no effect on the proportion of ammonia in the equilibrium mixture. If the catalyst wasn't used, the reaction would be so slow that virtually no ammonia would be produced.

## Uses of ammonia

A high proportion of ammonia is used to make fertilisers. Most of the ammonia used to make nitric acid eventually ends up in fertilisers as well.

Organic gardening using poultry manure.

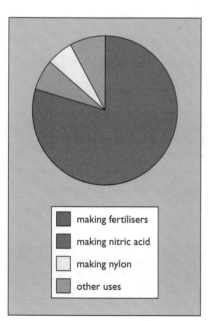

- making fertilisers
- making nitric acid
- making nylon
- other uses

## Making nitric acid

Ammonia is mixed with air and passed over a red-hot platinum–rhodium catalyst. The ammonia is oxidised by oxygen in the air to nitrogen monoxide, NO.

$$4NH_3(g) + 5O_2(g) \rightarrow 4NO(g) + 6H_2O(g)$$

On cooling, the nitrogen monoxide reacts with more oxygen from the air and is oxidised further to nitrogen dioxide, $NO_2$.

$$2NO(g) + O_2(g) \rightarrow 2NO_2(g)$$

This is finally absorbed in water in the presence of still more oxygen to give nitric acid.

$$2H_2O(l) + 4NO_2(g) + O_2(g) \rightarrow 4HNO_3(aq)$$

## Making fertilisers

Common nitrogen-containing fertilisers include ammonium nitrate and ammonium sulphate. These can be made by neutralisation reactions involving ammonia and either nitric acid or sulphuric acid.

$$NH_3(g) + HNO_3(aq) \rightarrow NH_4NO_3(aq)$$

$$2NH_3(g) + H_2SO_4(aq) \rightarrow (NH_4)_2SO_4(aq)$$

## Using fertilisers

Plants use carbon dioxide from the air and water from the soil to produce the carbohydrates which make up much of their bulk. But they also need other elements like nitrogen to produce proteins, or phosphorus to make DNA, and they have to get these from compounds in the soil.

These nutrients are removed from the soil by the plants, and are also washed away by heavy rain. They have to be regularly replaced to keep the soil fertile.

## Inorganic versus organic fertilisers

Fertilisers produced chemically are known as **inorganic fertilisers**. Manure and compost are **organic fertilisers**. Each can provide the nutrients that plants need to grow successfully.

Some organic chicken manure pellets have the composition shown on the right.

Notice the strange way that the percentages of the various elements are typically shown in fertilisers. Fertilisers don't contain $P_2O_5$, $K_2O$, etc. In reality they contain compounds like ammonium phosphate and potassium nitrate.

Bulky organic fertilisers have the advantage that they improve soil structure as well as adding nutrients. They improve the moisture-holding capacity of sandy soils, and break up heavy clay soils.

The nutrients are present as complicated molecules which have to be broken down by soil bacteria. This means that they are released gradually over a long period of time, and are less likely than inorganic fertilisers to be washed out of the soil by rain.

*Crops like wheat need a high input of fertilisers to give a good yield.*

| Typical Analysis | |
|---|---|
| Nitrogen as N | 4.5% |
| Phosphorus as $P_2O_5$ | 3.5% |
| Potassium as $K_2O$ | 2.5% |
| Magnesium as MgO | 1.0% |
| Sulphur as $SO_3$ | 0.5% |
| Calcium as CaO | 9.0% |

Plus full range of trace elements including Iron, Magnesium, Copper, Molybdenum and Zinc.

You need large quantities of bulky organic fertilisers to provide enough nutrients. It is easy to see how small mixed farms with animals as well as crops could use them. It isn't so easy to see how a huge farm growing only grain could do the same thing in a way that made economic sense.

The chart shows the results of research carried out at the Rothamsted Agricultural Experimental Station in the UK over a 5 year period. You can see that the unfertilised plot has very low yields compared with the others.

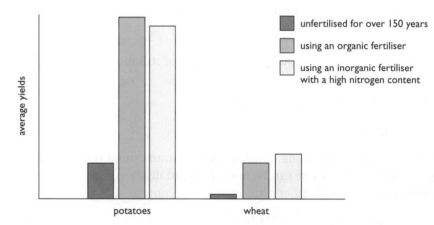

The wheat did better with the inorganic fertiliser – the potatoes with the organic one.

The advantages of inorganic fertilisers are that they are cheap, easy to use and very fast acting. For example, plants take up nitrogen as nitrate ions in solution, $NO_3^-(aq)$. Nitrate fertilisers provide that directly. Ammonium salts are rapidly converted into nitrate ions by soil bacteria.

Inorganic fertilisers do, however, have major environmental disadvantages.

- Over the long term, if little or no organic matter is returned to the soil, the soil structure will break down with the risk of wind or water erosion.

- Inorganic fertilisers use large amounts of energy, both in their manufacture and transport. This can lead to pollution and global warming.

- Inorganic fertilisers are very soluble in water and can pollute water courses. Phosphates and nitrates from fertilisers cause very rapid growth of water plants (**eutrophication**). When these plants die, their decay uses up all the oxygen dissolved in the water, causing the death of all the other water life.

- Nitrates can also pollute water supplies. Because all nitrates are soluble, it is very difficult to remove them from drinking water. There is an association between a rare disease in babies ('**blue baby syndrome**') and nitrates in drinking water, and there is a theoretical (but unproven) possibility of a link between high nitrate levels and some cancers.

## Activity 1: Fertilisers

### Investigative skills

P

Plan an experiment to see if ammonium sulphate fertiliser makes red peas or beans, or a similar quick-growing plant, grow better.

7. What are the raw materials for making ammonia by the Haber Process?
8. What is the effect of the catalyst in the Haber Process?
9. State two uses of ammonia.

# The manufacture and uses of sulphuric acid

### Sources of sulphur

The starting material for the manufacture of sulphuric acid is the element sulphur. This yellow solid is released by volcanoes and was mined by the ancient Romans in Sicily, near Mount Etna. Large quantities of sulphur are found in underground deposits in the USA. The sulphur is extracted by the Frasch Process which uses superheated (very hot) steam. The steam is pumped underground where it melts the sulphur, but not the surrounding rock. The liquid sulphur is pumped to ground level where it cools and sets solid. The sulphur produced this way is extremely pure and suitable for immediate use without further treatment.

Many metallic minerals are sulphides, for example, the major ores of lead and zinc. Natural gas often contains traces of hydrogen sulphide gas, a poisonous gas that smells of rotten eggs. When the natural gas is purified, sulphur can be extracted from the hydrogen sulphide.

### The Contact Process

At the heart of this process is a reversible reaction in which sulphur dioxide is converted to sulphur trioxide, but first you have to produce sulphur dioxide.

### Stage 1: making sulphur dioxide

Either burn sulphur in air:

$$S(s) + O_2(g) \rightarrow SO_2(g)$$

or heat sulphide ores strongly in air:

$$4FeS_2(s) + 11O_2(g) \rightarrow 2Fe_2O_3(s) + 8SO_2(g)$$

### Stage 2: making sulphur trioxide

Now the sulphur dioxide is converted into sulphur trioxide using an excess of air from the previous processes.

$$2SO_2(g) + O_2(g) \rightleftharpoons 2SO_3(g) \quad \Delta H = -196 \text{ kJ mol}^{-1}$$

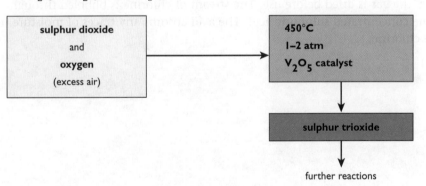

> Notice that the forward reaction is an exothermic change.

In this reaction an excess of oxygen is used, because it is important to make sure that as much sulphur dioxide as possible is converted into sulphur trioxide. Having sulphur dioxide left over at the end of the reaction is wasteful, and could cause possibly dangerous pollution.

### Investigative skills

P

Plan an experiment to find by how much you can dilute sulphuric acid before it stops showing typical acidic reactions (for example, reactions with metals or carbonates).

Because the forward reaction is exothermic, there would be a higher percentage conversion of sulphur dioxide into sulphur trioxide at a low temperature. However, at a low temperature the rate of reaction would be very slow. 450°C is a compromise. Even so, there is about a 99.5% conversion.

There are three gas molecules on the left-hand side of the equation, but only two on the right. Reactions in which the numbers of gas molecules decrease are favoured by high pressures. In this case, though, the conversion is so good at low pressures that it isn't economically worthwhile to use higher ones.

The catalyst, vanadium(V) oxide, has no effect on the percentage conversion, but helps to speed up the reaction. Without the catalyst the reaction would be extremely slow.

### Stage 3: making the sulphuric acid

In principle, you can react sulphur trioxide with water to make sulphuric acid. In practice, this produces an uncontrollable fog of concentrated sulphuric acid. Instead, the sulphur trioxide is absorbed in concentrated sulphuric acid to give **fuming sulphuric acid** (also called **oleum**).

$$H_2SO_4(l) + SO_3(g) \rightarrow H_2S_2O_7(l)$$

This is converted into twice as much concentrated sulphuric acid by careful addition of water.

$$H_2S_2O_7(l) + H_2O(l) \rightarrow 2H_2SO_4(l)$$

### Uses of sulphuric acid

Sulphuric acid has a wide range of uses throughout the chemical industry. The highest single use is in making fertilisers (including ammonium sulphate and 'superphosphate' – essentially a mixture of calcium phosphate and calcium sulphate).

It is also used in the manufacture of detergents, and as car battery acid.

### Sulphuric acid as a drying agent

In the laboratory preparation of chlorine gas from concentrated hydrochloric acid, the gas is dried before use. The stream of chlorine is bubbled through some concentrated sulphuric acid. The acid absorbs any traces of moisture in the chlorine.

# End of Chapter Checklist

**In this chapter you have learnt that:**

- some chemical reactions can go backwards as well as forwards and are called reversible reactions

- in a reversible reaction the forward and reverse reactions can go at the same rate which is called an equilibrium

- in a dynamic equilibrium, the forward and reverse reactions are continuous

- Le Chatelier's Principle helps us to predict what will happen in a reversible reaction, for example when changing the concentration, temperature or pressure

- ammonia is made from hydrogen and nitrogen and can be used to prepare nitric acid and fertilisers

- sulphuric acid is made from sulphur and is used to make fertilisers, detergents and lead-acid car batteries.

# Questions

1. When you heat copper sulphate crystals, the colour change is:

    A   white to blue
    B   no change
    C   blue to white
    D   blue to black                                              (1)

2. When a reversible reaction begins:

    A   the forward reaction is faster
    B   the reverse reaction is faster
    C   the forward and reverse reactions go at the same rate
    D   the reverse reaction slows down                            (1)

3. The reaction at the centre of the Haber Process for the manufacture of ammonia is:

    $$N_2(g) + 3H_2(g) \rightleftharpoons 2NH_3(g) \qquad \Delta H = -92 \text{ kJ mol}^{-1}$$

    a) What would happen to the percentage of ammonia in the equilibrium mixture and to the rate of the reaction if you:

    i)   increased the temperature
    ii)  increased the pressure
    iii) added a catalyst?                                         (6)

    b) In the light of your answer to aii), explain why ammonia plants usually operate with pressures of about 200 atmospheres.                              (2)

    c) State the sources of the nitrogen and the hydrogen used in the Haber Process.                              (2)

4. Sulphuric acid is:

    A   made from sulphur
    B   made without a catalyst
    C   a weak acid
    D   the same as oleum                                          (1)

5. a) What is the formula for sulphuric acid?                     (1)

    b) Explain the difference between sulphur dioxide and sulphur trioxide, and how one is turned into the other.                                                  (3)

    c) What is oleum and how is it used?                           (3)

6. a) State and explain Le Chatelier's Principle.                 (5)

    b) Apply the principle to the Haber Process to explain the use of:

    i) high pressure and ii) the temperature used.               (4)

    c) State one use of ammonia.                                   (1)

7. a) Explain how the fertiliser ammonium sulphate could be made on a large scale.                              (4)

    b) What are fertilisers?                                       (2)

    c) What are organic fertilisers?                              (2)

    d) Describe one environmental problem caused by the use of fertilisers.                                   (2)

8. a) Write out and explain the main equations for the Contact Process.                                      (5)

    b) What is a catalyst? Why is it useful?                       (2)

    c) What is a drying agent?                                     (1)

    d) What evidence is there that copper sulphate crystals contain water?                                   (2)

## Chapter 16: Electrolysis

**When you have completed this chapter, you will be able to:**

- explain how electrolysis works
- explain the different behaviour of dilute and concentrated solutions
- predict the products of electrolysis
- write equations to show electrode reactions
- understand how electrolysis is used in industry
- explain how electroplating is carried out.

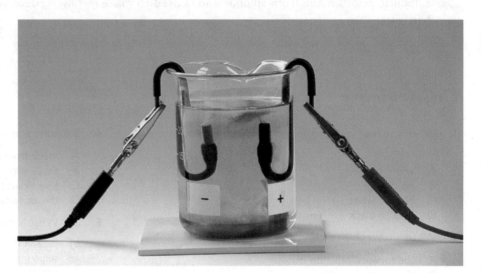

*Part of an electric circuit involving conducting metals, carbon and a solution of potassium iodide.*

The photograph shows what happens if you connect a solution of potassium iodide into a simple electrical circuit. If you look at what is happening in the solution, you can see obvious signs of chemical change. Some coloured substance is being produced at the positive carbon electrode, and a gas is being given off at the negative electrode. This chapter explores the effect of electricity on chemical compounds.

## Some important background

### Conductors and non-conductors

There is a reason why electrical wiring has metal on the inside but plastic on the outside, but never the other way round. Some materials conduct electricity and some do not. Those that don't conduct are called **insulators**.

### The conduction of electricity by metals and carbon

In a metal or carbon, electricity is simply a flow of electrons. The movement of the electrons doesn't produce any chemical change in the metal or the carbon.

Metals and carbon contain mobile electrons, and it is these that move. That's equally true even for a liquid metal like mercury. In an electrical circuit, you can think of a battery or a power pack as an 'electron pump', pushing the electrons through the various bits of metal or carbon.

### Passing electricity through compounds – electrolysis

Hardly any solid compounds conduct electricity. On the other hand, lots of compounds will conduct electricity when they are molten or when they are dissolved in water. All of these show signs of a chemical reaction while they are conducting.

#### *Defining electrolysis*

**Electrolysis** is a chemical change caused by passing an electric current through a compound which is either molten or in solution.

The reason that some compounds undergo electrolysis will be explored in the rest of this chapter.

#### *Some other important words*

An **electrolyte** is a substance which undergoes electrolysis. Electrolytes all contain ions. As you will see shortly, the movement of the ions is responsible for both the conduction of electricity and the chemical changes that take place.

In a solid electrolyte, the ions aren't free to move, and so nothing happens.

The electricity is passed into and out of the electrolyte via two **electrodes**. Carbon is frequently used for electrodes because it conducts electricity and is chemically fairly inert. Platinum is also fairly inert and can be used instead of carbon. Various other metals are sometimes used as well.

The positive electrode is called the **anode**. The negative electrode is called the **cathode**.

> Remember **PANC** – positive anode, negative cathode.

## The electrochemical series

This series is another way to compare metals. Each metal is placed in a circuit and combined with a standard electrode called the hydrogen electrode. The voltages produced give the following order, known as the **electrochemical series**.

| Metal | Voltage/volts |
|---|---|
| K | −2.92 |
| Ca | −2.87 |
| Na | −2.71 |
| Mg | −2.37 |
| Al | −1.66 |
| Zn | −0.76 |
| Fe | −0.44 |
| Pb | −0.13 |
| H (not a metal) | 0.00 |
| Cu | +0.34 |
| Au | +1.50 |

The most reactive metals are at the top. Note that calcium and sodium have changed places when compared to the chemical reactivity series of metals. The reasons for this are beyond the scope of this book.

## The electrolysis of molten compounds

### Electrolysing molten anhydrous lead bromide, $PbBr_2$

The power supply can be a 6 volt battery or a power pack. It doesn't matter which.

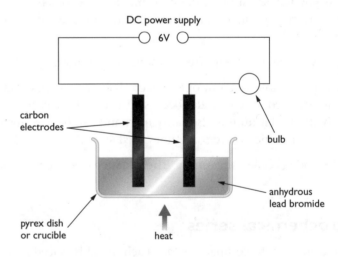

Nothing at all happens while the lead bromide is solid, but as soon as it melts:

- The bulb lights up, showing that electrons are flowing through it.

- There is bubbling around the electrode (the anode) connected to the positive terminal of the power source as bromine is given off.

- Nothing seems to be happening at the electrode (the cathode) connected to the negative terminal of the power source, but afterwards metallic lead is found underneath it.

- When you stop heating and the lead bromide solidifies again, everything stops – there is no more bubbling and the bulb goes out.

## Explaining what is happening

Lead bromide is an ionic compound. The solid consists of a giant structure of lead ions and bromide ions packed regularly in a crystal lattice. It doesn't have any mobile electrons, and the ions are locked tightly in the lattice and aren't free to move. The solid lead bromide doesn't conduct electricity.

As soon as the solid melts, the ions *do* become free to move around, and it is this movement which enables the electrons to flow in the external circuit. This is how it works...

As soon as you connect the power source, it pumps any mobile electrons away from the left-hand electrode towards the right-hand one as we've drawn it in the diagram.

The excess of electrons on the right-hand electrode makes it negatively charged. The left-hand electrode is positively charged because it is short of electrons. There is a limit to how many extra electrons the 'pump' can squeeze into the negative electrode because of the repulsion by the electrons already there.

Things change when the lead bromide melts, and the ions become free to move.

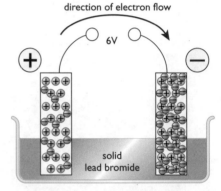

*Showing what happens when the lead bromide melts. The bulb also lights up.*

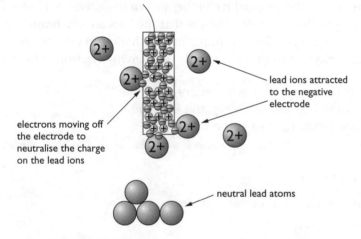

electrons moving off the electrode to neutralise the charge on the lead ions

lead ions attracted to the negative electrode

neutral lead atoms

*Showing what happens at the cathode.*

In this (and the next) diagram, the lead ions and bromide ions are drawn much bigger than the remainder of the atoms in the electrode (the green positively charged circles which represent the nuclei and the non-mobile electrons). This is so it is easier to see what is happening.

The positive lead ions are attracted to the cathode. When they get there, each lead ion picks up two electrons from the electrode and forms neutral lead atoms. These fall to the bottom of the container as liquid lead.

$$Pb^{2+}(l) + 2e^- \rightarrow Pb(l)$$

This leaves spaces in the electrode that more electrons can move into. The power source pumps new electrons along the wire to fill those spaces.

Bromide ions are attracted to the positive anode. When they get there, the extra electron which makes the bromide ion negatively charged moves onto the electrode because this electrode is short of electrons.

The loss of the extra electron turns each bromide ion into a bromine atom. These join in pairs to make bromine molecules. Overall:

$$2Br^-(l) \rightarrow Br_2(g) + 2e^-$$

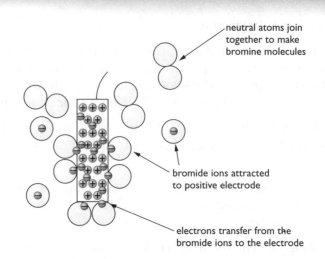

neutral atoms join together to make bromine molecules

bromide ions attracted to positive electrode

electrons transfer from the bromide ions to the electrode

*Showing what happens at the anode.*

The new electrons on the electrode are pumped away by the power source to help fill the spaces being created at the cathode. Because electrons are flowing in the external circuit, the bulb lights up.

Electrons can flow in the external circuit because of the chemical changes to the ions arriving at the electrodes. We say that the ions are **discharged** at the electrodes. Discharging an ion simply means that it loses its charge – either giving up electron(s) to the electrode or receiving electron(s) from it.

> 1. What is meant by the word 'electrolyte'?
> 2. What name is given to the negative electrode?
> 3. Why are metals formed at the cathode?

Remember OIL RIG.

### *Electrolysis and redox*

The lead ions gain electrons at the cathode:

$$Pb^{2+}(l) + 2e^- \rightarrow Pb(l)$$

Gain of electrons is reduction. The lead ions are reduced to lead atoms.

The bromide ions lose electrons at the anode.

$$2Br^-(l) \rightarrow Br_2(g) + 2e^-$$

Loss of electrons is oxidation. Bromide ions are oxidised to bromine molecules.

In any example of electrolysis (whether molten or in solution), reduction happens at the cathode and oxidation happens at the anode.

### The electrolysis of other molten substances

In each case, the positive ions are attracted to the negative cathode, where they are discharged by gaining electrons. Positive ions are known as **cations** because they are attracted to the cathode. The negative ions move to the anode, where they are discharged by giving electrons to the electrode. Negative ions are known as **anions**.

Remember that the cathode is negative and that a cation is attracted to it – so must be positive.

Not all ionic compounds can be electrolysed molten. Some break up into simpler substances before their melting point. For example, copper(II) carbonate breaks into copper(II) oxide and carbon dioxide even on gentle heating. It is impossible to melt it.

Other ionic compounds have such high melting points that it isn't possible to melt them in the lab, although it can be done industrially. For example, it is difficult to keep sodium chloride molten in the lab because its melting point is 801°C. If you could keep it molten you would get sodium at the cathode and chlorine at the anode. Sodium is manufactured by electrolysing molten sodium chloride.

At the cathode: $Na^+(l) + e^- \rightarrow Na(l)$     Molten sodium is produced.

At the anode: $2Cl^-(l) \rightarrow Cl_2(g) + 2e^-$   Chlorine gas is produced.

### Which ions are discharged from a molten substance during electrolysis?

Reactive metals lose electrons easily and are less likely to be discharged in electrolysis. The discharge reaction involves the gain of electrons:

metal ion + electron/electrons → metal atom

Metals that are low in the electrochemical series gain electrons more easily than those near the top. For example, copper is more likely to be discharged than magnesium from a mixture of both.

The concentration of the ions in solution also makes a difference. Ions present in high concentrations may be discharged more easily than others that are lower in the electrochemical series.

Finally, the nature of the electrodes can change the products of electrolysis. In the electrolysis of copper sulphate, carbon electrodes give oxygen at the anode but copper anodes dissolve (see pages 164 and 165).

## The electrolysis of aqueous solutions

### The electrolysis of sodium chloride solution

You might have thought that you would get the same products if you electrolysed sodium chloride solution as if you electrolysed it molten. You would be wrong!

Although chlorine is formed at the anode as you might expect, hydrogen is produced at the cathode rather than sodium. The hydrogen at the cathode is coming from the water.

Water is a very **weak electrolyte**. It ionises very slightly to give hydrogen ions and hydroxide ions.

$H_2O(l) \rightleftharpoons H^+(aq) + OH^-(aq)$

Whenever you have water present, you have to consider these ions *as well as* the ions in the compound you are electrolysing.

# Activity 1: Electrolysis of aqueous solutions

**You will need:**

- low voltage power supply (about 6 V)
- carbon electrodes
- test-tubes to cover them (see diagram)
- glass tube for solution
- dilute solutions of sulphuric acid, sodium chloride (concentrated sodium chloride solution gives mostly chlorine at the anode)
- splints

**Carry out the following:**

1. Fill the glass tube with one of the solutions.

2. Fill the test-tubes with the same solution and place in position over the electrodes (see diagram).

3. Connect the power supply and look for evidence of changes at the electrodes. Does anything collect in the test tubes?

4. Test any gases produced at the electrodes with both a glowing splint (oxygen test) and burning splint (hydrogen test).

5. Identify the products of electrolysis. Repeat with the other dilute solution.

The reason that you don't appear to get as much chlorine as you do hydrogen is that chlorine is slightly soluble and some of it dissolves in the water present.

**Electron transfer in electrolysis**

Electrolysis produces new materials at the electrodes. It happens because electrons move around the completed circuit. The electrons that are lost by ions at the anode then move around the circuit and are gained by ions from the cathode. This will help when you try to balance electron equations for the changes at the anode and cathode.

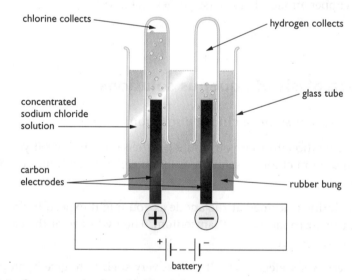

Electrolysis of concentrated sodium chloride solution.

## At the cathode

The solution contains $Na^+(aq)$ and $H^+(aq)$, and these are both attracted to the negative cathode. The $H^+(aq)$ gets discharged because it is much easier to persuade a hydrogen ion to accept an electron than it is a sodium ion. Each hydrogen atom formed combines with another one to make a hydrogen molecule.

For positive ions the lower an element is in the Reactivity Series, the more easily it will accept an electron.

$$2H^+(aq) + 2e^- \rightarrow H_2(g)$$

Remember that the hydrogen ions come from water molecules splitting up. Each time a water molecule ionises, it also produces a hydroxide ion. There is a build up of these in the solution around the cathode.

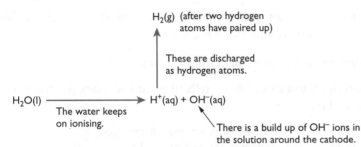

These hydroxide ions make the solution strongly alkaline in the region around the cathode. Because of the presence of the sodium ions attracted to the cathode, you can think of the electrolysis as also forming sodium hydroxide solution.

### At the anode

$Cl^-(aq)$ and $OH^-(aq)$ are both attracted by the positive anode. The hydroxide ion is *slightly* easier to discharge than the chloride ion is, but there isn't that much difference. There are far, far more chloride ions present in the solution, and so it is mainly these which get discharged.

$$2Cl^-(aq) \rightarrow Cl_2(g) + 2e^-$$

### Electrolysis of dilute sodium chloride solution

In a dilute solution there are fewer chloride ions and a different product is made. Water ionises a little to give both hydrogen ions ($H^+$) and hydroxide ions ($OH^-$). In a dilute solution it is the hydroxide ions that are discharged at the electrode. Hydroxide ions turn into water and oxygen gas. At the anode (+) oxygen gas bubbles off.

### The electrolysis of some other solutions using carbon electrodes

| | Cathode | | Anode | |
| | Product | Equation | Product | Equation |
|---|---|---|---|---|
| KI(aq) | hydrogen | $2H^+(aq) + 2e^- \rightarrow H_2(g)$ | iodine | $2I^-(aq) \rightarrow I_2(aq) + 2e^-$ |
| MgBr$_2$(aq) | hydrogen | $2H^+(aq) + 2e^- \rightarrow H_2(g)$ | bromine | $2Br^-(aq) \rightarrow Br_2(aq) + 2e^-$ |
| H$_2$SO$_4$(aq) | hydrogen | $2H^+(aq) + 2e^- \rightarrow H_2(g)$ | oxygen | $4OH^-(aq) \rightarrow 2H_2O(l) + O_2(g) + 4e^-$ |
| CuSO$_4$(aq) | copper | $Cu^{2+}(aq) + 2e^- \rightarrow Cu(s)$ | oxygen | $4OH^-(aq) \rightarrow 2H_2O(l) + O_2(g) + 4e^-$ |
| HCl(aq) (concentrated) | hydrogen | $2H^+(aq) + 2e^- \rightarrow H_2(g)$ | chlorine | $2Cl^-(aq) \rightarrow Cl_2(aq) + 2e^-$ |

In the electrolysis of water, a small amount of dilute sulphuric acid is added to allow a current to flow. The sulphate ions are too stable to be discharged. Instead, you get oxygen from discharge of hydroxide ions from the water. Twice as much hydrogen is produced as oxygen.

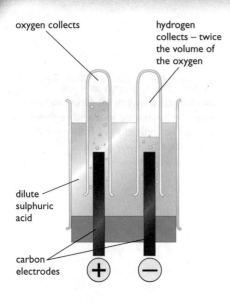

oxygen collects

hydrogen collects – twice the volume of the oxygen

dilute sulphuric acid

carbon electrodes

$+$ $-$

This also applies to using any other *inert* electrodes – like platinum, for example. As you will see below, some metal electrodes are changed during electrolysis.

Look at the sulphuric acid equations in the table. For every four electrons which flow around the circuit, you would get one molecule of oxygen. But four electrons would produce two molecules of hydrogen.

You get twice the number of molecules of hydrogen as of oxygen. Twice the number of molecules occupy twice the volume.

### Position of ions in the electrochemical series

The position of an ion in the series tells us what the likely products of electrolysis will be.

| Carbonate ion | |
| Nitrate ion | these ions are not discharged, we get oxygen instead |
| Sulphate ion | |

| Hydroxide ion from the water | |
| Chloride | halogens are discharged (when concentrated) and hydroxide ions stay in solution |
| Bromide | |
| Iodide | |

For metals, the lower the metal is in the series, the more likely it is to be discharged.

(*Note*: calcium and sodium exchange places in the electrochemical series when compared to the reactivity series of metals – the theory is beyond the scope of this book.)

### Summary of the electrolysis of solutions using carbon electrodes

- If the metal is high in the Reactivity Series, hydrogen is produced instead of the metal.

- If the metal is below hydrogen in the Reactivity Series, the metal is produced.

- If you have reasonably concentrated solutions of halides (chlorides, bromides or iodides), the halogen (chlorine, bromine or iodine) is produced. With other common negative ions, oxygen is produced.

This leaves the problem of what you get if you have a moderately reactive metal like zinc, for example. Reasonably concentrated solutions will give you the metal. Very dilute solutions will give you mainly hydrogen. In between, you will get both.

4. Why does the electrolysis of salt water produce hydrogen rather than sodium?
5. Why does the electrolysis of dilute sulphuric acid produce hydrogen and oxygen in the ratio 2:1?
6. What would be the products of the electrolysis of molten potassium chloride?

### The electrolysis of copper(II) sulphate solution using copper electrodes

If you use metal electrodes rather than carbon, different things can happen at the anode, unless the metal is extremely unreactive – like platinum.

Remember that positive ions are turned into atoms by taking electrons from the cathode. Electrons are pumped around the circuit from the anode to replace them. The reactions at the anode act as a source of these electrons.

So far, these electrons have come from negative ions giving up electrons to the anode, but there is another possibility. They can come from atoms in the electrode itself if that is an easier process.

A copper atom breaks away from the electrode forming a copper(II) ion, leaving its electrons behind on the electrode. Those electrons can then be pumped away by the power source around the circuit to the cathode.

$$Cu(s) \rightarrow Cu^{2+}(aq) + 2e^-$$

For every copper(II) ion that breaks away from the anode, two electrons are made available to pump around the circuit. That's exactly what you need to discharge a copper(II) ion arriving at the cathode.

$$Cu^{2+}(aq) + 2e^- \rightarrow Cu(s)$$

The overall effect is that:

- the anode loses mass

- the cathode gains exactly the same mass

- the number of copper(II) ions in the solution doesn't change at all. For every one that is discharged at the cathode, another one goes into solution at the anode.

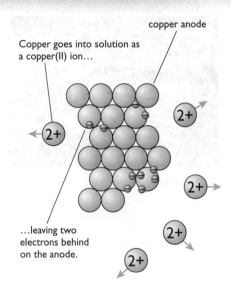
Copper goes into solution as a copper(II) ion...

copper anode

...leaving two electrons behind on the anode.

## Industrial electrolysis

The copper extracted by heating the ore is not pure enough to use as an electrical conductor. This impure copper is purified using electrolysis.

### Purifying copper

The purification of copper uses the electrolysis of copper(II) sulphate solution. The anode is impure copper and the cathode is made of pure copper. The copper from the anode goes into solution as $Cu^{2+}$ ions.

$$Cu(s) \rightarrow Cu^{2+}(aq) + 2e^-$$

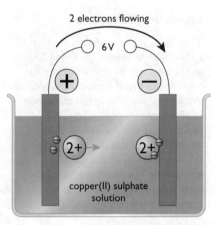
*Electrolysing copper(II) sulphate solution using copper electrodes.*

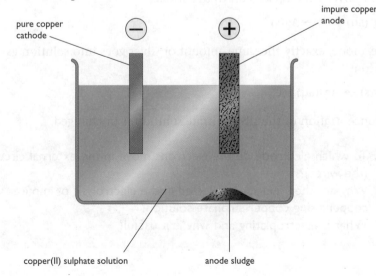

pure copper cathode

impure copper anode

copper(II) sulphate solution

anode sludge

*Copper refining.*

At the cathode, $Cu^{2+}$ ions are deposited as copper.

$$Cu^{2+}(aq) + 2e^- \rightarrow Cu(s)$$

The impure copper anode gradually disappears, and pure copper is plated on to the cathode.

### What happens to the impurities?

There are three sorts of impurities to think about: unreactive material left from the copper ore, metals below copper in the Reactivity Series, and metals above it.

Any metal in the impure anode which is lower than copper in the Reactivity Series (like silver) doesn't go into solution as ions. It remains as a metal and falls to the bottom of the cell with other unreactive material left from the ore. This is the 'anode sludge'.

Metals higher than copper in the Reactivity Series *do* go into solution as their ions. They then remain in the solution. They won't be discharged when they get to the cathode as long as their concentration doesn't get too high. The copper(II) sulphate solution has to be topped up to avoid that happening.

### Electroplating

Electroplating is a process very similar to the purification of copper. Suppose you wanted to plate copper onto steel. The steel object would be made the cathode in the electrolysis of copper(II) sulphate solution using a copper anode.

Copper would be transferred from the anode to the cathode, forming a layer of copper over the steel. The thickness and strength of the layer can be controlled by choice of the current and the time it is allowed to flow.

Silver, gold, nickel or chromium can all be plated on to other metals in a similar way.

To silver plate an object, it is hung in a solution of silver nitrate and is made the cathode in an electric circuit. The anode is a piece of silver.

At the cathode, silver ions are discharged by picking up an electron to form metallic silver which plates on to the article.

$$Ag^+(aq) + e^- \rightarrow Ag(s)$$

At the anode, exactly the same amount of silver goes into solution as silver ions.

$$Ag(s) \rightarrow Ag^+(aq) + e^-$$

The concentration of the silver nitrate solution is unchanged.

### Anodising aluminium

Aluminium metal is always covered in a thin layer of aluminium oxide. This oxide layer can be made thicker to protect the metal from corrosion. The aluminium object is made the anode of an electrolytic cell. Oxygen is produced at the anode and this combines with the metal to give a thicker oxide layer. It is called **anodising**. The anodised metal can absorb coloured dyes and give attractive metallic finishes.

*Silver plate. Silver can easily be electroplated onto other metals. This is much cheaper than using solid silver.*

7. To which electrode do the electrons move in the external circuit (the wires)?
8. Why are the impurities collected in the electrolysis of impure copper using copper sulphate solution?
9. What is electroplating and why is it useful?

## Uses of copper

Uses of copper include:

- Electrical wiring, because it is a good conductor of electricity and is easily drawn into wires.

- Domestic plumbing, because it doesn't react with water, and is easily bent into shape.

- Making brass. Brass is an alloy of copper and zinc in various proportions. Brass is harder than either copper or zinc individually.

- Coinage. 'Silver' coins are made from an alloy of copper and nickel ('cupronickel'), containing 75% copper and 25% nickel (except for 20p coins which are 84% copper and 16% nickel. If you look closely, you should be able to see the difference).

> 10. Describe two uses of copper metal.
> 11. What is the difference between brass and copper?

## Industrial electrolysis of brine

### Manufacture of chlorine, hydrogen and sodium hydroxide

Chlorine and hydrogen are two useful materials that can be prepared from ordinary salt, sodium chloride.

The salt solution is purified to remove ions other than sodium and chloride ions and is then electrolysed to produce three useful chemicals – sodium hydroxide, chlorine and hydrogen. The electrolysis can be carried out in a **membrane cell**.

The cell is designed to keep the products apart. If chlorine comes into contact with sodium hydroxide solution, it reacts to make bleach – a mixture of sodium chloride and sodium chlorate(I) solution. If chlorine comes into contact with hydrogen, it produces a mixture which would explode violently on exposure to sunlight or heat.

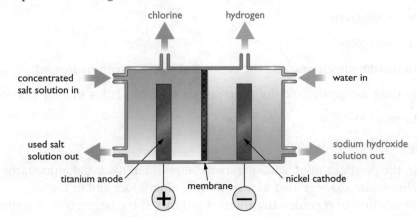

### *Explaining what's happening*

At the anode, chloride ions are discharged to produce **chlorine** gas.

$$2Cl^-(aq) \rightarrow Cl_2(g) + 2e^-$$

The membrane is designed so that only sodium ions can pass through it. The positive sodium ions are attracted into the right-hand compartment by the negatively charged electrode.

It is too difficult to discharge sodium ions, so hydrogen ions from the water are discharged instead to produce **hydrogen** gas.

$$2H^+(aq) + 2e^- \rightarrow H_2(g)$$

More and more water keeps splitting up to replace the hydrogen ions as soon as they are discharged. Each time a water molecule splits up it produces a hydroxide ion as well. This means that there will be a build up of sodium ions and hydroxide ions in the right-hand compartment – **sodium hydroxide solution** is formed.

### Manufacture of bleach

Bleaches contain several different bleaching agents. One example is hypochlorous acid and its salts. The acid itself is unstable and decomposes easily. The sodium salt is more stable and is usually called sodium hypochlorite, NaOCl (also called sodium chlorate(I)). The compound is prepared by electrolysing a cold solution of sodium chloride, common salt. When the products of this reaction are allowed to mix, bleach is produced.

sodium + chlorine → sodium + sodium + water
hydroxide            chloride   hypochlorite

### Uses of the products

Many of the products we use every day rely on the properties of these non-metals. Here are some examples.

Hydrogen is used:

- in the manufacture of ammonia
- in the manufacture of margarine
- as a fuel for the future.

Chlorine is used:

- to sterilise water
- to make bleach
- to make hydrochloric acid (by controlled reaction with hydrogen)
- to make antiseptics and disinfectants (for example, TCP – trichlorophenol)
- to make PVC.

Sodium hydroxide is used:

- in the purification of bauxite to make aluminium oxide ('alumina'): the aluminium oxide is used to manufacture aluminium and in the production of ceramics; the alumina industry is the largest user of sodium hydroxide in Australia.

You might feel that this is a slightly daunting list. Don't panic! You will come across the great majority of it elsewhere in the course. Otherwise, questions are quite likely to be of the form 'State one use for sodium hydroxide'. The worst case is that you could lose 1 mark if you've completely forgotten this list.

- to make soap by reacting the alkali with natural oils and fats, such as palm oil or olive oil. The reaction produces both soap and glycerol.

- to make paper from wood pulp, the strong alkali breaks down the fibrous structure of the wood

- to make the synthetic fibre rayon, once known as artificial silk, or the packaging material cellophane.

## Showing that ions move during electrolysis

So far, we have explained electrolysis in terms of the movement and discharge of ions. Is there any proof that they actually do move? You can show it simply, using substances with coloured ions.

Potassium manganate(VII) contains colourless $K^+$ ions and dark purple $MnO_4^-$ ions.

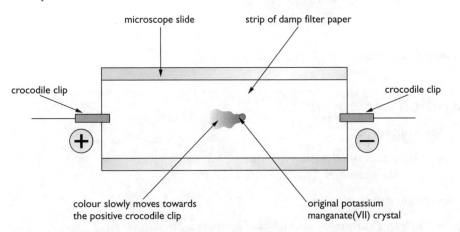

microscope slide  strip of damp filter paper

crocodile clip  crocodile clip

colour slowly moves towards the positive crocodile clip

original potassium manganate(VII) crystal

If a crystal of potassium manganate(VII) is put on a piece of damp filter paper and connected into a circuit, the purple colour spreads noticeably towards the positive end over the course of half an hour or so. This is because the purple $MnO^-_4$ ions are attracted to the positive electrode.

If you had a strongly coloured positive ion, it would obviously move the other way.

> **12.** Name one ion which has a strong colour.

### Teacher demonstration: Moving ions

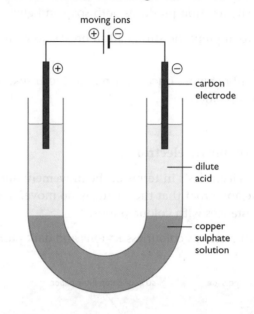

*U-tube demonstration.*

## Activity 2: U-tube demonstration

**Investigative skills**

A

**You will need:**

- U-tube half filled with a mixture of saturated sucrose and copper sulphate solutions. This mixture is very dense.

- dilute sulphuric acid

- droppers, carbon electrodes and a power source (6 volts)

**Carry out the following:**

1. Assemble the apparatus as shown, carefully adding the acid to float upon the blue solution.

2. Switch on the power and wait several minutes.

3. Look for evidence that the blue copper ions are moving towards one of the electrodes.

4. Explain what you see in terms of the charge on the electrode and the charge on the copper ions.

5. What would happen if you were to change over the connectors and reverse the current flow?

# End of Chapter Checklist

**In this chapter you have learnt that:**

- splitting up chemicals using electricity is called electrolysis

- the positive electrode is the anode and the negative electrode is the cathode

- in dilute solutions, the ions from water may be discharged

- the electrochemical series lets us predict the products of electrolysis

- copper and some other metals can be purified by electrolysis

- electroplating may be used to cover one metal with a layer of another one.

# Questions

1. Splitting a substance using electricity is called:

   A   electroplating
   B   conduction
   C   thermal decomposition
   D   electrolysis                                    (1)

2. A negative ion will be attracted to:

   A   the anode
   B   either electrode
   C   the cathode
   D   the top                                         (1)

3. One metal that is purified on a large scale by electrolysis is:

   A   iron
   B   copper
   C   gold
   D   platinum                                        (1)

4. Oxygen always collects at the anode so the ion must have a charge that is:

   A   large
   B   positive
   C   small
   D   negative                                        (1)

5. Which of the following statements is/are true?

   I     electrolysis uses d.c.
   II    electrolysis uses a.c.
   III   electrolysis works with solutions
   IV    electrolysis works with solids

   A   I and III
   B   II and III
   C   III and IV
   D   II and IV                                       (1)

6. *a)* Draw a labelled circuit diagram for the electrolysis of molten lead bromide.                    (4)

   *b)* Why must the lead bromide be molten rather than solid?                                           (2)

   *c)* Describe what you would see at each electrode.   (4)

7. *a)* Why is copper purified using electrolysis?       (2)

   *b)* What is the electrolyte used?                   (1)

   *c)* What happens at the anode and the cathode with copper electrodes?                              (4)

   *d)* Why are the impurities useful?                  (2)

   *e)* give one use for copper.                        (1)

8. *a)* Draw a labelled diagram for the electrolysis of dilute sulphuric acid, allowing you to collect any gases produced at the electrodes.              (5)

   *b)* *i)*   What gases are produced?

       *ii)*  What tests would you perform to identify them?                                         (4)

   *c)* What happens to the sulphate ion?              (1)

9. *a)* Draw diagrams to show the reactions at each electrode in the electrolysis of molten zinc bromide.        (6)

   *b)* It is difficult to collect the bromine. Why is this?   (2)

   *c)* *i)*   What is an external circuit?

       *ii)*  What carries the current in it?          (2)

10. *a)* Why is the electrochemical series useful in electrolysis?                                     (4)

    *b)* Draw a table to show the products of electrolysis of i) dilute sodium chloride, and ii) concentrated sodium chloride.                              (4)

    *c)* How could you identify each of the products?  (2)

**11.** Say what is formed at the cathode and at the anode during the electrolysis of the following substances. Assume that carbon electrodes were used each time. You don't need to write electrode equations.

  a)  molten zinc chloride (1)

  b)  molten lead(II) bromide (1)

  c)  sodium iodide solution (1)

  d)  molten sodium iodide (1)

  e)  copper(II) chloride solution (1)

  f)  dilute hydrochloric acid (1)

  g)  magnesium sulphate solution (1)

  h)  sodium hydroxide solution. (1)

**12.** Some solid potassium iodide was placed in a glass evaporating basin. Two carbon electrodes were inserted and connected to a 12 volt d.c. power source and a light bulb. The potassium iodide was heated. As soon as the potassium iodide was molten, the bulb came on. Purple fumes were seen coming from the positive electrode, and lilac flashes were seen around the negative one.

  a)  Explain why the bulb didn't come on until the potassium iodide melted. (2)

  b)  What name is given to the positive electrode? (1)

  c)  Name the purple fumes seen at the positive electrode, and write the electrode equation for their formation. (2)

  d)  The lilac flashes seen around the negative electrode are caused by the potassium which is formed. The potassium burns with a lilac flame. Write the electrode equation for the formation of the potassium. (2)

  e)  What differences would you expect to observe if you used molten sodium bromide instead of potassium iodide? (2)

  f)  Write the electrode equations for the reactions occurring during the electrolysis of molten sodium bromide. (2)

**13.** For each of the following electrolytes: *i)* write the cathode equation; *ii)* write the anode equation; *iii)* say what has been oxidised and what has been reduced.

  a)  molten lead(II) bromide using carbon electrodes (3)

  b)  sodium chloride solution using carbon electrodes (3)

  c)  calcium bromide solution using carbon electrodes (3)

  d)  copper(II) sulphate solution using carbon electrodes (3)

  e)  copper(II) sulphate solution using copper electrodes (3)

  f)  molten magnesium iodide using carbon electrodes (3)

  g)  dilute hydrochloric acid using carbon electrodes (3)

  h)  silver nitrate solution using silver electrodes. (3)

# Chapter 17: Electrolysis Calculations

> **When you have completed this chapter, you will be able to:**
> - carry out calculations involving coulombs and faradays
> - explain the importance of the Faraday Constant
> - explain electrode equations
> - calculate volumes of gases produced during electrolysis
> - calculate the masses of solid products in electrolysis.

## How to interpret electrode equations

### Moles of electrons

Magnesium is manufactured by electrolysing molten magnesium chloride. Magnesium is produced at the cathode (the negative electrode) and chlorine at the anode (the positive electrode). The electrode equations are:

$$Mg^{2+}(l) + 2e^- \rightarrow Mg(l)$$

$$2Cl^-(l) \rightarrow Cl_2(g) + 2e^-$$

In terms of moles, you can say:

- 1 mole of $Mg^{2+}$ ions gains 2 moles of electrons and produces 1 mole of magnesium, Mg.

- 2 moles of $Cl^-$ ions form 1 mole of chlorine, $Cl_2$, and release 2 moles of electrons.

When you are doing calculations, you just read $e^-$ as '**1 mole of electrons**'.

### Quantities of electricity

#### Coulombs

The coulomb is a measure of quantity of electricity. 1 coulomb is the quantity of electricity which passes if 1 ampere (amp) flows for 1 second.

> You may find this written in symbols using **Q** for the quantity of electricity, **I** for the current in amps, and **t** for the time in seconds. **Q = I × t.**

> **number of coulombs = current in amps × time in sec**

So, if 2 amps flows for 20 minutes, you can calculate the quantity of electricity (not forgetting to convert the time into seconds) as:

$$\text{Quantity of electricity} = 2 \times 20 \times 60 \text{ coulombs}$$
$$= 2400 \text{ coulombs}$$

#### The Faraday Constant

A flow of electricity is a flow of electrons. **1 faraday** is the quantity of electricity which represents 1 mole of electrons passing a particular point in the circuit – in other words, approximately $6 \times 10^{23}$ electrons.

> 1 faraday is more accurately quoted as 96 500 coulombs.

> **1 faraday = 96 500 coulombs (or approximately 96 000 C)**

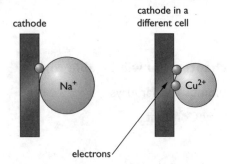

*Each sodium ion needs one electron from the cathode to neutralise its charge. Each copper ion needs twice that number.*

1. What is a coulomb?
2. What is the value of the Faraday Constant and what does it mean?
3. What unit of time must you use when calculating the numbers of coulombs?

### Interpreting electrode equations

In electrolysis calculations you are usually only interested in the quantity of electricity and the mass or volume of the product. For example:

$$Na^+(l) + e^- \rightarrow Na(l)$$

1 mole of sodium, Na, is produced by the flow of 1 mole of electrons (= 1 faraday).

$$Cu^{2+}(aq) + 2e^- \rightarrow Cu(s)$$

1 mole of copper, Cu, is produced by the flow of 2 moles of electrons (= 2 faradays). It takes twice as much electricity to produce a mole of copper as it does a mole of sodium. That's because the $Cu^{2+}$ ion carries twice the charge, and needs twice as many electrons to neutralise it.

$$2Cl^-(l) \rightarrow Cl_2(g) + 2e^-$$

1 mole of chlorine, $Cl_2$, is produced when 2 moles of electrons (= 2 faradays) flow around the circuit.

When 1 mole of chlorine, $Cl_2$, is produced at the anode, 2 moles of electrons are released. This means that the production of 1 mole of chlorine takes place when 2 moles of electrons flow in the circuit.

## Some sample calculations

### Electrolysing copper(II) sulphate solution

What mass of copper is deposited on the cathode during the electrolysis of copper(II) sulphate solution if 0.15 amps flows for 10 minutes?

The electrode equation is:

$$Cu^{2+}(aq) + 2e^- \rightarrow Cu(s)$$

($A_r$: Cu = 64. 1 faraday = 96 000 coulombs)

Start by working out the number of coulombs:

$$\begin{aligned} \text{number of coulombs} &= \text{amps} \times \text{time in seconds} \\ &= 0.15 \times 10 \times 60 \\ &= 90 \end{aligned}$$

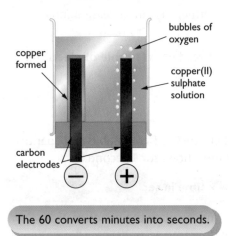

The 60 converts minutes into seconds.

Now work from the equation:

$$Cu^{2+}(aq) + 2e^- \rightarrow Cu(s)$$

2 moles of electrons give 1 mole of copper, Cu

$2 \times 96\,000$ coulombs give 64 g of copper

192 000 coulombs give 64 g of copper

90 coulombs give $\dfrac{90}{192\,000} \times 64$ g = 0.03 g

If you aren't happy with the last line, work out what 1 coulomb produces by dividing 64 by 192 000, and then multiply by 90.

## A calculation involving gases

During the electrolysis of dilute sulphuric acid using unreactive platinum electrodes, hydrogen is released at the cathode and oxygen at the anode. Calculate the volumes of hydrogen and oxygen produced (measured at room temperature and pressure) if 1.0 amp flows for 20 minutes. The electrode equations are:

$$2H^+(aq) + 2e^- \rightarrow H_2(g)$$

$$4OH^-(aq) \rightarrow 2H_2O(l) + O_2(g) + 4e^-$$

(The molar volume of a gas = 24 000 cm$^3$ at rtp. 1 faraday = 96 000 coulombs)

Again start by working out the number of coulombs:

$$\begin{aligned}\text{number of coulombs} &= \text{amps} \times \text{time in seconds}\\ &= 1.0 \times 20 \times 60\\ &= 1200\end{aligned}$$

### Calculating the volume of hydrogen

$$2H^+(aq) + 2e^- \rightarrow H_2(g)$$

2 moles of electrons give 1 mole of hydrogen, $H_2$
2 × 96 000 coulombs give 24 000 cm$^3$ of hydrogen at rtp
192 000 coulombs give 24 000 cm$^3$ of hydrogen at rtp

1200 coulombs give $\dfrac{1200}{192\,000} \times 24\,000$ cm$^3$ = 150 cm$^3$

### Calculating the volume of oxygen

$$4OH^-(aq) \rightarrow 2H_2O(l) + O_2(g) + 4e^-$$

A flow of 4 moles of electrons produces 1 mole of oxygen, $O_2$
4 × 96 000 coulombs produces 24 000 cm$^3$ of oxygen
384 000 coulombs produces 24 000 cm$^3$ of oxygen

1200 coulombs produces $\dfrac{1200}{384\,000} \times 24\,000$ cm$^3$ = 75.0 cm$^3$

Therefore, 150 cm$^3$ of hydrogen and 75.0 cm$^3$ of oxygen are produced.

> 4. What is meant by room temperature and pressure (rtp)?
> 5. When collecting gases, why must we know the temperature and pressure?
> 6. What is the problem with collecting chlorine gas over water?

## A reversed calculation

How long would it take to deposit 0.500 g of silver on the cathode during the electrolysis of silver(I) nitrate solution using a current of 0.250 amp? The cathode equation is:

$$Ag^+(aq) + e^- \rightarrow Ag(s)$$

($A_r$: Ag = 108. 1 faraday = 96 000 coulombs)

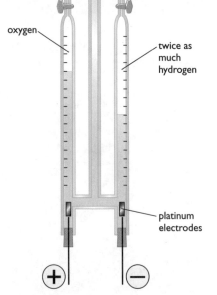

dilute sulphuric acid

oxygen

twice as much hydrogen

platinum electrodes

*Apparatus for electrolysing dilute sulphuric acid and measuring the volume of gases produced.*

### Activity 1: What affects the volume of gas produced?

**Investigative skills**

P

Plan and design an experiment to see if the volume of oxygen produced by the electrolysis of sulphuric acid varies with the size of the current. Draw a diagram of the apparatus you would use. Explain how you would use your results.

If you need help: Divide 96 000 by 108 to find out how many coulombs you need for 1 g of silver. Then multiply by 0.5 to find how many you need for 0.5 g.

1 mole of electrons gives 1 mole of silver, Ag
96 000 coulombs give 108 g of silver

To produce 0.500 g of silver you would need $\dfrac{0.500}{108} \times 96\,000 = 444.4$ coulombs.

$$\text{number of coulombs} = \text{amps} \times \text{time in seconds}$$
$$444.4 = 0.250 \times t$$
$$t = \frac{444.4}{0.250}$$
$$= 1780 \text{ sec}$$

The time needed to deposit 0.500 g of silver is 1780 seconds.

### Electrolysing more than one solution

Suppose you have two solutions connected together in series, so that the same quantity of electricity flows through both.

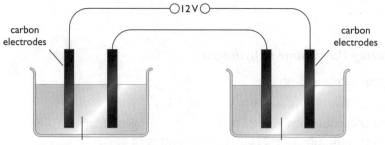

At the end of the electrolysis, it was found that 2.07 g of lead had been deposited on the cathode in the left-hand beaker.

(a) Calculate the quantity of electricity that passed during the experiment.

(b) Calculate the mass of silver that was deposited on the cathode in the right-hand beaker. ($A_r$S: Pb = 207; Ag = 108. 1 faraday = 96 000 coulombs)

### Calculation part (a)

$$Pb^{2+}(aq) + 2e^- \rightarrow Pb(s)$$

2 moles of electrons give 1 mole of lead, Pb

$2 \times 96\,000$ coulombs give 207 g of lead

If      192 000 coulombs give 207 g of lead

then    1920 coulombs give 2.07 g of lead

The quantity of electricity which passed = 1920 coulombs.

### Calculation part (b)

If 1920 coulombs passed through the beaker containing the lead(II) nitrate solution, then exactly the same amount passed through the rest of the circuit.

$$Ag^+(aq) + e^- \rightarrow Ag(s)$$

1 mole of electrons give 1 mole of silver, Ag
96 000 coulombs give 108 g of silver

1920 coulombs give $\dfrac{1920}{96\,000} \times 108$ g = 2.16 g

The mass of silver deposited is 2.16 g.

## An alternative way of solving part (b)

If you weren't asked to find the quantity of electricity in part (a), you could do part (b) much more easily without knowing anything at all about the Faraday Constant or even about coulombs.

Look again at the equations:

$$Pb^{2+}(aq) + 2e^- \rightarrow Pb(s)$$
$$Ag^+(aq) + e^- \rightarrow Ag(s)$$

If 2 moles of electrons flow, you will get 1 mole of lead and 2 moles of silver. However many electrons flow, you will always get twice as many moles of silver as of lead.

In this calculation, 2.07 g of lead were formed, which is 0.01 mol. You will therefore get 0.02 mol of silver = $0.02 \times 108$ g = 2.16 g.

### A similar example involving just one solution

During the electrolysis of concentrated copper(II) chloride solution, 3.2 g of copper was deposited at the cathode. What volume of chlorine (measured at rtp) would be formed at the anode? ($A_r$: Cu = 64. Molar volume = 24 000 cm$^3$ at rtp)

The electrode equations are:

$$Cu^{2+}(aq) + 2e^- \rightarrow Cu(s)$$
$$2Cl^-(aq) \rightarrow Cl_2(g) + 2e^-$$

Notice that for every 2 moles of electrons that flow, you will get 1 mole of Cu and 1 mole of chlorine, $Cl_2$. You are bound to get the same number of moles of each.

In this case, 3.2 g of copper is $\dfrac{3.2}{64}$ mol = 0.05 mol.

So you will also get 0.05 mol of chlorine, which is $0.05 \times 24\,000$ cm$^3$ at rtp = 1200 cm$^3$.

> 7. At which electrode would silver be deposited in the electrolysis of silver nitrate?
> 8. What is the connection between the type of charge on an ion and the electrode to which it moves?
> 9. Why do we use carbon electrodes rather than metal electrodes?

> How do you know which method to use? In an exam, look at the information you are given. If you are given a value for the Faraday Constant (96 000 coulombs), then you would be expected to use it. If you aren't given it, there must be another way of solving the problem.

> This time, you aren't given the Faraday Constant. There must be another way!

# End of Chapter Checklist

## In this chapter you have learnt that:

- quantity of electricity (in coulombs) = current (in amps) x time (in seconds)
- one mole of electrons is called one faraday
- at the electrodes, ions can either gain or lose electrons
- we can calculate the volumes of gases produced in electrolysis
- we can calculate the masses of metals deposited during electrolysis and also work out the size of ionic charges.

# Questions

1. One faraday is:

   A   the charge on the electron
   B   1 mole of electrons
   C   1 mole of protons
   D   a unit of mass                                              (1)

2. When using coulombs, time must be measured in:

   A   hours
   B   any time unit
   C   minutes
   D   seconds                                                     (1)

3. A cation is attracted to:

   A   the water
   B   the anode
   C   the cathode
   D   both electrodes                                             (1)

4. In the electrolysis of dilute sulphuric acid, there is more:

   A   oxygen collected than hydrogen
   B   water formed
   C   hydrogen collected than oxygen
   D   sulphate discharged                                         (1)

5. When measuring the volume of gases produced in electrolysis, we need to know all of these except:

   A   room temperature
   B   room pressure
   C   gas solubility
   D   exact size of the container                                 (1)

6. *a)* Draw a circuit diagram for the electrolysis of water (using dilute sulphuric acid).                       (4)

   *b)* *i)*   What are the products?

   *ii)*  What tests could you carry out to identify them?                                                        (4)

   *c)* What would you see at each electrode?                      (2)

7. *a)* Explain how you can calculate the number of coulombs used in the electrolysis of a solution.             (4)

   *b)* *i)*   What is a faraday?   *ii)*   Why is it useful?       (3)

   *c)* *i)*   Write an electrode equation for the conversion of 1 mole of potassium ions into the metal.

   *ii)*  How many electrons are needed?                           (3)

8. During the electrolysis of copper(II) sulphate solution using copper electrodes, copper from the anode goes into solution according to the equation:

   $$Cu(s) \rightarrow Cu^{2+}(aq) + 2e^-$$

   Calculate the loss in mass from the anode if a current of 0.50 amps flows for 1 hour.                          (8)

   ($A_r$: Cu = 64. 1 faraday = 96 000 coulombs)

9. During the electrolysis of lead(II) nitrate solution, lead is deposited at the cathode, and oxygen is released from the anode. If a current of 0.350 amp flows for 1000 seconds, calculate *a)* the mass of lead deposited, *b)* the volume of oxygen (measured at room temperature and pressure) produced. ($A_r$: Pb = 207. The molar volume of a gas is 24 000 cm$^3$ at rtp. 1 faraday = 96 000 coulombs)    (6)

   $$Pb^{2+}(aq) + 2e^- \rightarrow Pb(s)$$

   $$4OH^-(aq) \rightarrow 2H_2O(l) + O_2(g) + 4e^-$$

10. Some copper(II) sulphate solution was electrolysed using a pure copper cathode and an impure copper anode. Copper is lost from the anode and deposited on the cathode. Insoluble impurities in the anode form a sludge underneath the anode.

   Cathode equation:      $Cu^{2+}(aq) + 2e^- \rightarrow Cu(s)$

   Anode equation:        $Cu(s) \rightarrow Cu^{2+}(aq) + 2e^-$

   *a)* What mass of copper will be deposited on the cathode if 0.40 amps flows for 75 minutes? ($A_r$: Cu = 64. 1 faraday = 96 000 coulombs)                      (3)

*b)* If the anode was found to have lost 0.80 g during the experiment, calculate the percentage purity of the impure copper anode, assuming that only insoluble impurities were present. (3)

11. Aluminium is manufactured by electrolysing a solution of aluminium oxide, $Al_2O_3$, in molten cryolite. The electrode equation is:

$Al^{3+}(l) + 3e^- \rightarrow Al(l)$

A typical cell produces 1 tonne (1000 kg) of aluminium every 24 hours. What current (in amps) is needed to produce this amount of aluminium? ($A_r$S: Al = 27. 1 faraday = 96 000 coulombs) (6)

12. Two solutions were electrolysed in series using the apparatus on page 176. One beaker contained chromium(III) sulphate solution, and the other cobalt(II) sulphate solution. 0.295 g of cobalt was deposited on the cathode in the beaker containing cobalt(II) sulphate. The electrode equations are:

$Cr^{3+}(aq) + 3e^- \rightarrow Cr(s)$

$Co^{2+}(aq) + 2e^- \rightarrow Co(s)$

Calculate *a)* the quantity of electricity which flowed during the experiment, and *b)* the mass of chromium deposited on the cathode in the other beaker. ($A_r$S: Cr = 52; Co = 59. 1 faraday = 96 000 coulombs) (6)

13. Copper(II) sulphate solution and lead(II) nitrate solution were electrolysed in two beakers connected in series. If 0.64 g of copper was deposited at the cathode in one beaker, calculate the mass of lead deposited in the other one. ($A_r$S: Cu = 64; Pb = 207)

$Cu^{2+}(aq) + 2e^- \rightarrow Cu(s)$

$Pb^{2+}(aq) + 2e^- \rightarrow Pb(s)$ (8)

> **When you have completed this chapter, you will be able to:**
>
> - link the history of a metal to its place in the Reactivity Series
> - describe the extraction of aluminium, zinc and chromium
> - give examples of the uses of aluminium
> - understand how a blast furnace produces iron
> - explain the differences between iron and steel
> - understand why alloys are so useful.

## Extracting metals from their ores

### Minerals and ores

Most metals are found in the Earth's crust combined with other elements. The individual compounds are called **minerals**.

*Pyrite (iron pyrites), FeS$_2$*

*Magnetite, Fe$_3$O$_4$*

*Haematite, Fe$_2$O$_3$*

*Native gold on quartz.*

The photographs show samples of some iron-containing minerals, but they are normally found mixed with other unwanted minerals in rocks. An **ore** contains enough of the mineral for it to be worthwhile to extract the metal.

The price of a metal is affected by how common the ore is and how difficult it is to extract the metal from the ore.

A few very unreactive metals like gold are found **native**. That means that they exist naturally as the uncombined element. The photograph shows some native gold. Silver and copper are also sometimes found native – although much more rarely.

### Extracting the metal

Many ores are either oxides or compounds which are easily converted to oxides. Sulphides like sphalerite (zinc blende), ZnS, can be easily converted into an oxide by heating in air.

$$2ZnS(s) + 3O_2(g) \rightarrow 2ZnO(s) + 2SO_2(g)$$

To obtain the metal, you have to remove the oxygen – removal of oxygen is called reduction. Metals exist as positive ions in their ionic compounds, and to produce the metal, you would have to add electrons to the ion. Addition of electrons is also called reduction.

## Methods of extraction and the Reactivity Series

How a metal is extracted depends to a large extent on its position in the Reactivity Series. A manufacturer obviously wants to use the cheapest possible method of reducing an ore to the metal. There are two main economic factors to take into account:

- the cost of energy
- the cost of the reducing agent.

For metals up to zinc in the Reactivity Series, the cheapest method of reducing the ore is often to heat it with carbon or carbon monoxide. Carbon is cheap and can also be used as the source of heat. The extraction of iron is a good example of this.

Ores of metals higher than zinc in the Reactivity Series can't be reduced using carbon at reasonable temperatures.

Metals above zinc are usually produced by electrolysis. The metal ions are given electrons directly by the cathode. Unfortunately, the large amounts of electricity involved make this an expensive process – and so a metal like aluminium is much more expensive than one like iron.

Some metals, like titanium, are extracted by heating the compound with a more reactive metal. This is also bound to be an expensive method, because the more reactive metal itself will have had to be extracted by an expensive process first.

potassium

sodium

lithium

calcium

magnesium

aluminium

**(carbon)**

zinc

iron

copper

*A part of the Reactivity Series.*

### Metals in living systems

Metals are important in living systems. For example, the chlorophyll molecule contains magnesium. Without magnesium, photosynthesis would be impossible. Mammalian blood contains iron as part of the haemoglobin molecule. This forms the essential oxygen-carrying transport system in the body. Trace elements such as zinc, cobalt, iron and nickel are essential for a healthy diet.

Certain metals and their compounds are toxic, for example lead can cause damage to the nervous system.

## Activity 1: Displacement reactions

### Investigative skills

| O | A |
|---|---|

add metal (zinc, copper or magnesium)

copper sulphate solution

zinc sulphate solution

magnesium sulphate solution

**You will need:**

- test-tubes
- spatula
- solutions of copper sulphate, zinc sulphate, magnesium sulphate
- powdered metals: copper, zinc, magnesium

**Carry out the following:**

1. Half fill the test-tubes with each of the three solutions.

2. Add a small amount of zinc powder to each test-tube and look for any changes (colour, gas, new solid).

3. Repeat for the other two metals.

4. Tabulate your results and explain them in terms of the reactivity series.

### The Reactivity Series and the history of metal use

Why was the Bronze Age before the Iron Age? Why wasn't aluminium discovered until 1827?

Bronze is an alloy of copper and tin, both of which are low in the Reactivity Series. Both can be made easily from their ores by heating them with carbon. You can imagine the metals being found accidentally when charcoal (a form of impure carbon) in a fire came into contact with stones containing copper or tin ores.

Iron can also be made from its ores by heating them with carbon, but higher temperatures are needed. The iron produced is also more difficult to purify into a useful form than copper is. Iron therefore wasn't in common use until much later than bronze.

Because aluminium is above carbon in the Reactivity Series, it can't be made accidentally by heating aluminium oxide with carbon. It has to be extracted using electrolysis, and so it was difficult to get metallic aluminium before the discovery of electricity. This is true of all the metals from aluminium upwards in the Reactivity Series.

The method chosen to extract metals from their ores is determined by the position of the metal in the electrochemical series.

## Aluminium

### Extraction

Aluminium is the most common metal in the Earth's crust, making up 7.5% by mass. Its main ore is **bauxite** – a clay mineral which you can think of as impure aluminium oxide.

The bauxite is first treated to produce pure aluminium oxide.

Because aluminium is a fairly reactive metal it has to be extracted using electrolysis. Aluminium oxide, however, has a very high melting point and it isn't practical to electrolyse molten aluminium oxide.

Instead the aluminium oxide is dissolved in molten **cryolite**. This is another aluminium compound which melts at a more reasonable temperature. The electrolyte is a solution of aluminium oxide in molten cryolite at a temperature of about 1000°C.

*Bauxite – essentially impure aluminium oxide.*

*The cell room in an aluminium smelter.*

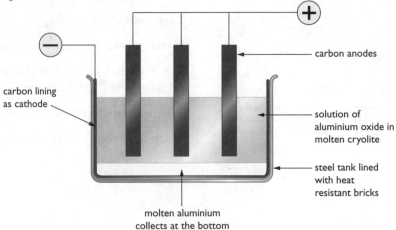

carbon anodes

carbon lining as cathode

solution of aluminium oxide in molten cryolite

steel tank lined with heat resistant bricks

molten aluminium collects at the bottom

The diagram on the previous page shows a very simplified view of the electrolysis cell. The molten aluminium is siphoned off from time to time, and fresh aluminium oxide is added to the cell. The cell operates at about 5–6 volts, but with currents of up to about 100 000 amps. The heat generated by the huge current keeps the electrolyte molten.

The two major costs of extracting aluminium are the cost of electricity and also that of replacing the anodes (positive electrodes) as they burn away. Aluminium smelters are sited where there is good access to low cost electricity. One example would be hydroelectric power.

### The chemistry of the process

Aluminium ions are attracted to the cathode and are reduced to aluminium by gaining electrons.

$$Al^{3+}(l) + 3e^- \rightarrow Al(l)$$

The molten aluminium produced sinks to the bottom of the cell.

The oxide ions are attracted to the anode and lose electrons to form oxygen gas.

$$2O^{2-}(l) \rightarrow O_2(g) + 4e^-$$

This creates a problem. Because of the high temperatures, the carbon anodes burn in the oxygen to form carbon dioxide. The anodes have to be replaced regularly and this adds to the expense of the process.

### Anodising aluminium

Given its position in the Reactivity Series, aluminium is a surprisingly unreactive metal under normal circumstances. Even apparently shiny aluminium is covered in a microscopically thin, but very strong, layer of aluminium oxide which prevents air and water getting at the aluminium and corroding it. This layer can be made even stronger by **anodising** the aluminium.

The aluminium is etched by treating it with sodium hydroxide solution. This removes existing aluminium oxide and produces the satin or matt finishes which are commonly seen in anodised aluminium.

The aluminium article is then made the anode in an electrolysis of dilute sulphuric acid. Oxygen is given off at the anode and this reacts with the aluminium to build up a thin film (about 0.02 mm thick) of aluminium oxide.

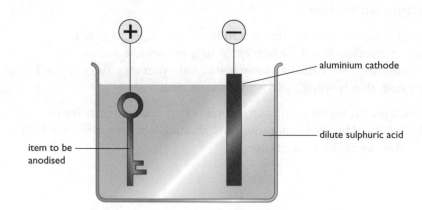

> **The alcoa process**
>
> An alternative method for the extraction of aluminium is the alcoa process. Alumina (aluminium oxide) is reacted with chlorine to produce aluminium chloride. The molten chloride is electrolysed in an inert atmosphere to produce aluminium metal. This method uses less electricity than the cryolite method but both aluminium chloride and chlorine are very corrosive and difficult to handle.

> The exact nature of the cathode is fairly irrelevant. Hydrogen would be evolved at the cathode, and the cathode itself wouldn't change. The anodising industry tends to use aluminium cathodes.

*Dyed anodised aluminium.*

Apart from a very thin barrier layer of aluminium oxide next to the aluminium, the film is porous, and will absorb dyes. Black, gold and bronze coloured dyes are typical of anodised aluminium. After the aluminium has been dyed, further treatment makes the surface completely non-porous.

### Summary

Anodising aluminium:

- makes it more corrosion resistant by strengthening the oxide film
- enables its surface to be dyed.

> 1. Why did it take so many years before we found out how to extract aluminium?
> 2. What is the name given to aluminium ore?
> 3. Why is anodising useful?

### Uses of aluminium

Pure aluminium isn't very strong, so aluminium alloys are normally used instead. The aluminium can be strengthened by mixing it with other elements such as silicon, copper or magnesium.

Aluminium's uses depend on its low density and strength (when alloyed), its ability to conduct electricity and heat, its appearance, and its ability to resist corrosion.

### Uses of aluminium alloys

Aluminium has a shiny appearance, resists corrosion, is light and a good conductor of heat.

Aluminium resists corrosion, is light and a good conductor of electricity. The aluminium in the cables is strengthened by a core of steel.

| Alloy | Other metals in addition to aluminium | Uses | Useful property |
|---|---|---|---|
| duralumin | copper, magnesium | aircraft structures | low density, strong |
| hypoeutectic | silicon, copper | car cylinder blocks | withstands high temperatures, hard |
| aluminium-magnesium | magnesium | marine castings (resists corrosion) | resists salt water corrosion |
| alclad | copper, magnesium, manganese | cladding for buildings | low density, resists corrosion |

### Recycling aluminium

Although aluminium is so common in the Earth's crust, it is only concentrated enough to be worth extracting in ores like bauxite. As with all other ores, these are a **finite resource** and will eventually run out – although in this case, that is going to take a very long time.

The main reason for recycling aluminium is that it is so expensive to produce. Recycling aluminium uses only about 5% of the energy needed to extract the same mass from its ore.

# Iron

## Extraction of iron using a blast furnace

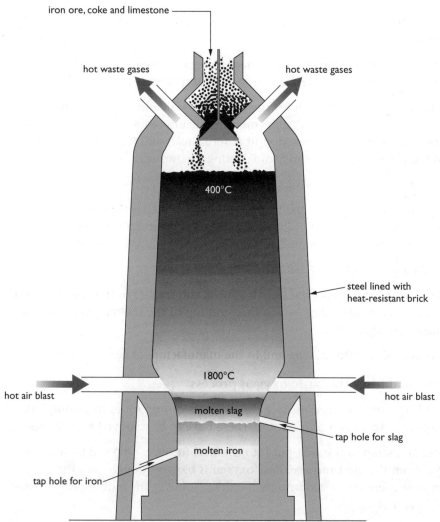

iron ore, coke and limestone

hot waste gases

hot waste gases

400°C

steel lined with
heat-resistant brick

1800°C

hot air blast

hot air blast

molten slag

tap hole for slag

molten iron

tap hole for iron

The hot waste gases at the top of the furnace are piped away and used to heat the air blast at the bottom.

Coke is impure carbon, and it burns in the hot air blast to form carbon dioxide. This is a strongly exothermic reaction.

$$C(s) + O_2(g) \rightarrow CO_2(g)$$

At the high temperatures in the furnace, the carbon dioxide is reduced by more carbon to give carbon monoxide.

$$CO_2(g) + C(s) \rightarrow 2CO(g)$$

It is the carbon monoxide which is the main reducing agent in the furnace – especially in the cooler parts. Assuming that the iron ore is haematite, $Fe_2O_3$:

$$Fe_2O_3(s) + 3CO(g) \rightarrow 2Fe(l) + 3CO_2(g)$$

The iron melts and flows to the bottom of the furnace where it can be tapped off.

An exothermic reaction is one which gives out heat.

In the hotter parts of the furnace, some of the iron oxide is also reduced by carbon itself.

$$Fe_2O_3(s) + 3C(s) \rightarrow 2Fe(l) + 3CO(g)$$

Notice that carbon monoxide, rather than carbon dioxide, is formed at these temperatures.

The limestone is added to the furnace to remove impurities in the ore which would otherwise clog the furnace with solid material.

> Thermal decomposition is splitting a compound into simpler bits using heat.

The furnace is hot enough for the limestone (calcium carbonate) to undergo thermal decomposition. It splits up into calcium oxide and carbon dioxide. This is an endothermic reaction (it absorbs heat) and it is important not to add too much limestone to avoid cooling the furnace.

$$CaCO_3(s) \rightarrow CaO(s) + CO_2(g)$$

> Calcium oxide is a basic oxide because it reacts with acids to form salts. Calcium silicate is a salt formed when calcium oxide and silicon dioxide react. Silicon dioxide is therefore described as an acidic oxide.

Calcium oxide is a basic oxide, and its function is to react with acidic oxides like silicon dioxide, $SiO_2$. Silicon dioxide occurs naturally as quartz, and is typical of the sort of impurities that need to be removed from the furnace.

$$CaO(s) + SiO_2(s) \rightarrow CaSiO_3(l)$$

The product is calcium silicate. This melts and trickles to the bottom of the furnace as a molten **slag**, which floats on top of the molten iron, and can be tapped off separately.

Slag is used in road making and in the manufacture of cement.

### Steel making – the basic oxygen process

The iron from the blast furnace contains lots of impurities including carbon, sulphur, silicon, and phosphorus. These have to be removed to make steel.

Recycled scrap iron is first put into the steel furnace, followed by molten iron from the blast furnace. Pure oxygen is blown through, and the impurities are converted into their oxides. Some of these disappear as gases ($CO_2$ and $SO_2$).

To remove the other inpurities, quicklime (calcium oxide) is added; this combines with them to make a slag which separates from the iron. For example, silicon dioxide would combine to make calcium silicate, exactly as in the blast furnace.

The slag floats on top of the iron and can be skimmed off.

Finally, small calculated amounts of other metals and non-metals are added to the pure iron to produce various different kinds of steel.

*Blowing oxygen through the impure molten iron.*

4. What four things are needed in a blast furnace for the extraction of iron?
5. How is the iron separated from slag?
6. What happens during a thermal decomposition?

## Properties and uses of the different kinds of iron

### Cast iron

Molten iron straight from the furnace can be cooled rapidly and solidified by running it into sand moulds. This is known as **pig iron**. If the pig iron is re-melted and cooled under controlled conditions, **cast iron** is formed. This is very impure iron, containing about 4% carbon as its main impurity.

Cast iron is very fluid when it is molten and doesn't shrink much when it solidifies, and that makes it ideal for making castings. Unfortunately, although cast iron is very hard, it is also very brittle – tending to shatter if it is hit hard. It is used for things like manhole covers, guttering and drainpipes, and cylinder blocks in car engines.

### Wrought iron

Pure iron is known as **wrought iron**. It was once used to make decorative gates and railings, but has now been largely replaced by mild steel. The purity of the iron makes it very easy to work because it is fairly soft, but the softness and lack of strength mean that it isn't useful for structural purposes.

### Mild steel

Mild steel is iron containing up to about 0.25% of carbon. This small amount of carbon increases the hardness and strength of the iron. It is used for (amongst other things) wire, nails, car bodies, ship building, girders and bridges.

*The first ever iron bridge (at Ironbridge in the UK) was made of cast iron.*

*Mild steel.*          *More mild steel.*

### High carbon steel

High carbon steel is iron containing up to 1.5% carbon. Increasing the carbon content makes the iron even harder, but at the same time it gets more brittle. High carbon steel is used for cutting tools and masonry nails. Masonry nails are designed to be hammered into concrete blocks or brickwork, where a mild steel nail would bend. If you miss-hit a masonry nail, it tends to fracture into two bits because of its increased brittleness.

### Stainless steel

Stainless steel is an alloy of iron with chromium and nickel. Chromium and nickel form strong oxide layers in the same way as aluminium, and these oxide layers protect the iron as well. Stainless steel is therefore very resistant to corrosion.

Obvious uses include kitchen sinks, saucepans, knives and forks, and gardening tools, but there are also major uses for it in the brewing, dairy and chemical industries where corrosion-resistant vessels are essential.

*Stainless steel in a modern winery.*

*A mild steel spade tarnishes – a stainless steel one doesn't.*

### Other alloy steels

There are lots of other specialist steels in which the iron is mixed with other elements. For example...

**Titanium steel** (iron alloyed with titanium) has the ability to withstand high temperatures and so is used in gas turbines and spacecraft, for example.

**Manganese steel** is iron alloyed with about 13% manganese and is intensely hard. It is used in rock-breaking machinery, for railway track subjected to very hard wear (for example, points), and for military helmets.

### A summary of types of steel

| Types of steel | Iron mixed with | Some uses | Useful property |
|---|---|---|---|
| mild steel | up to 0.25% carbon | nails, car bodies, ship building, girders | strong, malleable |
| high carbon steel | 0.25–1.5% carbon | cutting tools, masonry nails | very hard |
| stainless steel | chromium and nickel | cutlery, cooking utensils, kitchen sinks | resists corrosion, strong |
| manganese steel | manganese | rock-breaking machinery, military helmets | very hard and strong |

## Rusting

Iron rusts in the presence of oxygen and water. Rusting is accelerated in the presence of electrolytes like salt. **Warning!** Many metals corrode, but it is only the corrosion of iron that is referred to as *rusting*.

The formula of rust is $Fe_2O_3 \cdot xH_2O$, where x is a variable number. It simply behaves as a mixture of iron(III) oxide and water.

Forming this from iron is a surprisingly complicated process. The iron loses electrons to form iron(II) ions, $Fe^{2+}$, which are then oxidised by the air to iron(III) ions, $Fe^{3+}$. Reactions involving the water produce the actual rust.

### Preventing rusting by using barriers

The most obvious way of preventing rusting is to keep water and oxygen away from the iron. You can do this by painting it or coating it in oil or grease. Since water cannot get through the coating, rusting cannot begin. Plastic coated steel is also protected from corrosion as the plastic keeps out the water. If the barrier is damaged, for example when the paint cracks or some grease is wiped off, then corrosion will start. On old cars, you can observe that rusting continues underneath the damaged paint, making it blister and fall off.

### Preventing rusting by alloying the iron

We have already seen that alloying the iron with chromium and nickel to produce stainless steel prevents the iron from rusting. Even if the surface is scratched the stainless steel still won't rust. Unfortunately, stainless steel is expensive.

*Galvanised iron doesn't rust, even in constant contact with air and water.*

## Preventing rusting by using sacrificial metals

**Galvanised iron** is steel which is coated with a layer of zinc. As long as the zinc layer is unscratched, it serves as a barrier to air and water. However, the iron still doesn't rust even when the surface is broken.

Zinc is more reactive than iron, and so corrodes instead of the iron. During the process, the zinc loses electrons to form zinc ions.

$$Zn(s) \rightarrow Zn^{2+}(aq) + 2e^-$$

These electrons flow into the iron. Any iron atom which has lost electrons to form an ion immediately regains them. If the iron can't form ions, it can't rust.

Zinc blocks are attached to metal hulls or keels for the same reason. The corrosion of the more reactive zinc prevents the iron rusting. Such blocks are called **sacrificial anodes**.

Underground pipelines are also protected using sacrificial anodes. In this case, sacks containing lumps of magnesium are attached at intervals along the pipe. The very reactive magnesium corrodes in preference to the iron. The electrons produced as the magnesium forms its ions prevent ionisation of the iron.

*A sacrificial anode attached to a boat.*

## Activity 2: Corrosion of metals

**Investigative skills**

| O | A |
|---|---|

**You will need:**

- five test-tubes
- stoppers
- test-tube rack
- dropper
- metal samples: iron nails, aluminium foil pieces
- drying agent such as anhydrous calcium chloride
- vegetable oil
- salt

**Carry out the following:**

1. Set up the test-tubes according to the following pattern.

| Tube | 1 | 2 | 3 | 4 | 5 |
|------|---|---|---|---|---|
| **Contents** | water + nail | boiled water + nail + stopper (tube full of water) | boiled water + nail + 1 cm depth of oil on top (tube full of water) | dry tube, drying agent, nail and stopper | salt water + nail |
| **Conditions** | water + air | water but no air | water but no air | air but no water | salty water + air |

2. Boiling the water drives out any dissolved air (containing oxygen).
3. Leave for at least 5 days and look for signs of corrosion.
4. What conditions are needed for iron to rust?
5. Repeat the experiment, replacing the nails with aluminium foil.
6. How do the results compare?

## Activity 3: Corrosion race

### Investigative skills

P

Plan an experiment to compare how easily galvanised iron nails go rusty compared to ordinary iron nails. Make a prediction and give a reason for your choice.

*A zinc ore, sphalerite.*

7. Which metals make up stainless steel?
8. Give one use for mild steel.
9. What is galvanised iron? Why is it useful?

## Zinc

The major ore of zinc is zinc sulphide, ZnS. This can be converted into zinc oxide by heating in air. Zinc metal can be extracted from the oxide in two different ways.

### Electrolysis

About 85% of the zinc required each year is extracted this way.

The electrolyte is a solution of zinc sulphate, prepared from the oxide.

Zinc oxide + sulphuric acid → zinc sulphate + water

The anode is made of lead and the cathode of aluminium. Cathode reaction:

$$Zn^{2+}(aq) + 2e^- \rightarrow Zn(s)$$

The zinc metal builds up a layer around the cathode. When the cathode is removed, the zinc can be stripped off. The zinc is very pure indeed, 99.995% pure metal.

### Thermal extraction

This method accounts for about 15% of the annual zinc production. Coke is a form of carbon, as used in the blast furnace for making iron. Zinc oxide is heated with coke in a furnace.

Zinc oxide + carbon → zinc + carbon monoxide

Carbon monoxide is a good reducing agent. Once formed, this gas reduces more of the zinc oxide to the metal.

Zinc oxide + carbon monoxide → zinc + carbon dioxide

Zinc metal boils at about 900°C, a lower temperature than inside the furnace. The result is that zinc metal distils and can be condensed by cooling. The thermal process produces zinc that is 98.5% pure.

## Uses of zinc

| Alloy | Other metals in addition to zinc | Uses | Useful property |
|---|---|---|---|
| brass | copper | plumbing | resists corrosion |
| casting alloys | aluminium | castings for machines and toys | easy to form complicated shapes |
| zinc metal | none | galvanising (zinc coated steel) | protects steel from corrosion |

# Chromium

## Chromium extraction

The major ore of chromium is chromite, $FeCr_2O_4$. Some chromium is extracted directly from this ore to produce a mixture (alloy) of iron and chromium. The alloy is used to prepare stainless steel. If pure chromium is needed, the chromite must first be converted into chromium oxide, $Cr_2O_3$. The chromium metal is extracted by heating the oxide with aluminium. Aluminium is higher in the reactivity series than chromium. Aluminium can remove the oxygen from chromium oxide. We say that the aluminium is acting as a reducing agent.

Chromium oxide + aluminium → chromium + aluminium oxide

This is a spectacular reaction that gives out lots of heat (exothermic reaction). It is called the thermite reaction.

## Uses of chromium

| Alloy | Other metals in addition to chromium | Uses | Useful property |
|---|---|---|---|
| ferrochrome | iron | stainless steel | does not rust |
| chrome-vanadium | vanadium | drill bits | very hard |
| chromium metal | none | electroplating (coating other metals) | resistant to corrosion |

zinc oxide ZnO

hot coke (carbon)

zinc vapour

zinc vapour is condensed

zinc oxide + carbon monoxide
→ zinc + carbon dioxide
$ZnO + CO →$
$Zn + CO_2$

air          air

*Zinc furnace.*

*Stainless steel objects.*

# End of Chapter Checklist

**In this chapter you have learnt that:**

- metals low in the Reactivity Series were discovered first
- aluminium is extracted from bauxite by electrolysis
- zinc can be extracted by electrolysis or heating
- chromium can be extracted by the thermite process
- iron is extracted from iron oxide in a blast furnace
- steel alloys are stronger and less brittle than cast iron
- alloys have many useful properties.

# Questions

1. Bauxite is the ore of:

   A  lead
   B  titanium
   C  aluminium
   D  iron                                                    (1)

2. Aluminium is extracted from bauxite using:

   A  a blast furnace
   B  heating
   C  panning
   D  electrolysis                                            (1)

3. The materials used in a blast furnace are iron ore, limestone, coke and:

   A  oxygen
   B  bauxite
   C  air
   D  nickel                                                  (1)

4. Most car bodies are made using:

   A  stainless steel
   B  mild steel
   C  manganese steel
   D  titanium steel                                          (1)

5. Which of the following statements is/are true?

   I    lead has a high density
   II   lead is used in car batteries
   III  lead can be a health hazard
   IV   lead is a reactive metal

   A  I and II
   B  I, II and III
   C  III and IV
   D  I and IV                                                (1)

6. *a)* Why was copper discovered before aluminium?
       Explain your answer.                                  (5)

   *b)* Give *one* use each for copper and aluminium.        (2)

   *c)* What elements are immediately above and below
       copper in the Reactivity Series?                      (2)

   *d)* What is an alloy?                                     (1)

7. *a)* Explain how aluminium is extracted from its ore.
       Include a labelled diagram in your explanation.       (5)

   *b)* What are the name and formula of aluminium ore?      (2)

   *c)* Give the electron equation for the cathode reaction
       in the extraction of aluminium.                       (1)

   *d)* What is meant by anodising?                          (2)

8. *a)* Give *three* uses of aluminium metal and explain why
       it is chosen for these uses.                          (6)

   *b)* Why do we recycle aluminium?                         (2)

   *c)* Name *two* food-linked products that use aluminium.  (2)

9. *a)* What are the four materials needed in a blast
       furnace?                                              (4)

   *b)* How is the iron removed from the furnace?            (1)

   *c)* What is slag?                                         (1)

   *d)* What is the difference between cast iron and mild
       steel?                                                (4)

10. *a)* What is stainless steel?                            (2)

    *b)* Give *two* uses of stainless steel and explain why it is
        chosen for these uses.                               (4)

    *c)* Why don't we make all cars from stainless steel?    (2)

    *d)* Describe the use of *one* other steel alloy.        (2)

11. Give *two* uses for each of the following metals. In each case explain why the metal is particularly suitable for that purpose.

a) aluminium, b) iron (as mild steel), c) titanium, d) copper. (8)

12. a) Name the ore from which aluminium is extracted. (1)

b) Aluminium is manufactured using electrolysis. Carbon electrodes are used. Describe the nature of the electrolyte. (2)

c) At which electrode is the aluminium produced? (2)

d) Write the electrode equation for the formation of the aluminium. Is this an example of oxidation or reduction? (2)

e) Oxygen gas is formed at the other electrode. Explain why that causes a problem. (2)

f) Aluminium is frequently *anodised*.

i) Describe how you would anodise a piece of sheet aluminium. (3)

ii) Give *two* reasons why a manufacturer might choose to anodise an aluminium product. (2)

g) Aluminium alloys are used in aircraft construction.

i) What property of aluminium makes it particularly suitable for this purpose? (1)

ii) Why are aluminium alloys used in preference to pure aluminium? (2)

13. The following reactions take place in a blast furnace:

A $C(s) + O_2(g) \rightarrow CO_2(g)$ $\Delta H = -394 \text{ kJ mol}^{-1}$

B $CO_2(g) + C(s) \rightarrow 2CO(g)$ $\Delta H = +172 \text{ kJ mol}^{-1}$

C $Fe_2O_3(s) + 3CO(g) \rightarrow 2Fe(l) + 3CO_2(g)$

$\Delta H = +4 \text{ kJ mol}^{-1}$

D $CaCO_3(s) \rightarrow CaO(s) + CO_2(g)$ $\Delta H = +178 \text{ kJ mol}^{-1}$

a) Which one of these reactions provides the heat to maintain the temperature of the furnace? (2)

b) What materials are put into the furnace to provide sources of i) carbon, ii) oxygen, iii) iron(III) oxide, iv) calcium carbonate? (4)

c) What types of reaction are reactions B and C? Explain your answers. (2)

d) The calcium oxide produced in reaction D takes part in the formation of slag.

i) Give a chemical name for slag. (1)

ii) Write an equation to show its formation. (2)

iii) State one use for slag. (1)

e) Some iron is also produced by reaction between iron(III) oxide and carbon. Balance the following equation:

$Fe_2O_3(s) + C(s) \rightarrow Fe(l) + CO(g)$ (2)

14. a) Cast iron or pig iron from the bottom of the blast furnace contains an important impurity which limits its usefulness.

i) What is the impurity, and approximately what percentage of the cast iron does it make up? (2)

ii) What effect does this impurity have on the properties of cast iron which limits its usefulness? (2)

b) Describe the composition of stainless steel, and explain why it resists corrosion. State *one* use for stainless steel. (2)

c) Car bodies used to be made from mild steel which was then painted. In more recent cars, the steel is galvanised before it is painted. Describe and explain the effect that this has on the life of the car. (2)

15. Explain the reasons behind each of the following statements.

a) Although aluminium is fairly high in the Reactivity Series, it is slow to corrode. (2)

b) Overhead power cables are made of aluminium with a steel core. (2)

c) Blocks of zinc are attached to the hulls of steel ships. (2)

d) Gold is found native (uncombined) in nature, and yet gold is a very expensive metal. (2)

e) It is a good idea to recycle aluminium. (1)

# End of Section Questions

1. In an experiment to investigate the rate of decomposition of hydrogen peroxide solution in the presence of manganese(IV) oxide, 10 cm³ of hydrogen peroxide solution was mixed with 30 cm³ of water, and 0.2 g of manganese(IV) oxide was added. The volume of oxygen evolved was measured at 60 second intervals. The results of this experiment are recorded in the table below.

| Time (sec) | 0 | 60 | 120 | 180 | 240 | 300 |
|---|---|---|---|---|---|---|
| Vol (cm³) | 0 | 30 | 48 | 57 | 60 | 60 |

a) Balance the equation for the decomposition of hydrogen peroxide, including all the appropriate state symbols.

$$H_2O_2 \rightarrow H_2O + O_2$$

(2 marks)

b) Copy and complete the diagram to show how the volume of oxygen might have been measured.

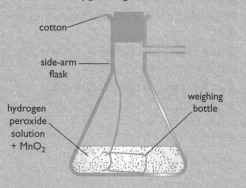

(2 marks)

c) Plot a graph of the results on a piece of graph paper, with time on the horizontal axis and volume of oxygen on the vertical axis. (4 marks)

d) Use your graph to find out how long it took to produce 50 cm³ of oxygen. (1 mark)

e) Explain why the graph becomes horizontal after 240 seconds. (2 marks)

f) Suppose the experiment had been repeated using the same quantities of everything but with the

reaction flask immersed in ice. Sketch the graph you would expect to get. Use the same grid as in part c). Label the new graph **F**. (2 marks)

g) On the same grid as in c) and f), sketch the graph you would expect to get if you repeated the experiment at the original temperature using 5 cm³ of hydrogen peroxide solution, 35 cm³ of water and 0.2 g of manganese(IV) oxide. Label this graph **G**. (2 marks)

**Total 15 marks**

2. Sodium chloride solution was electrolysed using the apparatus in the diagram.

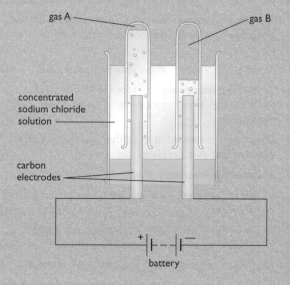

a) Name i) gas A, ii) gas B. (2 marks)

b) Describe how you would test for i) gas A, ii) gas B. (4 marks)

c) Suppose the sodium chloride solution was replaced by potassium iodide solution. What differences would you observe (if any) at i) the positive electrode, ii) the negative electrode? (3 marks)

d) The same apparatus was used to electrolyse another solution, C. A brown solid was formed on the negative electrode, and an orange solution around the positive one. Suggest a possible identity for solution C. (2 marks)

**Total 11 marks**

3. Reactions can be described as (amongst other things): neutralisation, precipitation, redox, thermal decomposition. Decide which of these types of reaction each of the following equations represents.

a) $Zn(s) + CuO(s) \rightarrow ZnO(s) + Cu(s)$ (1 mark)

b) $ZnCO_3(s) \rightarrow ZnO(s) + CO_2(g)$ (1 mark)

c) $ZnCO_3(s) + H_2SO_4(aq) \rightarrow$
$$ZnSO_4(aq) + CO_2(g) + H_2O(l)$$
(1 mark)

d) $Zn^{2+}(aq) + CO_3{}^{2-}(aq) \rightarrow ZnCO_3(s)$ (1 mark)

e) $Zn(s) + Pb^{2+}(aq) \rightarrow Zn^{2+}(aq) + Pb(s)$ (1 mark)

f) $Zn(s) + 2HCl(aq) \rightarrow ZnCl_2(aq) + H_2(g)$ (1 mark)

**Total 6 marks**

# Chapter 19: The Periodic Table – Introduction

**When you have completed this chapter, you will be able to:**

- understand the history of the Periodic Table
- explain the differences between groups and periods
- describe the physical properties of metals
- describe the reactions of metals with oxygen, water and acids
- compare the action of heat on metallic nitrates, carbonates and hydroxides
- describe the uses of the noble gases.

## The search for patterns

### Johann Wolfgang Döbereiner

> At this point in the history of Chemistry, atomic weight was measured relative to the mass of a hydrogen atom. For example, a sodium atom weighed 23 times as much as a hydrogen atom.

In work beginning in 1817, Döbereiner investigated some groups of three elements with similar chemical properties – for example, lithium, sodium and potassium. He noticed that the atomic weight (what we would now call the relative atomic mass) of the middle one was the average of the other two:

lithium: 7    sodium: 23    potassium: 39

$$23 = (7 + 39) \div 2$$

He found similar patterns with calcium, strontium and barium; with sulphur, selenium and tellurium; and with chlorine, bromine and iodine. These became known as Döbereiner's triads.

At the time, this was no more than a curiosity, but the real advances some 50 years later were again based on patterns in atomic weights.

### John Newlands

> For the moment we shall continue to use the term 'atomic weight' rather than relative atomic mass, because that's what it was called at the time.

In 1864–65, Newlands noticed that by arranging the known elements into atomic weight order, elements with similar properties recurred at regular intervals. Every eighth element was similar. He proposed a law which said that 'The properties of the elements are a periodic function of their atomic weights'.

By drawing an analogy with music, his law became known as the 'Law of Octaves'.

| 1 | 2 | 3 | 4 | 5 | 6 | 7 |
|---|---|---|---|---|---|---|
| H | Li | Be | B | C | N | O |
| F | Na | Mg | Al | Si | P | S |
| Cl | K | Ca | Cr | Ti | Mn | Fe |

This works up to a point, but it soon goes wrong. For example, there is no similarity at all between the metal manganese, Mn, and the non-metals nitrogen and phosphorus in Newlands' sixth group.

Newlands' ideas were greeted with ridicule at the time. One of his critics suggested that he might as well have arranged his elements in alphabetical order. With hindsight his theory was bound to fail.

There were lots of undiscovered elements, and some of the known ones had been given the wrong atomic weights. We also know now that the patterns get more complicated when you get beyond calcium.

### Dimitri Mendeleev

In 1869, only a few years after Newlands, Mendeleev used the same idea but improved it almost beyond recognition.

He arranged the elements in atomic weight order, but accepted that the pattern was going to be more complicated than Newlands suggested. Where necessary he left gaps in the table to be filled by as yet undiscovered elements. He also had the confidence to challenge existing values for atomic weight where they forced an element into the wrong position in his Periodic Table.

And he was right! One of Mendeleev's early triumphs was to leave a gap underneath silicon in the table for a new element which he called 'ekasilicon'. He predicted what the properties of this element and its compounds would be. When what we now call germanium was isolated in 1886, it proved to have almost exactly the properties that Mendeleev predicted.

### Part of Mendeleev's Table

Dimitri Mendeleev. You may find various spellings of 'Mendeleev'. Don't worry about this. They are all attempts at converting a Russian name into English.

| | | Group | | | | | | | |
|---|---|---|---|---|---|---|---|---|---|
| | | 1 | 2 | 3 | 4 | 5 | 6 | 7 | 8 |
| Period 1 | | H | | | | | | | |
| Period 2 | | Li | Be | B | C | N | O | F | |
| Period 3 | | Na | Mg | Al | Si | P | S | Cl | |
| Period 4 | | K | Ca | * | Ti | V | Cr | Mn | Fe Co Ni |
| | | Cu | Zn | * | * | As | Se | Br | |
| Period 5 | | Rb | Sr | Y | Zr | Nb | Mo | * | Ru Rh Pd |
| | | Ag | Cd | In | Sn | Sb | Te | I | |

You might notice that the noble gases (helium, neon, argon and the rest) are missing – they weren't known at this time. When argon was discovered in 1894, it was obvious that it didn't fit in anywhere in the existing table.

Argon was put into a new group – Group 0. This suggested that there must be other members of this group, and helium, neon, krypton and xenon were quickly found over the next four years.

There were some residual problems in Mendeleev's Table which had to be left for later. The atomic weight order wasn't *exactly* followed.

For example, tellurium, Te, has a slightly higher atomic weight than iodine – but iodine obviously has to go where it is in the Table because its properties are very similar to chlorine and bromine.

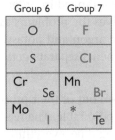

| Group 6 | Group 7 |
|---------|---------|
| O | F |
| S | Cl |
| Cr / Se | Mn / Br |
| Mo / I | * / Te |

correct order in terms
of atomic weight

| Group 6 | Group 7 |
|---------|---------|
| O | F |
| S | Cl |
| Cr / Se | Mn / Br |
| Mo / Te | * / I |

Mendeleev swapped these so that their
properties matched the rest of the group.

Another similar problem occurs when you come to slot argon into the Table.

Argon's atomic weight is slightly higher than potassium's. Putting the elements in their correct order of increasing atomic weight would put argon into the same group as lithium and sodium, and potassium in with gases like neon. That's obviously wrong!

The need for sub-groups (starting in Period 4) is also awkward. For example, lithium, sodium, potassium and rubidium are described as being in Group 1a; copper and silver are in Group 1b. There's not much resemblance between the two sub-groups.

These problems are completely overcome in the modern version of the Periodic Table.

| Group 0 | Group 1 |
|---------|---------|
| He | Li |
| Ne | Na |
| K | Ar |

*This is what the table would look like if potassium and argon were put in the order of their atomic weights.*

1. When did Mendeleev publish his periodic table?
2. What was the problem that Mendeleev found with argon and potassium?
3. How were the elements arranged in the first periodic table?

## The modern Periodic Table

| | 1 | 2 | ← Groups → | | | | | | | | | | | 3 | 4 | 5 | 6 | 7 | 0 |
|---|---|---|---|---|---|---|---|---|---|---|---|---|---|---|---|---|---|---|---|
| 1 | | | | | | | H | | | | | | | | | | | | | He |
| 2 | Li | Be | | | | | | | | | | | | | B | C | N | O | F | Ne |
| 3 | Na | Mg | | | | transition elements | | | | | | | | | Al | Si | P | S | Cl | Ar |
| 4 | K | Ca | Sc | Ti | V | Cr | Mn | Fe | Co | Ni | Cu | Zn | Ga | Ge | As | Se | Br | Kr |
| 5 | Rb | Sr | Y | Zr | Nb | Mo | Tc | Ru | Rh | Pd | Ag | Cd | In | Sn | Sb | Te | I | Xe |
| 6 | Cs | Ba | La | Hf | Ta | W | Re | Os | Ir | Pt | Au | Hg | Tl | Pb | Bi | Po | At | Rn |
| 7 | Fr | Ra | Ac | | | | | | | | | | | | | | | |

inner transition elements

| | Ce | Pr | Nd | Pm | Sm | Eu | Gd | Tb | Dy | Ho | Er | Tm | Yb | Lu |
|---|---|---|---|---|---|---|---|---|---|---|---|---|---|---|
| | Th | Pa | U | Np | Pu | Am | Cm | Bk | Cf | Es | Fm | Md | No | Lr |

Periods

*The Periodic Table.*

In the modern Periodic Table, the elements are arranged in order of atomic number – not of relative atomic mass. That solves the problems associated with the positions of tellurium and iodine, and argon and potassium. Stretching the Table to keep the transition elements separate gets rid of the problem of having sub-groups.

The inner transition elements are usually dropped out of their proper places, and written separately at the bottom of the Periodic Table. The reason for this isn't very subtle. If you put them where they should be, everything has to be drawn slightly smaller to fit on the page. That makes it more difficult to read.

*Full width Periodic Table.*

Most Periodic Tables stop at the end of the second inner transition series. Some continue beyond that to include a small number of more recently discovered elements.

New elements are still being discovered, although they are highly radioactive and have incredibly short existences. Knowing where they are in the Periodic Table, however, means that you can make good predictions about what their properties would be.

### The Periodic Table and atomic structure

Remember that the atomic number counts the number of protons in the atoms of the element, and therefore the number of electrons in the neutral atom. Elements in the same group have the same number of electrons in their outer energy levels (shells). That governs how they react and means that they are likely to have similar chemical properties.

There is, however, a change in properties (sometimes gradual, sometimes quite rapid) from the top to the bottom of a group. You will find examples later in this chapter.

### Metals and non-metals in the Periodic Table

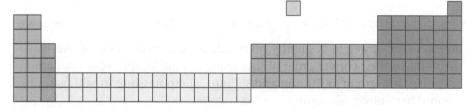

Although the division into metals and non-metals is shown as clear-cut, in practice there is a lot of uncertainty on the dividing line. For example, arsenic, As, has properties of both metals and non-metals. Even tin, Sn, is close enough to the boundary that it is possible to find some non-metallic character hidden away in its properties.

# Characteristics of metals

### What is a typical metal?

Almost all metals have certain properties in common:

- they conduct heat well (e.g. calcium)
- they conduct electricity well (e.g. magnesium or copper)
- they are shiny (clean surface); this is known as metallic lustre (e.g. polished calcium or aluminium)
- they are malleable, i.e. they can be rolled into thin sheets (e.g. magnesium ribbon or copper foil)
- they are ductile, i.e. they can be pulled into thin wires (e.g. copper, silver or gold)
- they are sonorous, i.e. they make a sound when hit (e.g. copper or iron).

Some metals have unusual properties. For example:

- mercury is a liquid at room temperature
- sodium floats on water.

We need to examine the physical and chemical properties of metals to find their similarities and differences. Most of the elements in the Periodic Table are metals.

### Physical properties of metals

| Metal | Symbol | Melting point/°C | Boiling point/°C | Thermal conductivity/ W m$^{-1}$ K$^{-1}$ | Density/ g cm$^{-3}$ | Colour |
|---|---|---|---|---|---|---|
| aluminium | Al | 660 | 2 450 | 237 | 2.69 | silvery |
| calcium | Ca | 850 | 1 490 | 200 | 1.55 | silvery |
| copper | Cu | 1 083 | 2 567 | 401 | 8.95 | brown |
| iron | Fe | 1 528 | 2 750 | 80 | 7.87 | silvery |
| magnesium | Mg | 650 | 1 100 | 156 | 1.74 | silvery |
| sodium | Na | 98 | 885 | 141 | 0.97 | silvery |
| zinc | Zn | 419 | 907 | 116 | 7.1 | silvery |

4. Water has a density of 1 g cm$^{-3}$. Which of the metals in the table float on water?
5. Which metal is the best conductor of heat?
6. Which metal has the lowest boiling point?

## Chemical properties of metals

Metals react with oxygen to form oxides. Some very reactive metals, such as sodium, are stored under oil to keep oxygen and moisture away. Iron rusts when exposed to both oxygen and moisture. Some metals, such as sodium, burn brightly in oxygen gas.

| Metal | Reaction with oxygen | Formula of oxide |
|---|---|---|
| sodium | burns rapidly, white oxide | $Na_2O$ |
| calcium | burns less rapidly, white oxide | $CaO$ |
| magnesium | ribbon burns well, white oxide | $MgO$ |
| aluminium | powder burns well, white oxide | $Al_2O_3$ |
| zinc | powder burns, white oxide | $ZnO$ |
| iron | powder or wire burns, black oxide | $Fe_3O_4$ |
| copper | slow darkening on heating, black oxide | $CuO$ |

### Activity 1: Reaction of iron with oxygen

**Investigative skills**

**P**

Plan an experiment to find out if iron wool weighs more or less after burning in oxygen gas. Explain what apparatus you would use and what measurements you would need to take.

Example equations:

$$\text{magnesium} + \text{oxygen} \rightarrow \text{magnesium oxide}$$
$$2Mg(s) + O_2(g) \rightarrow 2MgO(s)$$

$$\text{copper} + \text{oxygen} \rightarrow \text{copper oxide}$$
$$2Cu(s) + O_2(g) \rightarrow 2CuO(s)$$

Metals react with water and steam to give oxides or hydroxides.

| Metal | Reaction with water |
|---|---|
| sodium | rapid, produces sodium hydroxide and hydrogen gas |
| calcium | quite rapid, produces calcium hydroxide and hydrogen gas |
| magnesium | slow with water, rapid with steam producing magnesium oxide and hydrogen gas |
| aluminium | no reaction, even with steam – metal is protected by oxide coating |
| zinc | slow in steam, produces zinc oxide and hydrogen gas |
| iron | very slow (rusting); needs oxygen in addition to water |
| copper | no reaction at all |

Example equations:

$$\text{sodium} + \text{water} \rightarrow \text{sodium hydroxide} + \text{hydrogen}$$
$$2Na(s) + 2H_2O(aq) \rightarrow 2NaOH(aq) + H_2(g)$$

$$\text{magnesium} + \text{steam} \rightarrow \text{magnesium oxide} + \text{hydrogen}$$
$$2Mg(s) + H_2O(aq) \rightarrow 2MgO(s) + H_2(g)$$

## Activity 2: Metals and water

### Investigative skills

| O | A |
|---|---|

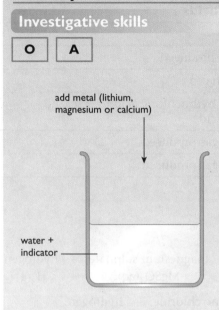

add metal (lithium, magnesium or calcium)

water + indicator

**You will need:**

- safety screen
- sandpaper
- 100 cm³ beakers
- metals: lithium, calcium, magnesium
- indicator such as phenolphthalein

**Carry out the following:**

1. Half fill a beaker with cold water and add three drops of indicator.

2. Add a very small (2 mm³) piece of lithium. View from behind the safety screen.

3. Observe any changes (gas evolved, movement, colour change).

4. Repeat using calcium and magnesium, each rubbed with sandpaper to give a clean surface.

5. Tabulate your results and put the metals in order of reactivity.

---

7. Why is copper used for water pipes?
8. Name two metals that react with steam.
9. Give the chemical formulae for three metal oxides.

---

Metals react with acids to form a salt and hydrogen gas. Hydrochloric acid gives chloride salts and sulphuric acid gives sulphate salts.

## Activity 3: Metals and acids

### Investigative skills

| O | A |
|---|---|

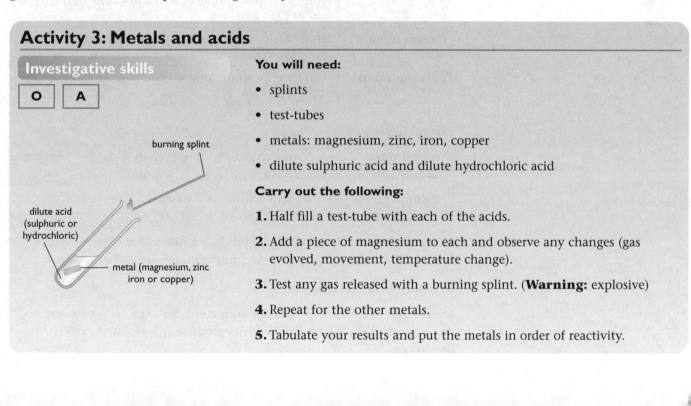

burning splint

dilute acid (sulphuric or hydrochloric)

metal (magnesium, zinc iron or copper)

**You will need:**

- splints
- test-tubes
- metals: magnesium, zinc, iron, copper
- dilute sulphuric acid and dilute hydrochloric acid

**Carry out the following:**

1. Half fill a test-tube with each of the acids.

2. Add a piece of magnesium to each and observe any changes (gas evolved, movement, temperature change).

3. Test any gas released with a burning splint. (**Warning:** explosive)

4. Repeat for the other metals.

5. Tabulate your results and put the metals in order of reactivity.

*Summary of the reactions of metals with acids*

| Metal | Reaction with dilute acids |
|---|---|
| sodium | violent, very rapid |
| calcium | very rapid, hydrogen gas produced |
| magnesium | rapid, hydrogen gas produced |
| aluminium | needs moderately strong hydrochloric acid to react, no reaction with sulphuric acid |
| zinc | slow reaction, hydrogen gas produced |
| iron | slow reaction, hydrogen gas produced |
| copper | no reaction at all |

Example equations:

magnesium  +  sulphuric acid  →  magnesium sulphate  +  hydrogen
$Mg(s)$    +    $H_2SO_4(aq)$    →       $MgSO_4(aq)$        +     $H_2(g)$

zinc  +  hydrochloric acid  →  zinc chloride  +  hydrogen
$Zn(s)$ +      $2HCl(aq)$       →       $ZnCl_2(aq)$     +      $H_2(g)$

## Action of heat on metal compounds

Many metal compounds are changed by heat. We call it **thermal decomposition**.

### Action of heat on nitrates

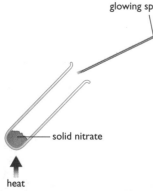

glowing splint

solid nitrate

heat

| Nitrate | Changes and products |
|---|---|
| copper nitrate | blue to black; produces copper oxide, nitrogen dioxide and oxygen gas |
| lead nitrate | white to yellow, produces lead oxide, nitrogen dioxide and oxygen gas |
| sodium nitrate | white, melts, white solid on cooling, produces nitrite and oxygen gas |
| potassium nitrate | white, melts, white solid on cooling, produces nitrite and oxygen gas |

Example equations:

copper nitrate  →  copper oxide  +  nitrogen dioxide  +  oxygen
$2Cu(NO_3)_2(s)$   →      $2CuO(s)$       +        $4NO_2(g)$        +      $O_2(g)$

sodium nitrate  →  sodium nitrite  +  oxygen
$2NaNO_3(s)$    →       $2NaNO_2(s)$      +      $O_2(g)$

Metals high in the Reactivity Series produce a nitrite and oxygen. Metals lower in the Series produce an oxide, nitrogen dioxide and oxygen gas.

### Action of heat on carbonates

Metals high in the Reactivity Series do not change – the carbonates are not decomposed. Metals lower in the Series produce an oxide and carbon dioxide gas.

Example equation:

calcium carbonate   →   calcium oxide   +   carbon dioxide

$$CaCO_3(s) \rightarrow CaO(s) + CO_2(g)$$

Most carbonates of metals lower in the Reactivity Series than calcium behave in the same way.

### Action of heat on hydroxides

Metals at the top of the Reactivity Series have hydroxides which are unchanged by heat, such as potassium and sodium. Metals lower in the Series are decomposed to an oxide and water.

Example equation:

copper hydroxide   →   copper oxide   +   water

$$Cu(OH)_2(s) \rightarrow CuO(s) + H_2O(l)$$

## Group 0 – The noble gases

### Physical properties

The noble gases are all colourless gases. Radon, at the bottom of the Group, is radioactive. Argon makes up almost 1% of the air. Helium is the second lightest gas (after hydrogen). Radon is nearly 8 times denser than air.

All the gases are **monatomic**. That means that their molecules consist of single atoms.

Their densities and boiling points illustrate typical patterns (trends) in physical properties as you go down a Group in the Periodic Table.

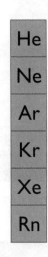

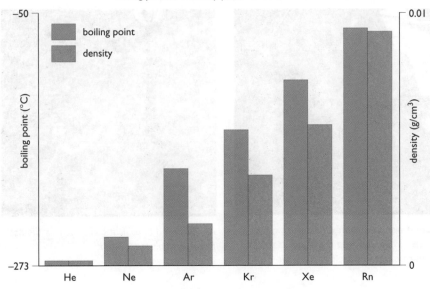

*Boiling point and density patterns in the noble gases.*

The density increases as the atoms get heavier.

*Some of the noble gases are used in advertising lighting. Neon, for example, produces a bright red colour.*

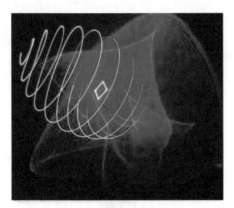

*A light show using krypton and argon lasers.*

The boiling points also increase as you go down the Group. This is because the attractions between one molecule and its neighbours get stronger as the atoms get bigger. More energy is needed to break the stronger attractions. In helium, these intermolecular attractions are very, very weak. Very little energy is needed to break these attractions, and so helium's boiling point is very low.

## Chemical reactivity

The noble gases don't form stable ions, and so don't produce ionic compounds. They are reluctant to form covalent bonds because in most cases it costs too much energy to rearrange the full energy levels to produce the single electrons that an atom needs if it is to form simple covalent bonds by sharing electrons. That means that these gases are generally unreactive.

Until the 1960s scientists thought that the noble gases were completely unreactive. Then they found that they could make xenon combine with fluorine just by heating the two together. After that a number of other compounds of both xenon and krypton were found – mainly combined with fluorine and oxygen.

## Summarising the main features of the Group 0 elements

Group 0 elements:

- are colourless, monatomic gases
- become denser as you go down the Group
- have higher boiling points as you go down the Group
- are generally unreactive.

## Uses of the noble gases

*Helium is used in weather balloons because it is less dense than air, and safer than hydrogen.*

*Helium is also used in mixtures with oxygen for deep sea diving. For deep diving, this is safer than using oxygen by itself, or using air.*

*Argon is used to fill light bulbs because the hot filament won't react with it.*

# End of Chapter Checklist

**In this chapter you have learnt that:**

- many people were involved in the development of the Periodic Table
- groups contain similar elements
- periods of elements are rows across the table. Each period starts with metals and ends with non-metals
- metals react with oxygen to form oxides
- metals react with water or steam to give hydroxides or oxides
- metals react with acids to give a salt and hydrogen gas
- metals near the top of the Reactivity Series, such as sodium, are very reactive
- metals near the bottom of the Series, such as copper, are unreactive
- nitrates, carbonates and hydroxides of metals low in the Reactivity Series tend to decompose on heating
- noble gases have many uses.

# Questions

1. The modern Periodic Table arranges elements in order of:

   A   atomic mass      C   atomic number
   B   atomic weight     D   colour     (1)

2. Elements in the same group are:
   A   similar       C   metals
   B   different     D   non-metals     (1)

3. Metals react with acids to form a salt and:

   A   water        C   carbon dioxide
   B   hydrogen     D   precipitate     (1)

4. The word describing the appearance of a metallic surface is:

   A   shine        C   lustre
   B   oxide       D   reflection     (1)

5. Which of the following statements is/are true?

   I    all metals conduct heat
   II    all metals are solids
   III    all metals sink in water
   IV    most metals float on water

   A   I, II, III and IV     C   I
   B   I and II         D   III     (1)

6. *a)* Draw the apparatus needed to collect the gas evolved when magnesium reacts with sulphuric acid.     (4)

   *b)* What is the gas? What test would you carry out to identify it?     (2)

   *c)* How rapid will this reaction be? Explain your answer.     (2)

   *d)* Write word and balanced symbol equations for the reaction.     (2)

7. *a)* How could you compare the thermal conductivities of copper and iron?     (6)

   *b)* Why do iron saucepans sometimes have a copper base?     (2)

   *c)* Which metal boils below 1000°C? Why might this be a problem in a furnace?     (2)

8. *a)* Plot line graphs to show the variation in melting point and boiling point of the metals in the table on page 199.     (5)

   *b)* What patterns do you find?     (3)

   *c)* What does this tell you about the strength of metallic bonds in different metals?     (2)

9. *a)* Draw a table to show the reactions of three *metals* with oxygen and dilute acid.     (5)

   *b)* Potassium is in the same group as sodium. Predict the reaction of potassium with water and name the products.     (3)

   *c)* Aluminium is a reactive metal but is normally coated in an unreactive layer of oxide. How could you show this?     (2)

# Chapter 20: The Group 1 and 2 Metals

**When you have completed this chapter, you will be able to:**

- identify the alkali metals of Group 1
- understand the reactions of the alkali metals
- identify the metals of Group 2
- understand the reactions of Group 2 metals
- make predictions about other elements and compounds in Groups 1 and 2.

## Group 1 – the alkali metals

This Group contains the familiar reactive metals sodium and potassium, as well as some less common ones.

Francium at the bottom of the Group is radioactive. One of its isotopes is produced during the radioactive decay of uranium-235, but is extremely short-lived. Once you know about the rest of Group 1 you can predict what francium would be like, but you can't realistically observe its properties. We will make those predictions later.

### Physical properties

Nobody is expecting you to remember these values! They are here just to show the patterns. You *will* be expected to know those patterns.

| | Melting point/ °C | Boiling point/ °C | Density/ g cm³ |
|---|---|---|---|
| Li | 181 | 1 342 | 0.53 |
| Na | 98 | 883 | 0.97 |
| K | 63 | 760 | 0.86 |
| Rb | 39 | 686 | 1.53 |
| Cs | 29 | 669 | 1.88 |

Notice that the melting and boiling points of the elements are very low for metals, and get lower as you go down the Group.

Their densities tend to increase – although not as tidily as the noble gases. Lithium, sodium and potassium are all less dense than water, and so will float on it.

The metals are also very soft and are easily cut with a knife, becoming softer as you go down the Group. They are shiny and silver when freshly cut, but tarnish within seconds on exposure to air.

### Storage and handling

All these metals are extremely reactive, and get more reactive as you go down the Group. They all react quickly with air to form oxides, and anywhere between rapidly and violently with water to form strongly alkaline solutions of the metal hydroxides.

To stop them reacting with oxygen or water vapour in the air, lithium, sodium and potassium are stored under oil. If you look carefully at the photograph, you will see traces of bubbles in the beaker containing the sodium. There must have been a tiny amount of water present in the oil that the sodium was placed in.

Rubidium and caesium are so reactive that they have to be stored in sealed glass tubes to stop any possibility of oxygen getting at them.

Great care must be taken not to touch any of these metals with bare fingers. There could be enough sweat on your skin to give a reaction producing lots of heat and a very corrosive metal hydroxide.

*Lithium, sodium and potassium have to be kept in oil to stop them reacting with oxygen in the air.*

### The reactions with water

All of these metals react with water to produce a metal hydroxide and hydrogen.

> **metal + cold water → metal hydroxide + hydrogen**

The main difference between the reactions is how fast they happen. The reaction between sodium and water is typical.

### *Sodium*

$$2Na(s) + 2H_2O(l) \rightarrow 2NaOH(aq) + H_2(g)$$

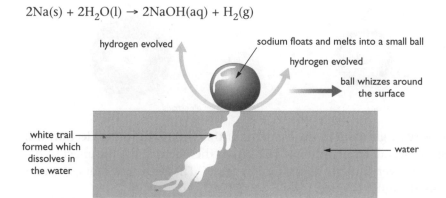

The sodium floats because it is less dense than water. It melts because its melting point is low and lots of heat is produced by the reaction. Because the hydrogen isn't given off symmetrically around the ball, the sodium is pushed around the surface of the water.

The white trail formed is the sodium hydroxide which dissolves to make a strongly alkaline solution.

### *Lithium*

$$2Li(s) + 2H_2O(l) \rightarrow 2LiOH(aq) + H_2(g)$$

The reaction is very similar to sodium's reaction, except that it is slower. Lithium's melting point is higher and the heat isn't produced so quickly, so the lithium doesn't melt.

Strictly speaking, most of the time the sodium is reacting, it is present as molten sodium – not solid sodium. Writing (l) for the state symbol, though, has the potential for confusing an examiner, and is probably best avoided!

## Potassium

$$2K(s) + 2H_2O(l) \rightarrow 2KOH(aq) + H_2(g)$$

Potassium's reaction is faster than sodium's. Enough heat is produced to ignite the hydrogen which burns with a lilac flame. The reaction often ends with the potassium spitting around.

As you go down the Group, the metals become more reactive.

### Explaining the increase in reactivity

In all of these reactions, the metal atoms are losing electrons and forming metal ions in solution. For example:

$$Na(s) \rightarrow Na^+(aq) + e^-$$

The electrons released by the metal are gained by the water molecules, producing hydroxide ions and hydrogen.

$$2H_2O(l) + 2e^- \rightarrow 2OH^-(aq) + H_2(g)$$

The differences between the reactions depend in part on how easily the outer electron of the metal is lost in each case. That depends on how strongly it is attracted to the nucleus in the original atom. Remember that the nucleus of an atom is positive because it contains protons, and so attracts the negative electrons.

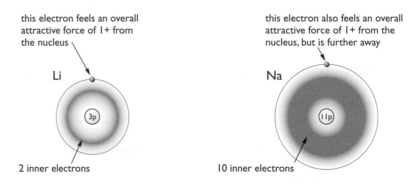

In every single atom in the elements of this Group, the outer electron will feel an overall attractive force of 1+ from the nucleus, but the effect of the force falls very quickly as distance increases. The bigger the atom, the more easily the outer electron is lost.

1. What is the order of reactivity of the metals sodium, lithium and potassium?
2. Describe one reaction that suggests that these metals all belong in the same Group in the Periodic Table.
3. Why does Group 1 have the name 'alkali metals'?

## Compounds of the alkali metals

All Group 1 metal ions are colourless. That means that their compounds will be colourless or white unless they are combined with a coloured negative ion. Potassium dichromate(VI) is orange, for example, because the dichromate(VI) ion is orange.

*Potassium dichromate(VI) solution is orange because the dichromate(VI) ion is orange.*

The compounds are typical ionic solids and are mostly soluble in water. Some lithium compounds (like lithium carbonate, for example) are rather less soluble.

You don't normally come across bottles containing the oxides of the metals in the lab. They react instantly with water (including water vapour in the air) to form hydroxides. For example:

$$Na_2O(s) + H_2O(l) \rightarrow 2NaOH(aq)$$

The hydroxides all form strongly alkaline solutions. This is the origin of the name 'alkali metals'.

| Metal | Hydroxide | Chloride | Sulphate | Nitrate | Carbonate |
|---|---|---|---|---|---|
| Lithium | LiOH | LiCl | $Li_2SO_4$ | $LiNO_3$ | $Li_2CO_3$ |
| Sodium | NaOH | NaCl | $Na_2SO_4$ | $NaNO_3$ | $Na_2CO_3$ |
| Potassium | KOH | KCl | $K_2SO_4$ | $KNO_3$ | $K_2CO_3$ |

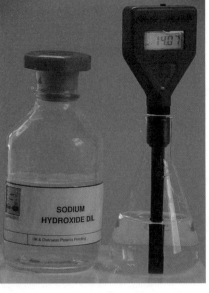

Sodium hydroxide solution is strongly alkaline.

The hydroxides are strong bases and form solutions which are alkaline with very high pH values.

The chlorides and sulphates are colourless salts which dissolve easily in water giving neutral solutions. Sodium chloride is the most common salt in sea water.

The nitrates of sodium and potassium decompose on heating to give the nitrite salt and oxygen gas.

$$2NaNO_3(s) \xrightarrow{\text{heat}} 2NaNO_2(s) + O_2(g)$$

All of the metal nitrates behave the same except lithium. The first member of a Group often has unusual properties. The nitrate and carbonate of lithium behave the same as Group 2 nitrates and carbonates such as magnesium (see below).

The carbonates of sodium and potassium are soluble and are not decomposed by heating in a Bunsen flame. Again, lithium carbonate is unusual and behaves like the Group 2 carbonates.

Sodium and potassium carbonate both dissolve in water to give alkaline solutions. Sodium carbonate is sold as washing soda and is a good degreasing agent.

4. Name two potassium compounds whose solutions would be alkaline.
5. What pH range would you expect for a solution of rubidium hydroxide?
6. What is the formula of caesium bromide?

# Activity 1: Reactions of some alkali metal compounds

**You will need:**

- test-tubes and holder
- splints
- dilute hydrochloric acid and dropper
- solid carbonates and nitrates of sodium and potassium
- spatula

**Carry out the following:**

1. Place a little of each carbonate into separate tubes and add about 3 cm$^3$ of the acid.

2. Observe any changes.

3. Test the gas with a burning splint.

4. Heat a little of each solid nitrate in a separate tube. (**Care:** dangerous fumes)

5. Observe any changes.

6. Test the gas with a glowing splint.

7. Write a summary table of your results.

# Activity 2: Identifying unknown compounds

You are provided with solutions of sodium hydroxide, sodium carbonate, sodium chloride, hydrochloric acid and universal indicator. The solutions of sodium compounds are not labelled. How could you identify each of them?

Remember that francium is highly radioactive and short-lived. You couldn't actually observe any of these things.

## Summarising the main features of the Group 1 elements

Group 1 elements:

- are metals
- are soft, with melting points and densities which are very low for metals
- react rapidly with air to form coatings of the metal oxide
- react with water to produce the metal hydroxide and hydrogen
- increase in reactivity as you go down the Group
- form compounds in which the metal has a 1+ ion
- have oxides which react with water to produce soluble alkaline hydroxides
- have white compounds which dissolve to produce colourless solutions.

### Using these features to predict the properties of francium

If you followed the trends in Group 1, you could predict that:

- Francium is very soft with a melting point just above room temperature.
- Its density is probably just over 2 g/cm$^3$.
- Francium will be a silvery metal but will tarnish almost instantly in air.

- It would react violently with water to give francium hydroxide and hydrogen.

- Francium hydroxide solution will be strongly alkaline.

- Francium oxide will react with water to give francium hydroxide.

- Francium compounds are white and dissolve in water to give colourless solutions.

## Group 2 – magnesium and calcium

This Group is known as the alkaline earth metals since some of the elements occur widely in rocks and minerals. Calcium occurs naturally as calcium carbonate in chalk, limestone and marble as well as the mineral calcite. Magnesium carbonate occurs together with calcium carbonate in the rock dolomite. The element radium is strongly radioactive.

### Physical properties

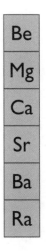

| Element | Melting point/ °C | Boiling point/ °C | Density/ g/cm$^3$ |
|---------|-------------------|-------------------|-------------------|
| Be | 1 280 | 2 970 | 1.85 |
| Mg | 650 | 1 100 | 1.74 |
| Ca | 850 | 1 490 | 1.55 |
| Sr | 760 | 1 380 | 2.6 |
| Ba | 710 | 1 640 | 3.5 |

Notice that the trends, ignoring the unusual first element beryllium, are that density increases down the Group, while boiling point shows a more complex pattern. The melting point values are low when compared to the transition metals which typically have melting points above 1 000°C.

The metals magnesium and calcium are shiny on freshly cut surfaces but react with air and moisture to give a dull appearance.

The Group 2 metals are chemically reactive, but much less so than the Group 1 alkali metals. The reactivity increases down the Group, as in Group 1.

### The reactions with water

Magnesium reacts extremely slowly with cold water but rapidly with steam.

# Activity 3: Reacting magnesium with water

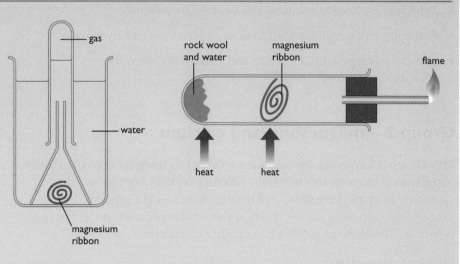

Magnesium + steam → magnesium oxide + hydrogen gas

$$Mg + H_2O \rightarrow MgO + H_2$$

**You will need:**

- beaker, test-tube, glass funnel (transparent)
- magnesium strip, cleaned with emery paper
- boiling tube, stopper and flame tube
- rock wool
- Bunsen burner

**Carry out the following:**

1. Place a coil of magnesium under the funnel, inside the beaker.

2. Fill the beaker with cold water.

3. Invert a water-filled test-tube over the funnel to collect any gas.

4. Leave for several days then look for evidence of hydrogen gas.

5. Place some wet rock wool in a boiling tube (the wool keeps the water in one place).

6. Fix a coil of cleaned magnesium in the centre of the tube.

7. Clamp the tube and heat the magnesium strongly, occasionally heating the rock wool to release some steam.

8. Observe the way the metal reacts with steam, glows and releases hydrogen. The hydrogen can be ignited at the end of the tube.

9. How does this compare with the reactions of Group 1 metals with water?

## Activity 4: Reaction of calcium with water

Calcium + water → calcium hydroxide + hydrogen

$$Ca + 2H_2O \rightarrow Ca(OH)_2 + H_2$$

**You will need:**

- calcium metal
- test-tubes, filter funnel and filter paper
- splint, drinking straw
- universal indicator solution or paper

**Carry out the following:**

1. One-third fill a test-tube with cold water and add a small piece of calcium.

2. Observe the changes, including the temperature.

3. Test the gas with a burning splint.

4. When the reaction is over, filter the mixture into a clean test-tube. Test the pH of a sample.

5. Gently blow through the solution using a straw. Note any changes.

6. The solution of calcium hydroxide is strongly alkaline and is also known as lime water.

## Compounds of magnesium and calcium

| metal | oxide | hydroxide | chloride | nitrate | carbonate |
|---|---|---|---|---|---|
| Mg | MgO | $Mg(OH)_2$ | $MgCl_2$ | $Mg(NO_3)_2$ | $MgCO_3$ |
| Ca | CaO | $Ca(OH)_2$ | $CaCl_2$ | $Ca(NO_3)_2$ | $CaCO_3$ |

*Limestone Alps.*

Both magnesium and calcium oxides are strong bases and can be prepared by burning the elements in air or by heating the carbonates.

$$2Mg + O_2 \rightarrow 2MgO \quad \text{white solid, high melting point}$$
$$2Ca + O_2 \rightarrow 2CaO \quad \text{white solid, high melting point}$$

Both magnesium and calcium hydroxides form alkaline solutions in water and both are strong bases. They can neutralise acids to give salts.

$$Mg(OH)_2 + 2HCl \rightarrow MgCl_2 + 2H_2O$$
$$Ca(OH)_2 + 2HCl \rightarrow CaCl_2 + 2H_2O$$

Both magnesium and calcium chlorides are colourless salts which crystallise from solution with water of crystallisation. Anhydrous calcium chloride is a useful drying agent for gases, except for ammonia with which it reacts.

The nitrates of magnesium and calcium can be made from the metals, the oxides or hydroxides by reaction with dilute nitric acid. Both nitrates decompose on heating to give brown fumes of nitrogen dioxide and oxygen gas.

$$2Mg(NO_3)_2 \xrightarrow{\text{heat}} 2MgO + 4NO_2 + O_2$$

$$2Ca(NO_3)_2 \xrightarrow{\text{heat}} 2CaO + 4NO_2 + O_2$$

The carbonates of magnesium and calcium occur widely in nature. They are decomposed by heating to give the corresponding oxides.

$$MgCO_3 \xrightarrow{\text{heat}} MgO + CO_2$$

$$CaCO_3 \xrightarrow{\text{heat}} CaO + CO_2$$

Calcium oxide is known as quicklime and can be converted into calcium hydroxide by adding water. This reaction is very exothermic.

$$CaO + H_2O \rightarrow Ca(OH)_2$$

Calcium hydroxide is known as slaked lime and is used in agriculture to neutralise acidic soils and so improve crop yields.

### Summarising the main features of the Group 2 elements

Group 2 elements magnesium and calcium:

- are metals
- have melting points and densities that are quite low for metals
- react with air, water or steam to form oxides or hydroxides
- show the same reactivity trend as Group 1, increasingly reactive as you go down the Group
- form compounds in which the metal has a +2 ion
- form white or colourless compounds which give colourless solutions.

7. Predict the physical properties of strontium by using the Group 2 trends.
8. Write the equation for the reaction of strontium with water.
9. The barium compound called barium sulphate occurs naturally as the mineral heavy spar. Why does it have this name?

# End of Chapter Checklist

## In this chapter you have learnt that:

- the alkali metals of Group 1 are very reactive
- the alkaline earth metals are reactive, but less than the alkali metals
- compounds of both Groups 1 and 2 are colourless
- the Group trends are similar, the metals become more reactive as you go down the Group
- you can predict the properties of other metals in these Groups by considering the Group trends.

## Questions

You will need to use the Periodic Table.

1. Answer the questions which follow using *only* the elements in this list:

   caesium, chlorine, molybdenum, neon, nickel, nitrogen, magnesium, strontium, tin

   a) Name an element which is in
      i) Group 2
      ii) the same period as silicon
      iii) the same group as phosphorus. (3)

   b) How many electrons are there in the outer levels of atoms of: *i)* strontium, *ii)* sodium, *iii)* calcium? (3)

   c) Divide the list of elements at the beginning of the question into metals and non-metals. (2)

   d) Name *i)* the most reactive metal, *ii)* the least reactive metal. (2)

2. This question concerns the chemistry of the elements Li, Na, K, Rb and Cs on the extreme left-hand side of the Periodic Table. In each case, you should identify the substances represented by letters.

   a) **A** is the lightest of all metals. (1)

   b) When metal **B** is dropped onto water it melts into a small ball and rushes around the surface. A gas **C** is given off and this burns with a lilac flame. A white trail dissolves into the water to make a solution of **D**. (3)

   c) When metal **E** is heated in a green gas **F**, it burns with an orange flame and leaves a white solid product **G**. (3)

   d) Write equations for:
      i) the reaction of **B** with water (2)
      ii) the reaction between **E** and **F**. (2)

   e) What would you expect to see if solution **D** was tested with red and blue litmus paper? (2)

   f) Explain why **B** melts into a small ball when it is dropped onto water. (1)

3. Three metals are named X, Y and Z. X reacts with cold water, releasing a gas. Y reacts only with steam. Z reacts only with acids.

   a) Which gas is released when metals react with water? What test can you carry out to identify it? (2)

   b) Which is the most reactive metal of the three? Explain your answer. (3)

   c) Write a word equation for the reaction of Y with steam. (2)

   d) Suggest possible identities for X, Y and Z. (3)

4. Three metal nitrates are named P, Q and R. On heating: P melts and effervesces. The gas given off relights a glowing splint. Q goes from white to yellow and releases brown fumes. R goes from blue to black and releases brown fumes.

   a) Which nitrate contains a transition metal? Explain your answer. (2)

   b) Which nitrate probably contains a Group 1 metal? Explain your answer and write a word equation for the reaction. (4)

   c) What are the brown fumes produced by Q and R? What test could you carry out to identify them? (2)

   d) Suggest the identity of R. (2)

5. Two metal carbonates are named A and B. A is green in colour. On heating it turns black and releases a gas that turns lime water cloudy. B is white. On heating it glows and releases a gas which also turns lime water cloudy.

   a) Which carbonate is unlikely to be a transition metal compound? Explain your answer. (2)

   b) How do you test a gas with lime water? Why does it go cloudy? (3)

   c) Compound B is used to make cement and glass. Suggest the identity of B and give your reasons. (3)

   d) Name *one* other carbonate that is similar to B and explain why it is similar. (2)

## Chapter 21: Trends in Group 7, the Halogens

**When you have completed this chapter, you will be able to:**

- describe the colours and physical states of the halogens
- understand their reactions with hydrogen and metals
- explain why halogens are oxidising agents
- describe examples of displacement reactions
- write ionic equations to explain halogen behaviour
- summarise the main features of Group 7 elements
- understand the main trends in Period 3 elements.

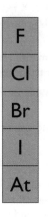

### Group 7 – the halogens

The name 'halogen' means 'salt-producing'. When they react with metals these elements produce a wide range of salts like calcium fluoride, sodium chloride, potassium iodide and so on.

The halogens are non-metallic elements with diatomic molecules – $F_2$, $Cl_2$ and so on. As the molecules get larger towards the bottom of the group, the melting and boiling points increase. Fluorine and chlorine are gases. Bromine is a liquid which turns to vapour very easily, and iodine is a solid.

Astatine is radioactive and is formed during the radioactive decay of other elements like uranium and thorium. Most of its isotopes are so unstable that their lives can be measured in seconds or fractions of a second.

|         | State  | Colours                          |
|---------|--------|----------------------------------|
| $F_2$   | gas    | yellow                           |
| $Cl_2$  | gas    | green                            |
| $Br_2$  | liquid | dark red liquid – red/brown vapour |
| $I_2$   | solid  | dark grey solid – purple vapour  |

Because the halogens are non-metals, they will be poor conductors of heat and electricity. When they are solid (for example, iodine at room temperature), their crystals will be brittle and crumbly.

### Safety

Fluorine is so dangerously reactive that you would never expect to come across it in a school lab.

Apart from any safety problems associated with the reactivity of the elements (especially fluorine and chlorine), all the elements have extremely poisonous vapours and have to be handled in a fume cupboard.

Liquid bromine is also very corrosive and great care has to be taken to keep it off the skin. It is a good idea to have a beaker of dilute sodium thiosulphate solution handy whenever you use bromine. This reacts at once with any bromine you might have got on your skin or the bench, without being particularly harmful itself.

Chlorine, bromine and iodine.

Iodine has a purple vapour.

## Activity 1: Iodine vapour

### Investigative skills

P

Iodine vapour is toxic. Plan an experiment to produce iodine vapour safely to see its colour. Sketch the apparatus you would need.

1. Which halogen is a green gas?
2. Which halogen is a solid at room temperature?
3. What does the word halogen mean?

## Activity 2: Preparation of chlorine gas. Demonstration – hazardous

### Investigative skills

O   A

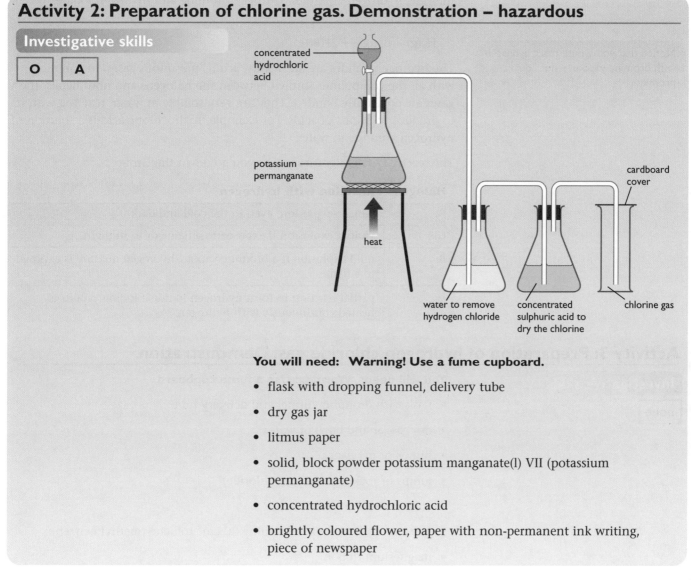

concentrated hydrochloric acid

potassium permanganate

heat

water to remove hydrogen chloride

concentrated sulphuric acid to dry the chlorine

cardboard cover

chlorine gas

**You will need: Warning! Use a fume cupboard.**

- flask with dropping funnel, delivery tube

- dry gas jar

- litmus paper

- solid, block powder potassium manganate(l) VII (potassium permanganate)

- concentrated hydrochloric acid

- brightly coloured flower, paper with non-permanent ink writing, piece of newspaper

**Carry out the following:**

1. Add a few grams of the solid manganate together with about 10 cm³ water.

2. Slowly add enough acid to start the reaction, look for the green colour of the gas.

3. Test the gas using damp blue litmus paper, held near the top of the jar. The chlorine gas turns it red and then bleaches it.

4. Drop the flower, writing and newspaper into the jar. Seal it and leave to stand. See which materials can be bleached by the gas. Try to explain the effect of chlorine on newspaper ink.

## Reactions with hydrogen

The halogens react with hydrogen to form **hydrogen halides** – hydrogen fluoride, hydrogen chloride, hydrogen bromide and hydrogen iodide. For example:

$$H_2(g) + Br_2(g) \rightarrow 2HBr(g)$$

The hydrogen halides are all steamy, acidic, poisonous gases. In common with all the compounds formed between the halogens and non-metals, the gases are covalently bonded. They are very soluble in water, reacting with it to produce solutions of acids. For example, hydrochloric acid is a solution of hydrogen chloride in water.

The reactivity falls significantly as you go down the Group.

> Notice that you would react hydrogen with bromine vapour – not liquid bromine.

| Halogen | Reaction with hydrogen |
|---------|------------------------|
| $F_2$ | violent explosion, even in the cold and dark |
| $Cl_2$ | violent explosion if exposed to a flame or to sunlight |
| $Br_2$ | mild explosion if a bromine vapour/hydrogen mixture is exposed to a flame |
| $I_2$ | partial reaction to form hydrogen iodide if iodine vapour is heated continuously with hydrogen |

## Activity 3: Preparation of hydrogen chloride gas. Demonstration

**Investigative skills**

**You will need:  Warning! Use a fume cupboard.**

- flask with dropping funnel and delivery tube

- dry gas jar and bowl of water

- litmus or pH paper

- lumps of rock salt (sodium chloride)

- concentrated sulphuric acid

- 100 cm³ conical flask containing 20 cm³ toluene (methyl benzene)

- magnesium ribbon

## Activity 3: Preparation of hydrogen chloride gas. Demonstration (continued)

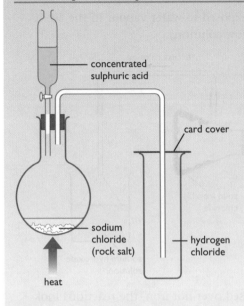

concentrated sulphuric acid

card cover

sodium chloride (rock salt)

hydrogen chloride

heat

**Carry out the following:**

1. Set up the apparatus as shown and add a small amount of acid to the rock salt. Warm very gently if necessary.

2. The dense colourless hydrogen chloride gas fumes in moist air.

3. Test the gas by placing a damp indicator paper at the top of the gas jar.

4. Add a little water. The gas dissolves in the water and forms hydrochloric acid. Add a piece of magnesium ribbon and observe any changes.

5. Collect a second jar or tube of gas and invert it in a bowl of water, open end downwards. The water rises as the gas dissolves. Remember that hydrogen chloride is a covalent gas but it ionises completely in water forming hydrochloric acid.

6. Now bubble the gas through the 20 cm$^3$ toluene, an organic solvent. After 5 minutes, test the solution by adding indicator paper followed by a small piece of magnesium. There are no acidic reactions in the absence of water.

## Properties

| Property of HCl gas | HCl in water | HCl in methylbenzene (toluene) |
|---|---|---|
| Solubility | extremely soluble | soluble |
| Add dry litmus | goes red | no change |
| Add Mg ribbon | effervescence, hydrogen gas | no change |
| Add limestone | effervescence, carbon dioxide gas | no change |

## Reactions with metals

### With iron

Iron(II) chloride can also be prepared by heating iron in a stream of dry hydrogen chloride gas.

$$Fe + 2HCl \rightarrow FeCl_2 + H_2$$

The halogens are strong oxidising agents and are able to oxidise the metal to its higher valency giving iron(III) chloride.

### Making anhydrous iron(III) chloride from iron and chlorine

This time, pure dry chlorine is passed over heated iron wool in the long glass tube. The iron burns and can produce a whole range of coloured products.

$$2Fe(s) + 3Cl_2(g) \rightarrow 2FeCl_3(s)$$

Pure anhydrous iron(III) chloride is actually black, but it reacts with any trace of water present to produce various orange and red compounds. There

*Iron(III) chloride crystals scraped out of the reaction tube.*

is usually enough water present in rubber bungs and tubing to affect the iron(III) chloride while it is still in the tube.

As soon as the iron(III) chloride is exposed to water vapour in the air, it reacts with it to form orange or yellow solutions.

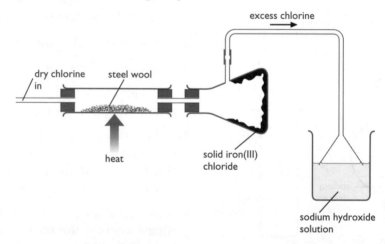

If bromine or iodine vapour are passed over hot iron, the reactions look similar to the chlorine reaction, but the reactivity again falls as you go down the group. For example, the iron burns more brightly with chlorine than it does with bromine vapour. If you used fluorine, the iron would burn even more brightly still. Notice that the less reactive iodine forms iron(II) iodide, $FeI_2$, rather than iron(III) iodide, $FeI_3$.

| Halogen | Product | Equation |
|---------|---------|----------|
| $F_2$ | iron(III) fluoride | $2Fe(s) + 3F_2(g) \rightarrow 2FeF_3(s)$ |
| $Cl_2$ | iron(III) chloride | $2Fe(s) + 3Cl_2(g) \rightarrow 2FeCl_3(s)$ |
| $Br_2$ | iron(III) bromide | $2Fe(s) + 3Br_2(g) \rightarrow 2FeBr_3(s)$ |
| $I_2$ | iron(II) iodide | $Fe(s) + I_2(g) \rightarrow FeI_2(s)$ |

## The reaction between sodium and chlorine

Sodium burns in chlorine with its typical orange flame to produce white, solid sodium chloride.

$$2Na(s) + Cl_2(g) \rightarrow 2NaCl(s)$$

Sodium chloride is, of course, an ionic solid. Typically, when the halogens react with metals from Groups 1 and 2, they form ions.

It is useful to look at this from the point of view of the sodium and of the chlorine separately by writing the ionic equation for the reaction:

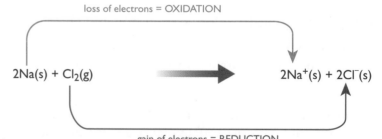

*Sodium burning in chlorine to produce sodium chloride.*

The sodium has lost electrons and so has been oxidised to sodium ions. This means that chlorine is acting as an **oxidising agent**. That is typical of the reactions of chlorine. It is a strong oxidising agent.

### Exploring another reaction of chlorine as an oxidising agent

If you add chlorine solution ('chlorine water') to iron(II) chloride solution, the very pale green solution becomes yellowish.

The iron(II) chloride has been converted into iron(III) chloride.

$$2FeCl_2(aq) + Cl_2(aq) \rightarrow 2FeCl_3(aq)$$

Adding some sodium hydroxide solution to a sample of the product gives an orange-brown precipitate, showing that the solution now contains iron(III) ions. If you still had iron(II) ions present, you would have got a green precipitate.

The ionic equation shows that this is another example of chlorine acting as an oxidising agent.

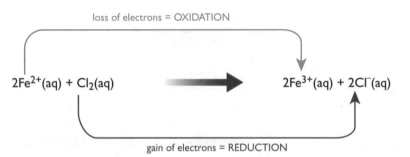

loss of electrons = OXIDATION

$$2Fe^{2+}(aq) + Cl_2(aq) \longrightarrow 2Fe^{3+}(aq) + 2Cl^-(aq)$$

gain of electrons = REDUCTION

*Testing the solution formed when chlorine reacts with iron(II) chloride solution.*

Any other iron(II) salt would be oxidised in exactly the same way.

This is the reason that iron(III) chloride is formed when you pass chlorine over hot iron. Chlorine is a powerful enough oxidising agent to remove three electrons from the iron and oxidise it all the way to iron(III) chloride.

Iodine, on the other hand, is a much weaker oxidising agent. It is powerful enough to remove two electrons from the iron to make iron(II) ions, but won't oxidise iron(II) ions onwards to iron(III) ions. This is why iron(II) iodide is formed when iodine vapour is passed over hot iron.

> 4. When sodium burns in chlorine, what is oxidised?
> 5. When iron burns in chlorine, what is the product?
> 6. Why is it dangerous to react hydrogen with chlorine?

### Displacement reactions involving the halogens

Just as you can use the Reactivity Series of metals to make sense of their displacement reactions, so you can also use a corresponding Reactivity Series for the halogens.

We shall concentrate on the three commonly used halogens, but the trend continues for the rest of the group.

### *Reacting chlorine with potassium bromide or potassium iodide solutions*

If you add chlorine solution to colourless potassium bromide solution, the solution becomes orange as bromine is formed.

$$2KBr(aq) + Cl_2(aq) \rightarrow 2KCl(aq) + Br_2(aq)$$

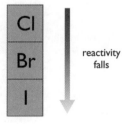

reactivity falls

The more reactive chlorine has displaced the less reactive bromine from potassium bromide.

Similarly, adding chlorine solution to potassium iodide solution gives a dark red solution of iodine. If an excess of chlorine is used, you may get a dark grey precipitate of iodine.

$$2KI(aq) + Cl_2(aq) \rightarrow 2KCl(aq) + I_2(aq \text{ or } s)$$

Both of these are redox reactions. In each case, the chlorine is acting as an oxidising agent.

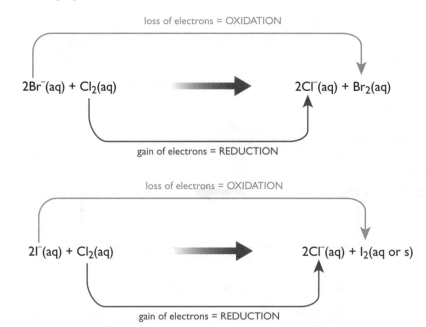

The potassium ions are spectator ions. The reaction would be the same with any soluble bromide or iodide.

### The reaction of bromine with potassium iodide solution

In exactly the same way, the more reactive bromine displaces the less reactive iodine from potassium iodide solution.

Adding bromine solution ('bromine water') to colourless potassium iodide solution gives a dark red solution of iodine (or a dark grey precipitate if an excess of bromine is added).

$$2KI(aq) + Br_2(aq) \rightarrow 2KBr(aq) + I_2(aq \text{ or } s)$$

The ionic equation shows bromine acting as an oxidising agent.

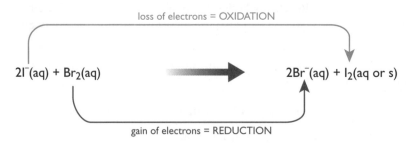

Small amounts of an organic solvent are sometimes added to these reactions. The bromine or iodine produced dissolve in the solvents and give strong colours which can make them easier to recognise.

## Activity 4: Halogen displacement

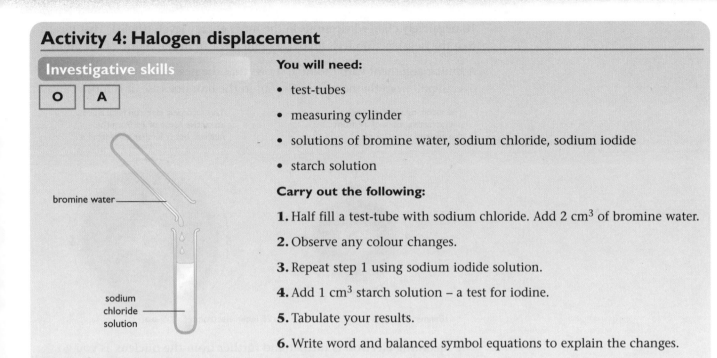

**Investigative skills**

| O | A |

bromine water

sodium chloride solution

**You will need:**

- test-tubes
- measuring cylinder
- solutions of bromine water, sodium chloride, sodium iodide
- starch solution

**Carry out the following:**

1. Half fill a test-tube with sodium chloride. Add 2 cm$^3$ of bromine water.
2. Observe any colour changes.
3. Repeat step 1 using sodium iodide solution.
4. Add 1 cm$^3$ starch solution – a test for iodine.
5. Tabulate your results.
6. Write word and balanced symbol equations to explain the changes.

### Summary of the properties of chlorine gas

| Solubility in water | soluble, about 2.5 volumes in 1 volume water, solution is acidic |
|---|---|
| Colour | green |
| Odour | choking, pungent (toxic) |
| Density | more dense than air |
| Reactivity | very reactive, combines directly with metals and non-metals |
| Oxidising action | powerful oxidising agent |
| Displacement | displaces halogens lower down in Group 7 |
| Test for the gas | damp blue litmus paper goes red and is then bleached |

## Explaining the trend in the reactivity of the halogens

As you go down the Group, the oxidising ability of the halogens falls.

When a halogen oxidises something, it does so by removing electrons from it.

Each halogen has the ability to oxidise the ions of those underneath it in the group, but not those above it. Chlorine can remove electrons from bromide or iodide ions, and bromine can remove electrons from iodide ions.

The reason has to do with how easily the atoms gain electrons to make their ions.

Chlorine is a strong oxidising agent because its atoms readily attract an extra electron to make chloride ions. Bromine is less successful at attracting electrons, and iodine is less successful still.

You have to consider the amount of attraction the incoming electron feels from the nucleus. In chlorine, there are 17 positively charged protons offset by the

10 negatively charged electrons in the inner energy levels (shells). That means that the new electron feels an overall pull of 7+ from the centre of the atom.

A similar argument with bromine shows that the new electron also feels an overall pull from the nucleus of 7+, but in the bromine case, it is further away.

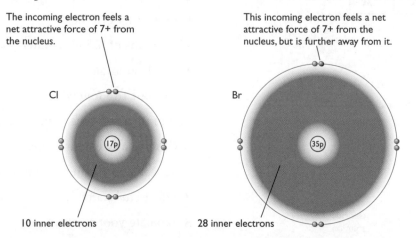

The incoming electron feels a net attractive force of 7+ from the nucleus.

This incoming electron feels a net attractive force of 7+ from the nucleus, but is further away from it.

10 inner electrons

28 inner electrons

The incoming electron is further and further from the nucleus as you go down the Group, and so it is less strongly attracted. That means the ion is less easily formed.

It also means that it is easier to remove the extra electron from, say, a bromide ion than from a chloride ion. Bromide ions are more easily oxidised than chloride ions. Iodide ions hold the extra electron weakly enough that they are very easily oxidised.

### Summarising the main features of the Group 7 elements

Group 7 elements:

- have diatomic molecules, $X_2$
- go from gases to liquid to solid as you go down the Group
- have coloured poisonous vapours
- form ionic salts with metals and molecular compounds with non-metals
- become less reactive towards the bottom of the Group
- are oxidising agents, with oxidising ability decreasing down the Group
- will displace elements lower down the Group from their salts.

> **7.** What is meant by displacement?
> **8.** Are the halogens oxidising or reducing agents?
> **9.** Which halogens exist as diatomic molecules?

## Uses of the halogens and their compounds

### Fluorine

Fluorides such as sodium fluoride are added to toothpaste or sometimes to drinking water supplies to help prevent tooth decay.

There is a lot of controversy about this. A study by the University of York in the UK in September 2000 showed that much of the research in this area wasn't very good. By comparing the results of lots of research projects they found that fluoridation of water supplies reduced dental decay by an average of about 15%. There was a useful effect, even for those people already using fluoride toothpaste.

However, there was a marked increase in dental fluorosis. This is a mottling of the teeth caused by absorption of fluorine. In about 1 in 8 people this was enough to be thought visually unattractive.

There was no evidence for any effects on bone structure or development, and no evidence that fluoridation causes cancer. On the other hand, the studies were largely short term, and there could be hidden long term effects.

Many people object strongly to having medication forced on them via the water supply.

### Chlorine

Chlorine is added to the water supply in very small quantities to kill bacteria.

Chlorine is also a powerful bleach. A solution of chlorine in water ('chlorine water') could be used as a bleach, but would be dangerous because of the free chlorine. Household bleach is a chlorine compound, sodium chlorate(I), NaClO.

### Iodine

A dilute solution of iodine in potassium iodide solution is used as an antiseptic.

### Silver halides

Three of the silver halides (silver chloride, silver bromide and silver iodide) are important in photography.

For example, when silver bromide is exposed to light, a redox reaction happens in which silver ions gain electrons from the bromide ions. This is called a **photochemical reaction** because it involves light.

Small amounts of metallic silver are formed.

$$2Ag^+Br^-(s) \rightarrow 2Ag(s) + Br_2(g)$$

The presence of the silver leaves a faint image on the photographic film or paper. This image is strengthened during the process of developing the film or paper.

X-rays and radiation from radioactive substances affect photographic film or paper in the same way.

### Predicting the properties of fluorine

The trend in reactivity within Group 7 is that the reactivity falls as you go down the Group, as the atomic number increases. Fluorine is above chlorine in the halogens of Group 7. Use your knowledge of halogen chemistry to answer these questions.

*Chlorine is used to sterilise water.*

If you are interested, you might like to do some research on what happens during the formation of the negative and the final print. You might also like to find out how colour images are formed.

10. Compare the reaction of fluorine and of iodine with hydrogen, stating the likely products in each case.
11. How will fluorine react with magnesium? Write both word and symbol equations.
12. Comment on this displacement reaction.
    Chlorine + sodium fluoride → fluorine + sodium chloride

# Trends in Period 3

In groups such as Group 7, the elements all have similar properties. The halogens of Group 7 may be similar but there is a trend in reactivity within the group. Fluorine is the most reactive and iodine the least. In a period, a horizontal row in the Periodic Table, the trends are different.

Period 3 is one example. It starts with sodium (a metal) and finishes with argon (a non-metal).

Period 3 trends.

| Elements | Na, Mg, Al | Si | P, S, Cl, A$_r$ |
|---|---|---|---|
| Metal/non-metal | metals | metalloid | non-metals |
| Structures | metallic | giant molecular | simple molecular |
| Bonding | strong metallic forces of attraction | covalent, very strong attraction | molecular, weak forces between molecules |
| Properties (melting and boiling points) | high | very high | low |
| Conductivity | good | poor | do not conduct |

Silicon in Group 4 is described as a metalloid, also called a semi-metal. The properties of metalloids are in between those of metals and non-metals.

Properties of Period 3 elements.

| Element | Atomic number | Properties |
|---|---|---|
| sodium | 11 | soft metal, very reactive, less dense than water (0.97 g cm$^{-3}$) |
| magnesium | 12 | harder metal, less reactive, low density (1.7 g cm$^{-3}$) |
| aluminium | 13 | similar hardness to magnesium, less reactive, low density (2.7 g cm$^{-3}$) |
| silicon | 14 | non-metal, brittle, not very reactive, density 2.3 g cm$^{-3}$ |
| phosphorus | 15 | non-metal, white allotrope reactive, red not reactive, density of white is 1.8 g cm$^{-3}$ |
| sulphur | 16 | non-metal, reasonably reactive, yellow solid, density 1.9 g cm$^{-3}$ |
| chlorine | 17 | non-metal, green gas, very reactive, density 0.003 g cm$^{-3}$ |
| argon | 18 | non-metal, colourless gas, completely unreactive, density 0.0016 g cm$^{-3}$ |

## Oxides and chlorides of elements in Period 3

Sodium and magnesium oxides are basic, as expected for metallic oxides. Aluminium oxide is **amphoteric**, it reacts with both acids and alkalis. The oxides of the non-metals are acidic but argon does not react with oxygen and so has no oxide.

Sodium and magnesium chlorides are ionic, typical of metallic chlorides. Aluminium chloride has both an ionic and covalent character. The chlorides of the non-metals are all covalent. Argon does not form a chloride.

# End of Chapter Checklist

**In this chapter you have learnt that:**

- each of the halogens gives a different coloured vapour

- halogens react with hydrogen to form hydrogen halides

- halogens react with metals to form metal halides such as iron chloride and sodium chloride, ordinary salt

- halogens are oxidising agents and are themselves reduced

- any halogen can displace another that is lower down in Group 7

- we can use ionic equations to show how halogens react

- halogen compounds have useful properties

- Period 3 is a typical period in the Periodic Table; it starts with metallic elements but finishes with non-metals.

# Questions

1. The halogen that is a solid at room temperature is:

   A   chlorine          B   fluorine
   C   iodine            D   bromine          (1)

2. The colour of iodine vapour is:

   A   green             B   yellow
   C   purple            D   black            (1)

3. All halogens are:

   A   reducing agents   B   oxidising agents
   C   very easily oxidised  D   gases        (1)

4. A molecule containing two similar atoms could be called:

   A   double
   B   normal
   C   small
   D   diatomic                               (1)

5. Which of the following statements is/are true?

   I    chlorine can displace bromine
   II   chlorine can displace fluorine
   III  bromine can displace chlorine
   IV   bromine can displace iodine

   A   I and II
   B   I and IV
   C   I
   D   III and IV                             (1)

6. a) Describe one use for iodine.          (2)

   b) Write a balanced symbol equation for the displacement of iodine from potassium iodide solution.                              (2)

7. a) Describe the way chlorine is used to ensure water is safe to drink.                  (5)

   b) Name two products made using chlorine.  (2)

   c) When certain waste materials containing chlorine are burned, they release chlorine into the air. Why is this a problem?                  (3)

8. a) i)   Describe the reaction of sodium with chlorine.

      ii)  How can this be done safely?       (5)

   b) Write a balanced symbol equation for the reaction.  (2)

   c) What has been oxidised in the reaction and what has been reduced?                   (2)

   d) What is the colour of the sodium flame?  (1)

9. a) Use balanced symbol equations to explain how halogens can displace each other in solution.   (5)

   b) Choose one example of displacement and describe the colour changes you would see.     (2)

   c) Explain the redox reaction between chlorine and bromide ions, using half-equations.   (3)

10. a) Adding fluorides to drinking water can reduce the amount of dental decay suffered by children. Should we add a medication to everyone's drinking water even if they do not need or want it? Explain your answer.                               (6)

    b) Describe an alternative way to provide people with fluoride to protect their own teeth.   (2)

    c) Explain why silver halides are important in photography.                          (2)

# Chapter 22: *The Transition Metals*

**When you have completed this chapter, you will be able to:**

- locate the transition metals in the Periodic Table
- explain the main features of the metals and their compounds
- describe the reactions of iron and copper
- understand how complex ions form
- know that transition metals and their compounds are useful as catalysts.

The bold type simply picks out some of the more familiar transition elements. You may not have heard of many of the inner transition elements except uranium and plutonium.

## Transition metals

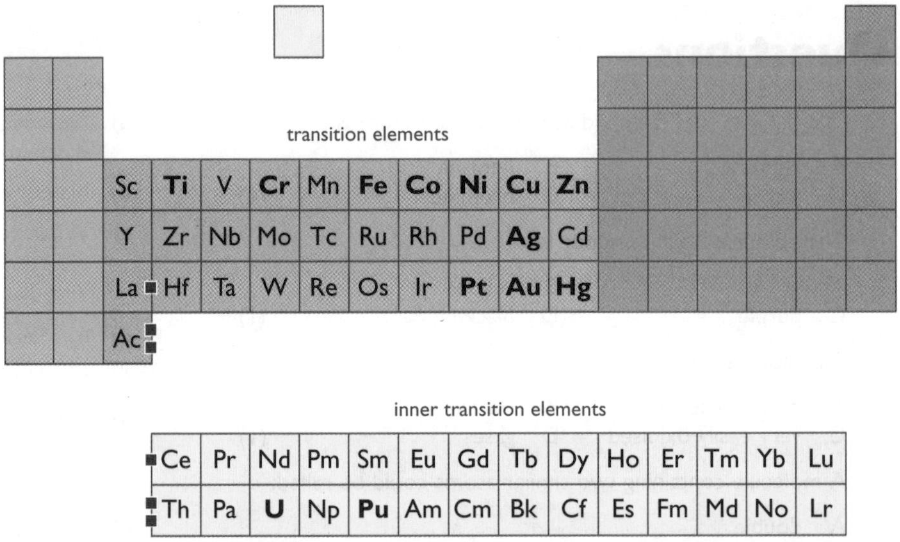

transition elements

| Sc | **Ti** | V | **Cr** | Mn | **Fe** | **Co** | **Ni** | **Cu** | **Zn** |
| Y | Zr | Nb | Mo | Tc | Ru | Rh | Pd | **Ag** | Cd |
| La | Hf | Ta | W | Re | Os | Ir | **Pt** | **Au** | **Hg** |
| Ac | | | | | | | | | |

inner transition elements

| Ce | Pr | Nd | Pm | Sm | Eu | Gd | Tb | Dy | Ho | Er | Tm | Yb | Lu |
| Th | Pa | **U** | Np | **Pu** | Am | Cm | Bk | Cf | Es | Fm | Md | No | Lr |

These are all typically metallic elements. They are good conductors of heat and electricity, workable, strong, and mostly with high densities. With the exception of liquid mercury, Hg, they have melting points which range from fairly high to very high.

They are much less reactive than the metals in Groups 1 and 2, and so they don't react as rapidly with air or water.

Because of their useful physical properties and relative lack of reactivity, several of the transition elements are important in everyday life.

1. Where are the transition elements in the Periodic Table?
2. What is the difference between a group and a period?
3. Where in the Periodic Table do we find non-metals?

You will also find information about the transition metals iron and copper in Chapter 19 which deals with trends in the Periodic Table.

The transition metals form a block of elements in the centre of the Periodic Table. There are ten elements in the first transition series, from scandium to zinc.

## Variable valency

The formulae of metal chlorides illustrate what we mean by variable valency. Sodium chloride is NaCl (Group 1), calcium chloride is CaCl2 (Group 2) and aluminium chloride is $AlCl_3$ (Group 3). The transition metals are unusual in showing more than one valency. Iron forms two chlorides $FeCl_2$ and $FeCl_3$ since the metal forms two kinds of ion, one with charge +2 and the other +3. Copper also forms two chlorides $CuCl_2$ and CuCl. This property of transition metals is what we refer to as variable valency. We write the valency using Roman numerals, for example copper(II) oxide is CuO and iron(III) chloride is $FeCl_3$.

> 4. What are the formulae of the bromides of copper?
> 5. Write the formulae of the two copper oxides containing metal ions with charges +1 and +2.
> 6. What is the valency of iron in the compound $Fe_2O_3$?

## Transition metals form coloured compounds

The majority of transition metal compounds are coloured. The photographs show some examples, both in and out of the lab.

Solutions of some common transition metal compounds.

The blue on this Yuan Dynasty vase was created using cobalt compounds in the glaze.

## Transition metals and their compounds are often useful catalysts

Examples you will come across:

- Iron in the manufacture of ammonia.

- Vanadium(V) oxide, $V_2O_5$, in the manufacture of sulphuric acid.

*Weathering on the copper covered spire of Truro Cathedral.*

- Platinum and rhodium in the manufacture of nitric acid.

- Platinum and palladium in catalytic converters for cars.

- Nickel in the manufacture of margarine.

- Manganese(IV) oxide, $MnO_2$, in the decomposition of hydrogen peroxide.

## Reactions of iron

Iron reacts with solutions of dilute acids to form iron(II) salts.

$$\text{Iron} + \text{sulphuric acid} \rightarrow \text{iron(II) sulphate} + \text{hydrogen}$$
$$\text{Fe} + \text{H}_2\text{SO}_4 \rightarrow \text{FeSO}_4 + \text{H}_2$$

Iron reacts slowly with water to form rust. With steam, the reaction is different. Hot iron reacts with steam to form hydrogen gas. This reaction is reversible.

$$3\text{Fe(S)} + 4\text{H}_2\text{O(g)} \underset{\text{reversible reaction}}{\rightleftharpoons} \text{Fe}_3\text{O}_4\text{(s)} + 4\text{H}_2\text{(g)}$$

This reaction can be used on an industrial scale to prepare hydrogen from steam and scrap iron.

### Hydroxides of iron

Iron forms two different hydroxides, each with its own special colour. They can both be prepared by precipitation since the hydroxides are both insoluble.

## Activity 1: Preparation of the hydroxides of iron

**Investigative skills**

A

**You will need:**

- test-tubes

- fresh solutions of iron(II) sulphate and iron(III) chloride

- evaporating basin

- dilute sodium hydroxide

**Carry out the following:**

1. Place 2 cm³ iron(II) sulphate in a tube and add 2 cm³ sodium hydroxide.

2. Observe the precipitate and pour the mixture into an evaporating basin and leave exposed to the air.

3. Repeat using iron(III) chloride. Compare the colours at the start and later.

4. Iron(II) hydroxide is white when pure but usually appears pale green. Iron(III) hydroxide is rusty brown. Use this information to explain your observations.

# Reactions of copper

Copper is very low in the reactivity series of metals. Copper does not react with dilute hydrochloric or sulphuric acids but it does react with the oxidising acid, nitric acid. With concentrated nitric acid, brown fumes of nitrogen dioxide are released. This gas is toxic.

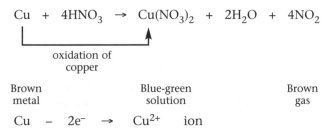

$$Cu + 4HNO_3 \rightarrow Cu(NO_3)_2 + 2H_2O + 4NO_2$$

| Brown metal | | Blue-green solution | | Brown gas |

$$Cu - 2e^- \rightarrow Cu^{2+} \quad ion$$

This is an example of a redox reaction: the copper is oxidised and the nitric acid is reduced.

## Compounds of copper

When copper is heated in a flame, it turns black with a coating of copper(II) oxide.

$$2Cu + O_2 \rightarrow 2CuO$$

This oxide can also be produced by heating the green copper(II) carbonate.

$$CuCO_3 \rightarrow CuO + CO_2$$

This is an example of thermal decomposition. Copper(II) oxide is a base – it reacts with acids to form salts.

$$CuO + H_2SO_4 \rightarrow CuSO_4 + H_2O$$

black solid — blue solution

$$Base + acid \rightarrow salt + water$$

Blue copper(II) sulphate solution reacts with alkalis such as sodium hydroxide or ammonia solution to form copper(II) hydroxide.

$$CuSO_4 + 2NaOH \rightarrow Cu(OH)_2 + Na_2SO_4$$

blue solution — blue-green gelatinous solid

$$CuSO_4 + NH_3(aq) \rightarrow Cu(OH)_2$$

ammonia solution — add ↓ more ammonia deep blue complex ion

*Malachite, a green copper ore.*

The ion is $[Cu(H_2O)_2 (NH_3)_4]^{2+}$

It is octahedral in shape with copper in the middle.

The formation of the deep blue complex ion with ammonia is used as a test for the presence of copper in solution. Other metals give green-blue solutions, for example nickel or cobalt, but only copper forms this coloured complex.

When the base copper(II) oxide reacts with other acids, it produces salts.

$$CuO + 2HCl \rightarrow CuCl_2 + H_2O$$

black — green

$$CuO + 2HNO_3 \rightarrow Cu(NO_3)_2 + H_2O$$

blue

## Activity 2: Chemistry of copper

**Investigative skills**

P

Plan an experiment to convert copper(II) carbonate into copper(II) oxide, then to copper(II) sulphate and finally produce a solution of a complex ion.

Both salts can be obtained as coloured crystals if the solutions are left to evaporate. Copper nitrate decomposes easily on heating, releasing brown fumes.

$$2Cu(NO_3)_2 \xrightarrow{\text{heat}} 2CuO + 4NO_2 + O_2$$

blue     black     brown gas     relights a glowing splint

### Copper(I) compounds

In addition to the green and blue copper(II) compounds, the metal also forms copper(I) compounds. For example, copper(I) oxide has the formula $Cu_2O$. It is a red-brown solid. Most of the important chemistry of copper involves copper(II) compounds.

# End of Chapter Checklist

**In this chapter you have learnt that:**

Transition elements:

- are the group of elements found in the centre of the Periodic Table
- are metals with a high melting point (except mercury)
- are hard, tough and strong
- are much less reactive than Group 1 metals
- have many uses in everyday life
- form coloured compounds
- are often useful catalysts
- form complex ions.

# Questions

1. Copy and complete:

| Metal | Valency | Name using Roman numerals | Colour of ion |
|-------|---------|---------------------------|---------------|
| Iron | +2 | | |
| Iron | +3 | | |
| Copper | +1 | | |
| Copper | +2 | | |

(8)

2. Copper roofs often turn green in the presence of air that contains carbon dioxide. Explain what compound may be forming on the metal surface. (4)

3. Describe the formation of a complex ion of copper starting from crystals of copper sulphate. (6)

4. a) What is meant by variable valency? (2)

   b) Give examples of three different coloured transition metal ions. (3)

   c) How can you test a blue solution to check if it contains copper(II) ions? (3)

5. Copper turns black on heating in a Bunsen flame. The black material reacts with dilute sulphuric acid to give a blue solution. Explain these changes and write word and symbol equations. (8)

6. All of these are characteristics of transition metals, except:

   A   coloured compounds
   B   active as catalysts
   C   very reactive metals
   D   variable valencies           (1)

## Chapter 23: Non-metals: Nitrogen, Oxygen, Sulphur and Hydrogen

**When you have completed this chapter, you will be able to:**

- describe the physical and chemical properties of non-metals
- carry out laboratory preparations of oxygen, hydrogen, ammonia and carbon dioxide
- describe how oxygen and nitrogen are extracted from air
- describe the properties of sulphur.

## Physical properties of non-metals

| Name | Symbol | Melting point/°C | Boiling point/°C | Thermal conductivity/ $W\ m^{-1}\ °C^{-1}$ | Density/ $g\ cm^{-3}$ | Colour |
|------|--------|------------------|------------------|---------------------------------------------|------------------------|--------|
| hydrogen | H | −259 | −253 | 0.181 | 0.00008 | none |
| nitrogen | N | −210 | −196 | 0.025 | 0.00117 | none |
| oxygen | O | −219 | −183 | 0.267 | 0.00133 | none |
| sulphur | S | 115 | 445 | 0.269 | 1.96 | yellow |
| chlorine | Cl | −101 | −34 | 0.009 | 0.00299 | green |

1. Which of the elements in the table are in the same group?
2. Which element is a solid at room temperature?
3. Which two elements have a colour?

## The present atmosphere

### Composition

The composition of the air is believed to have been largely unchanged for the past 200 million years. The approximate percentages (by volume) of the main gases present in *unpolluted, dry* air are:

It is important to realise that these figures apply only to *dry* air. Air can have anywhere between 0% and 4% of water vapour.

| nitrogen | 78.1% | about 4/5 |
|----------|-------|-----------|
| oxygen | 21.0% | about 1/5 |
| argon | 0.9% | |
| carbon dioxide | 0.03% | |

There are also very small amounts of the other noble gases in air.

## Showing that air contains about one fifth oxygen

This apparatus can be used to find the percentage of oxygen in the air:

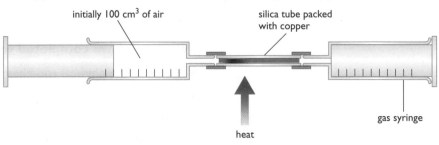

A silica tube looks like glass, but won't melt, however strongly you heat it with a Bunsen burner.

The apparatus initially contains 100 cm³ of air. This is pushed backwards and forwards over the heated copper, which turns black as copper(II) oxide is formed. The volume of air in the syringes falls as the oxygen is used up.

$$2Cu(s) + O_2(g) \rightarrow 2CuO(s)$$

As the copper reacts, the Bunsen burner is moved along the tube, so that it is always heating fresh copper.

Eventually, all the oxygen in the air is used up. The volume stops contracting and the copper stops turning black. On cooling, somewhere around 79 cm³ of gas is left in the syringes, showing that 21% has been used up.

Therefore the air contained 21% of oxygen.

## Extraction of oxygen and nitrogen from liquid air

When a compressed gas escapes from a balloon or an inflated tyre, the temperature falls, the expanding gas feels colder. This effect is used to cool air until it liquefies at about –200°C. By recycling the cooled air several times, the temperature continues to fall until liquid air is produced. Both water vapour and carbon dioxide must first be removed, since both turn into solids at low temperatures. The ice would block the pipes and stop the process continuing. The remaining gases are nitrogen, oxygen and traces of the noble gases, mainly argon.

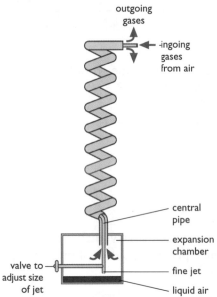

principles of the liquefaction of air

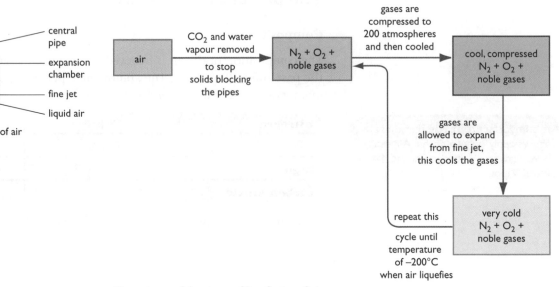

*Flow scheme of the stages of liquefaction of air.*

The liquid air can be separated into separate gases by fractional distillation. This is the same type of separation process as used to separate crude oil into useful hydrocarbon fractions. As the temperature of the liquid air rises from about –200°C, nitrogen is the first to boil at about –196°C, followed by oxygen at about –183°C. Careful distillation allows the noble gases to be separated as well.

## Activity 1: Laboratory preparation of oxygen

**Investigative skills**

O   A

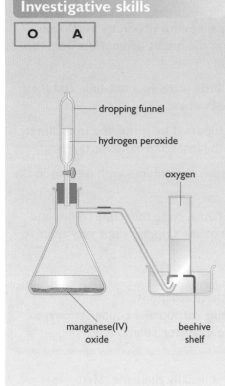

dropping funnel

hydrogen peroxide

oxygen

manganese(IV) oxide

beehive shelf

**You will need:**

- conical flask
- dropping funnel
- delivery tube
- trough
- beehive shelf or other support
- gas jars with lids
- measuring cylinder
- burning spoon
- tongs
- hydrogen peroxide 20 vol. strength
- manganese(IV) oxide
- magnesium ribbon
- sulphur powder
- indicator solution

**Carry out the following:**

1. Assemble the apparatus as shown.
2. Add about 1 g of the catalyst, manganese(IV) oxide, to the flask.
3. Add 25 cm$^3$ hydrogen peroxide.
4. Collect several jars or tubes of oxygen gas and test as follows:
   a) Place a glowing splint in the oxygen.
   b) Ignite a 3 cm piece of magnesium ribbon and transfer it to the oxygen. The product from the burning is magnesium oxide. Shake it with some water and test with indicator.
   c) Ignite 0.5 g sulphur in a flame, using a burning spoon or spatula. Transfer it to the oxygen. A gas will be produced. Shake the product with some water and test with indicator.
5. Tabulate your results.
6. Can you draw any general conclusion about metallic and non-metallic oxides when dissolved in water and tested with indicators?

### Forming oxides by burning elements in air

In addition to burning elements in pure oxygen, we can form oxides by simply heating them in air. For example:

Magnesium + oxygen → magnesium oxide

$$2Mg + O_2 \rightarrow 2MgO$$

sulphur + oxygen → sulphur dioxide

$$S + O_2 \rightarrow SO_2$$

# Activity 2: Burning elements in air

**You will need:**

- Bunsen burner and tongs
- Spatula, test-tubes
- Samples of magnesium, iron filings, iron wool, copper foil, carbon (charcoal) and universal indicator

**Carry out the following:**

1. Use tongs to burn a 3 cm piece of magnesium ribbon by heating in a hot gas flame. Do not look directly at the bright flame. How does this compare to burning in pure oxygen?

2. Shake the magnesium oxide with a little water in a test-tube and then add indicator. The solid metal oxide is a base.

3. Use tongs to burn iron wool in air, followed by sprinkling iron filings into a hot gas flame. Both produce the insoluble oxide $Fe_3O_4$.

4. Burn a piece of charcoal in a hot flame. It combines with oxygen in the air to form carbon dioxide gas.

5. Heat a piece of copper foil in a hot flame. Note the flame colour and the black coating of solid copper(II) oxide. Copper is not very reactive.

---

4. Write word equations for all of these reactions.
5. Write symbol equations for burning carbon and copper in oxygen.
6. Why doesn't iron oxide affect the indicator solution?

---

The natural gas used in Bunsen burners is usually methane. Methane is oxidised in air as it burns to give carbon dioxide and water vapour.

$$CH_4 + 2O_2 \rightarrow CO_2 + 2H_2O$$

This oxidation is very exothermic, which is why we use it to produce heat.

*Summary*

| Element | Metal | Non-metal | Acidic oxide | Basic oxide |
|---------|-------|-----------|--------------|-------------|
| sulphur | | yes | yes | |
| magnesium | yes | | | yes |
| copper | yes | | | yes |
| carbon | | yes | yes | |

---

7. What pattern do you notice about the acid-base properties of these oxides?
8. Phosphorus is a non-metal. Will its oxide be acidic or basic?
9. Why does fresh soda water have a sharp taste?

Sulphur dioxide reacts with water to form an acidic solution containing sulphurous acid, $H_2SO_3$. This is also called sulphuric(IV) acid. Ordinary sulphuric acid is sulphuric(VI) acid. If sulphur dioxide is bubbled into an alkali such as sodium hydroxide, it forms a hydrogen sulphite salt.

$$NaOH + SO_2 \rightarrow NaHSO_3 \text{ sodium hydrogen sulphite}$$

If more alkali is added, the sulphite salt is produced

$$NaHSO_3 + NaOH \rightarrow Na_2SO_3 + H_2O \text{ sodium sulphite formed}$$

The test for sulphur dioxide is described in Chapter 24, Analysis.

# Carbon dioxide

## Activity 3: Laboratory preparation of carbon dioxide

**Investigative skills**

O   A

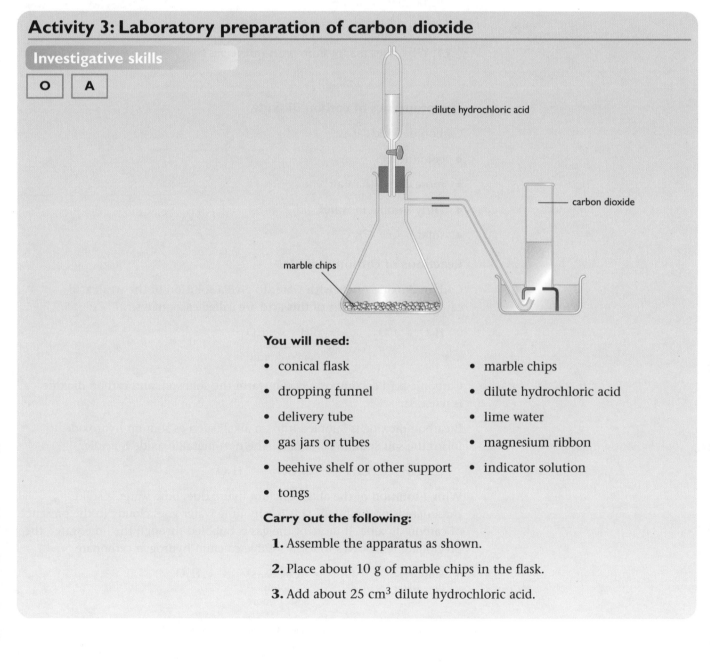

dilute hydrochloric acid

carbon dioxide

marble chips

**You will need:**

- conical flask
- dropping funnel
- delivery tube
- gas jars or tubes
- beehive shelf or other support
- tongs

- marble chips
- dilute hydrochloric acid
- lime water
- magnesium ribbon
- indicator solution

**Carry out the following:**

1. Assemble the apparatus as shown.
2. Place about 10 g of marble chips in the flask.
3. Add about 25 cm³ dilute hydrochloric acid.

## Activity 3: Laboratory preparation of carbon dioxide (continued)

**4.** Collect several jars or tubes of the gas. (Carbon dioxide is more dense than air.)

**5.** Bubble the gas through lime water and observe any changes.

**6.** Place a burning splint in the gas.

**7.** Shake a tube of gas with water and test with indicator.

**8.** Ignite some magnesium ribbon and transfer it to the gas. The word equation is:

magnesium + carbon dioxide → magnesium oxide + carbon

**9.** Tabulate your results.

**10.** What are the tests for carbon dioxide?

**11.** Why does magnesium burn in the gas?

### The properties of carbon dioxide

Carbon dioxide gas is:

- colourless
- more dense than air
- fairly soluble in water
- solid below –78°C

### Reactions of carbon dioxide

Carbon dioxide reacts with water to give a solution of the weak acid carbonic acid. The salts of this acid are called carbonates.

$$H_2O + CO_2 \underset{\text{reversible}}{\rightleftharpoons} \underbrace{2H^+ + CO_3^{2-}}_{\text{carbonic acid}}$$

Carbonic acid decomposes if you warm the solution, and carbon dioxide is released.

If carbon dioxide is bubbled into an alkali such as sodium hydroxide, it forms the salt sodium carbonate. The non-metallic oxide is acidic.

$$2NaOH + CO_2 \rightarrow Na_2CO_3 + H_2O$$

With a solution of the alkali calcium hydroxide, lime water, a solid carbonate is precipitated. This is why lime water goes cloudy in the presence of carbon dioxide. If more of the gas is bubbled through the lime water, the precipitate redissolves to form soluble calcium hydrogen carbonate.

$$\underset{\text{lime water}}{Ca(OH)_2} + CO_2 \rightarrow \underset{\text{cloudy}}{CaCO_3(s)} + H_2O$$

$$\text{add} \downarrow \text{more } CO_2$$

$$\underset{\text{soluble}}{Ca(HCO_3)_2(aq)}$$

## Uses of carbon dioxide

Carbon dioxide is a dense gas that extinguishes fires. Electrical fires should never be put out using water owing to the danger of electric shocks.

Carbon dioxide is used to make carbonated drinks, fizzy drinks. The gas is dissolved under pressure and is released when the bottle or can is opened.

*A carbon dioxide fire extinguisher in action.*

*A selection of fizzy drinks.*

The sharp taste of fizzy drinks is caused by carbonic acid. As the gas bubbles out when the drink is opened, drinks tend to taste sweeter.

## Nitrogen and oxygen

Nitrogen is a very unreactive gas. This is why it does not react with oxygen in the air under normal conditions. In thunder storms, the energy released by lightning is often enough to make the two gases react. The product is a mixture of nitrogen oxides.

The high temperatures and pressures in car engines can also make nitrogen and oxygen react. For example:

$$N_2 + O_2 \rightarrow 2NO \text{ nitrogen monoxide}$$

Followed by

$$2NO + O_2 \rightarrow 2NO_2 \text{ nitrogen dioxide}$$

Nitrogen dioxide is a brown acidic gas that dissolves in water to give an acidic solution. This is one of the causes of acid rain pollution.

### Nitrogen and food preservation

Liquid nitrogen is so cold that it can rapidly freeze fresh food to preserve it. Frozen food can be stored for weeks or months without going bad. Since nitrogen gas is so unreactive, it is used to fill food containers in place of air. Crisps and nuts are often sold in sealed packets filled with nitrogen. The absence of oxygen stops the food oxidising and reduces the chances of bacteria growing and spoiling the food.

## Ammonia

Under industrial conditions, nitrogen combines with hydrogen to form ammonia.

Ammonia is the only common alkaline gas. Ammonia solution is used as an alkali in titrations and for other reactions.

# Activity 4: Laboratory preparation of ammonia

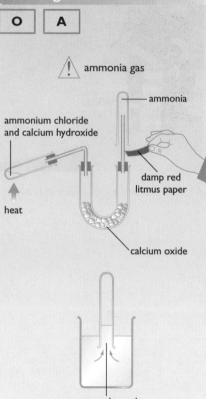

ammonia gas

ammonia

ammonium chloride
and calcium hydroxide

damp red
litmus paper

heat

calcium oxide

water replaces the
ammonia which dissolves
in the water

**You will need:**

- straight glass tube

- delivery tube

- U-tube

- test-tubes and stoppers

- beaker

- solid ammonium chloride and calcium hydroxide

- red litmus paper

- calcium oxide (for drying the gas)

**Carry out the following:**

1. Assemble the apparatus as shown.

2. Warm a mixture of about 2 g each of the two solids.

3. Collect several tubes of the gas as shown and seal with stoppers. (Ammonia is less dense than air so will rise upwards.)

4. Test the gas with moist red litmus paper.

5. Invert one tube of gas into a beaker of water (see diagram). What does this tell you about the solubility of ammonia?

6. Very carefully waft some of the gas towards your nose to recognise the smell.

7. Tabulate your results.

Ammonia is manufactured on an industrial scale by the Haber Process, and has several uses such as making nitric acid and fertilisers.

## Properties of ammonia gas

- colourless

- less dense than air

- very soluble in water, 770 volumes of ammonia dissolve in 1 volume of water at room temperature

- the only common alkaline gas

- ammonia solution is a weak alkali. Full details are given in Chapter 10, Acids, Bases and pH.

## Ammonium salts

Ammonia gas reacts with hydrogen chloride gas, both covalent, to form the ionic salt ammonium chloride.

$$NH_3(g) \ + \ HCl(g) \quad \rightleftharpoons \quad NH_4Cl(g)$$

covalent gases    reversible    with ions $NH_4^+$ and $Cl^-$

If solid ammonium chloride is warmed, it sublimes. It changes back to the two gases without first melting. Tests for both gases are described in Chapter 24, Analysis.

Ammonia also reacts with sulphuric acid to give the salt ammonium sulphate.

$$2NH_3(aq) \ + \ H_2SO_4(aq) \quad \rightarrow \quad (NH_4)_2SO_4$$

solutions react    ionic salt

The white crystalline salt can be obtained by evaporating the solution. Ammonium sulphate is an important fertiliser, as is ammonium nitrate.

### *Making explosives and fireworks*

Ammonia can be oxidised to form nitric acid. When the two are combined, they produce ammonium nitrate. This is a cheap oxygen-providing compound called an **oxygen carrier**. It is a major ingredient in many commercial explosives.

Small-scale explosives are sold as fireworks. Most contain gunpowder which is a mixture of potassium nitrate, sulphur and carbon in the correct proportions.

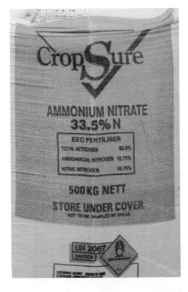

*Ammonium nitrate fertiliser pack label.*

# Sulphur

Sulphur is a yellow, solid element. It is released by many volcanoes (see Chapter 15, page 153). It exists in two different crystalline forms called the **allotropes** of sulphur as well as another form called plastic sulphur.

## Rhombic sulphur

This is the form of sulphur that is stable at room temperature. Rhombic crystals can be obtained by dissolving powdered sulphur in the toxic solvent carbon disulphide, followed by evaporation. The individual sulphur molecules are rings of eight atoms.

## Monoclinic sulphur

If sulphur is heated until it melts and is then cooled, long needle-shaped crystals form. These are monoclinic sulphur. At room temperature they gradually change back to the other allotrope, rhombic sulphur.

## Plastic sulphur

If hot molten sulphur is poured rapidly into cold water, it forms plastic sulphur, that looks a bit like chewing gum. The sulphur cools so quickly that

the structure of the $S_8$ molecule

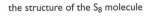

105°

the crystal shapes of (a) rhombic sulphur, (b) monoclinic sulphur

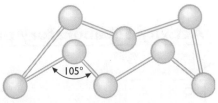

(a)

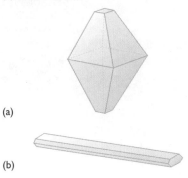

(b)

crystals cannot form. After some days, the flexible plastic sulphur sets hard as it turns back to rhombic.

### Reactions of sulphur

Sulphur burns in oxygen to give an acidic gas, sulphur dioxide.

$$\text{sulphur} + \text{oxygen} \rightarrow \text{sulphur dioxide}$$
$$S + O_2 \rightarrow SO_2$$

Many metals react with sulphur to form sulphides.

$$\text{iron} + \text{sulphur} \rightarrow \text{iron sulphide}$$
$$Fe + S \rightarrow FeS$$

Many of the most important metal ores are sulphides, such as galena and lead sulphide. Sulphur dioxide is used to make sulphuric acid.

## Hydrogen

In Chapter 9 on the Reactivity Series, we saw examples of the general reaction:

Metal + acid → a salt + hydrogen

In each case, the new salt takes its name from the acid used.

| Acid name | Salt name |
|---|---|
| sulphuric acid | sulphate |
| nitric | nitrate |
| hydrochloric | chloride |

The hydrogen gas released can be tested with a burning splint, the 'pop' test. The same reaction can be used to prepare a larger sample of hydrogen gas.

### Activity 5: Laboratory preparation of hydrogen gas

**Investigative skills**

| O | A |
|---|---|

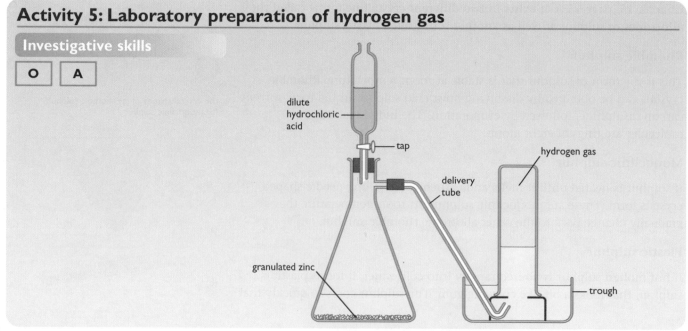

dilute hydrochloric acid

tap

hydrogen gas

delivery tube

granulated zinc

trough

## Activity 5: Laboratory preparation of hydrogen gas (continued)

**You will need:  Danger: explosive gas**

- flask fitted with a delivery tube and funnel

- trough, gas jar and support

- granulated zinc and dilute hydrochloric acid

- indicator solution

**Carry out the following: Warning! Use a fume cupboard**

1. Set up the apparatus as shown with an inverted water-filled gas jar in place

2. Add acid slowly until a steady stream of bubbles can be seen. Do not collect the gas immediately since it will be mixed with air from the flask. Hydrogen-air mixtures are explosive.

3. Collect several small gas jars or tubes of hydrogen.

4. Test one with indicator. Hydrogen is neutral.

5. Carefully ignite a small sample of hydrogen. Wear safety glasses. The gas explodes to form water, hydrogen oxide.

### Testing hydrogen oxide

If a jet of burning hydrogen is placed in contact with a cold surface, a colourless liquid condenses. This liquid is neutral, boils at 100°C under normal conditions and is pure water. Two volumes of hydrogen combine with one volume of oxygen. This explains the formula of water, $H_2O$.

$$2H_2 + O_2 \rightarrow 2H_2O$$

## Activity 6: Testing water

**Investigative skills**

P

Plan an experiment to determine which of the following unlabelled water samples is pure water.

- distilled water

- tap water

- sea water

- mineral water

Think about evaporation and what might be left over.

Under normal conditions, pure water boils at 100°C and has a density of $1 \text{ g/cm}^3$.

# End of Chapter Checklist

**In this chapter you have learnt that:**

● non-metals have different physical and chemical properties to metals

● you can prepare oxygen, carbon dioxide, hydrogen and ammonia gases in the laboratory

● ammonia can be oxidised to nitric acid

● ammonium nitrate is both a fertiliser and an explosive

● cooling air gives liquid air which can be separated

● sulphur forms allotropes.

# Questions

1. Which of the following is a solid, yellow, non-metallic element?

   A chlorine
   B sulphur
   C nitrogen
   D copper                                        (1)

2. Nitrogen and hydrogen combine to give:

   A nitrogen oxide
   B ammonia
   C hydrogen oxide
   D nitric acid                                    (1)

3. *a)* Draw and label the apparatus needed to prepare oxygen in the laboratory.            (5)

   *b)* Write word and balanced symbol equations for the reaction.                        (2)

   *c)* What is the test for oxygen?                 (1)

   *d)* Describe the reaction of burning sulphur with oxygen.                            (2)

4. *a)* Describe how ammonia is converted into nitric acid.                              (4)

   *b)* Give *two* uses each for ammonia and nitric acid.                               (4)

   *c)* What is unusual about the uses of ammonium nitrate?                            (2)

5. *a)* Describe the preparation of carbon dioxide from marble chips and dilute hydrochloric acid.   (5)

   *b)* Could you prepare carbon dioxide from sea shells and hydrochloric acid? (*Hint:* sea shells contain calcium carbonate.) Explain your answer.       (2)

   *c)* Describe *two* tests for carbon dioxide.      (2)

   *d)* Write a word equation for the reaction of magnesium with carbon dioxide.          (1)

6. Describe the allotropes of sulphur and sketch the crystal shapes.                      (8)

   What happens when sulphur burns in air?         (2)

7. Draw a table to compare the properties of hydrogen, ammonia, nitrogen and carbon dioxide gases.   (10)

8. Describe everyday uses of each:

   *a)* carbon dioxide                              (6)

   *b)* nitrogen                                    (2)

   *c)* oxygen                                      (2)

> **When you have completed this chapter, you will be able to:**
> - collect soluble and insoluble gases
> - test and identify gases
> - test liquids for their water content
> - carry out flame tests
> - test metal ions in solution
> - test for carbonates, sulphites, sulphates and halides
> - test for nitrates, acids and alkalis.

## Collecting and identifying gases

### Collecting gases

Gases can be collected:

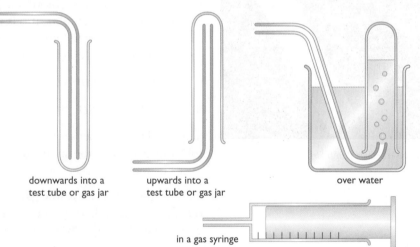

downwards into a
test tube or gas jar

upwards into a
test tube or gas jar

over water

in a gas syringe

Collecting into a gas syringe is fine if you want to measure the volume of the gas. To test the gas, though, you would have to transfer it to a test tube using one of the first three methods.

### Hazards associated with collecting the gases

Hydrogen forms explosive mixtures with air. Chlorine, sulphur dioxide, hydrogen chloride and ammonia are all poisonous. Sulphur dioxide can trigger asthma attacks.

### How to collect and test individual gases

#### Hydrogen, $H_2$

Hydrogen is less dense than air and is almost insoluble in water. Collect it over water or upwards into a test tube or gas jar.

Hydrogen pops when a lighted splint is held to the mouth of the tube. The hydrogen combines explosively with oxygen in the air to make water.

$$2H_2(g) + O_2(g) \rightarrow 2H_2O(l)$$

### Oxygen, $O_2$

Oxygen has almost the same density as air and is almost insoluble in water. You normally collect it over water.

Oxygen relights a glowing splint.

### Carbon dioxide, $CO_2$

Carbon dioxide is denser than air and can be collected downwards into a test tube or gas jar. It is only slightly soluble in water and so it can be collected over water as well.

Carbon dioxide turns lime water milky (or chalky). Lime water is calcium hydroxide solution. Carbon dioxide reacts with it to form a white precipitate of calcium carbonate.

$$Ca(OH)_2(aq) + CO_2(g) \rightarrow CaCO_3(s) + H_2O(l)$$

### Chlorine, $Cl_2$

Chlorine is denser than air and is usually collected downwards into a test tube or gas jar. Because chlorine is green it is easy to see when the tube or gas jar you are collecting it into is full. Chlorine is too soluble to collect it satisfactorily over water, but you can collect it over concentrated salt solution instead. It is less soluble in the salt solution.

Chlorine is a green gas which bleaches damp litmus paper.

Chlorine bleaches damp litmus paper.

### Sulphur dioxide, $SO_2$

Sulphur dioxide is denser than air, and can be collected downwards into a test tube or gas jar. It is too soluble to collect over cold water, but you can collect it over hot water. Gases become less soluble as the temperature increases.

Sulphur dioxide is an acidic gas, turning blue litmus paper red. You can pick out sulphur dioxide from other acidic gases because it turns potassium dichromate(VI) paper from orange to green.

### Hydrogen chloride, HCl

Hydrogen chloride is denser than air and is extremely soluble in water (to make hydrochloric acid). It must be collected downwards into a test tube or gas jar.

Hydrogen chloride is a steamy acidic gas. The steaminess is due to the hydrogen chloride reacting with water vapour in the air to form droplets of concentrated hydrochloric acid. It turns damp blue litmus paper red.

### Ammonia, $NH_3$

Ammonia is less dense than air and is extremely soluble in water. It can only be collected upwards into a test tube or gas jar.

Ammonia is the only alkaline gas that you will meet. It turns damp red litmus paper blue.

## Nitrogen dioxide, $NO_2$

Nitrogen dioxide is a poisonous, brown gas. The smell is a little like chlorine. It is soluble in water, producing an acidic solution. Nitrogen dioxide turns damp blue litmus paper red.

1. What is the test for carbon dioxide gas?
2. Name one gas that turns blue litmus paper red.
3. Which is the only common alkaline gas?
4. What is the effect of chlorine on damp litmus paper?

# Testing for water

## Using anhydrous copper(II) sulphate

Water turns white anhydrous copper(II) sulphate blue.

Anhydrous copper(II) sulphate lacks water of crystallisation and is white. Dropping water onto it replaces the water of crystallisation, and turns it blue.

$$CuSO_4(s) + 5H_2O(l) \rightarrow CuSO_4 \cdot 5H_2O(s)$$

This test works for anything which contains water. It does *not* show that the water is pure.

## Using cobalt chloride paper

Cobalt chloride paper is simply filter paper which has been dipped into cobalt(II) chloride solution and then dried thoroughly in a desiccator. A desiccator is a piece of glassware or a small cabinet which contains a tray of some substance which absorbs water.

In the absence of any water, the paper is blue. Adding water to it turns it pink.

Anything which contains water will turn cobalt chloride paper from blue to pink. The water doesn't have to be pure.

# Testing for ions

## Flame tests

A flame test is used to show the presence of certain metal ions in a compound. A platinum or nichrome wire is cleaned by dipping it into concentrated hydrochloric acid and then holding it in a hot Bunsen flame. This is repeated until the wire doesn't impart any colour to the flame.

The wire is dipped back into the acid, then into a tiny sample of the solid you are testing, and back into the flame.

Nichrome (a nickel–chromium alloy) is a cheap alternative to platinum. It does, however, always produce a faint yellow colour in the flame which you have to ignore.

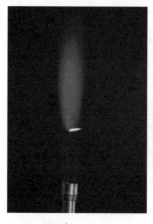

Red shows $Li^+$ ions.

Strong persistent yellow-orange shows $Na^+$ ions.

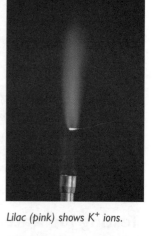

Lilac (pink) shows $K^+$ ions.

Orange-red ('brick red') shows $Ca^{2+}$ ions.

Blue-green shows $Cu^{2+}$ ions.

Pale green shows $Ba^{2+}$ ions.

The calcium flame test often has so much orange in it that it can be confused with sodium. The orange-red often appears for quite a short time only.

5. Describe one test for water.
6. Which metal ion shows a blue-green flame colour?

### Testing for positive ions using sodium hydroxide solution

Of the common hydroxides, only sodium, potassium and ammonium hydroxides dissolve in water. Most metal hydroxides are insoluble. This means that if you add sodium hydroxide solution to a solution containing the metal ions, you will get a precipitate of the metal hydroxide.

The colour of these precipitates can help you to identify the metal ion.

#### Blue precipitate

This shows the presence of copper(II) ions. The precipitate is copper(II) hydroxide.

$$Cu^{2+}(aq) + 2OH^-(aq) \rightarrow Cu(OH)_2(s)$$

Any copper(II) salt in solution will react with sodium hydroxide solution in this way. For example, with copper(II) sulphate solution, the full equation would read

$$CuSO_4(aq) + 2NaOH(aq) \rightarrow Cu(OH)_2(s) + Na_2SO_4(aq)$$

The blue precipitate of copper(II) hydroxide.

## An orange-brown precipitate

This shows the presence of iron(III) ions. The precipitate is iron(III) hydroxide.

$$Fe^{3+}(aq) + 3OH^-(aq) \rightarrow Fe(OH)_3(s)$$

Any iron(III) compound in solution will give this precipitate. An example full equation might be

$$FeCl_3(aq) + 3NaOH(aq) \rightarrow Fe(OH)_3(s) + 3NaCl(aq)$$

Notice how much more complicated the full equations for these reactions are. They also obscure what is going on. Use ionic equations for precipitation reactions wherever possible.

## A green precipitate

This shows the presence of iron(II) ions. The precipitate is iron(II) hydroxide.

$$Fe^{2+}(aq) + 2OH^-(aq) \rightarrow Fe(OH)_2(s)$$

This could be the result of reacting, say, iron(II) sulphate solution with sodium hydroxide solution:

$$FeSO_4(aq) + 2NaOH(aq) \rightarrow Fe(OH)_2(s) + Na_2SO_4(aq)$$

The green precipitate darkens on standing and turns orange around the top of the tube. This is due to the iron(II) hydroxide being oxidised to iron(III) hydroxide by the oxygen in the air.

## White precipitates

White precipitates are more common and therefore less informative. The examples you may need are:

- *A white precipitate which doesn't dissolve when you add excess sodium hydroxide solution.*

  This shows the presence of either magnesium or calcium ions in solution. The precipitate may be either magnesium hydroxide or calcium hydroxide.

  $$Mg^{2+}(aq) + 2OH^-(aq) \rightarrow Mg(OH)_2(s)$$

  or $\quad Ca^{2+}(aq) + 2OH^-(aq) \rightarrow Ca(OH)_2(s)$

  You could tell which you had by doing a flame test on the original compound. Calcium gives an orange-red colour, whereas magnesium has no flame colour.

- *A white precipitate which dissolves when you add excess sodium hydroxide solution.*

  This shows the presence of aluminium ions in the solution. The white precipitate is aluminium hydroxide.

  $$Al^{3+}(aq) + 3OH^-(aq) \rightarrow Al(OH)_3(s)$$

  The precipitate dissolves because the aluminium hydroxide reacts with excess hydroxide ions to give $Al(OH)_4^-$ ions – called tetrahydroxoaluminate ions.

The orange-brown precipitate of iron(III) hydroxide.

The green precipitate of iron(II) hydroxide.

Precipitate formed from magnesium sulphate solution and sodium hydroxide solution.

*Copper(II) carbonate turns black on heating.*

$$Al(OH)_3(s) + OH^-(aq) \rightarrow Al(OH)_4^-(aq)$$

The full equation for this is:

$$Al(OH)_3(s) + NaOH(aq) \rightarrow NaAl(OH)_4(aq)$$

The compound formed is called sodium tetrahydroxoaluminate.

Another white precipitate which is soluble in excess sodium hydroxide is zinc. Zinc hydroxide is an **amphoteric** material, this means that it reacts both with acids and alkalis. This test will not distinguish between aluminium and zinc ions in solution since they react in the same way.

### No precipitate, but a smell of ammonia

This shows the presence of an ammonium salt. Sodium hydroxide solution reacts with ammonium salts (either solid or in solution) to produce ammonia gas. In the cold, there is just enough ammonia gas produced for you to be able to smell it. If you warm it, you can test the gas coming off with a piece of damp red litmus paper. Ammonia is alkaline and turns the litmus paper blue.

$$NH_4^+(s \text{ or } aq) + OH^-(aq) \rightarrow NH_3(g) + H_2O(l)$$

A typical full equation might be:

$$NH_4Cl(s) + NaOH(aq) \rightarrow NaCl(aq) + NH_3(g) + H_2O(l)$$

## Testing for carbonates and sulphites

### Testing for carbonates by heating them

Most carbonates split up to give the metal oxide and carbon dioxide when you heat them. This is a good example of **thermal decomposition** – breaking something up by heating it. You can test the carbon dioxide given off by passing it through lime water, and there may be helpful colour changes as well.

For example, copper(II) carbonate is a green powder which decomposes on heating to produce black copper(II) oxide.

$$CuCO_3(s) \rightarrow CuO(s) + CO_2(g)$$

Zinc carbonate is a white powder which decomposes on heating to give zinc oxide, which is yellow when it is hot, but turns back to white on cooling.

$$ZnCO_3(s) \rightarrow ZnO(s) + CO_2(g)$$

### Testing for carbonates using dilute acids

If you add a dilute acid to a solid carbonate, carbon dioxide is produced in the cold. It is probably best to use dilute nitric acid. Some acid–carbonate combinations can produce an insoluble salt which coats the solid carbonate and stops the reaction, but this doesn't happen if you use nitric acid because all nitrates are soluble.

Add a little dilute nitric acid, look for bubbles of gas produced in the cold, and test the gas with lime water.

The ionic equation shows any carbonate reacting with any acid.

$$CO_3^{2-}(s) + 2H^+(aq) \rightarrow CO_2(g) + H_2O(l)$$

For example, using zinc carbonate and dilute nitric acid:

$$ZnCO_3(s) + 2HNO_3(aq) \rightarrow Zn(NO_3)_2(aq) + CO_2(g) + H_2O(l)$$

### Testing for sulphites using dilute acids

Sulphites contain the ion $SO_3^{2-}$, and have formulae just like carbonates. Carbonates react with dilute acids to give off $CO_2$. Sulphites react with dilute acids to give off $SO_2$.

$$SO_3^{2-}(s) + 2H^+(aq) \rightarrow SO_2(g) + H_2O(l)$$

The only difference is that sulphites usually need warming with the acid before you produce enough sulphur dioxide to test adequately. Remember that sulphur dioxide is an acidic gas which turns potassium dichromate(VI) paper from orange to green.

The full equation would be exactly like the carbonate equation – replacing the carbon by sulphur.

### Testing for sulphates

Make a solution of your suspected sulphate, add enough dilute *hydrochloric* acid to make it acidic and then add some barium *chloride* solution. A sulphate will produce a white precipitate of barium sulphate.

$$Ba^{2+}(aq) + SO_4^{2-}(aq) \rightarrow BaSO_4(s)$$

You acidify the solution to destroy other compounds which might also produce white precipitates when you add the barium chloride solution. For example, if you didn't add acid, you would also get a white precipitate if there was a carbonate present because barium carbonate is also white and insoluble. The acid reacts with and removes the carbonate ions.

You could equally well use *nitric* acid and barium *nitrate* solution. You must *never* acidify the solution with sulphuric acid because sulphuric acid contains sulphate ions. If you add those, you are bound to get a precipitate of barium sulphate, whatever else is present.

> **7.** Which metal ion gives an orange-brown precipitate with sodium hydroxide?
> **8.** How can you distinguish between magnesium and calcium ions?
> **9.** How do carbonates react with acids?

### Testing for chlorides, bromides and iodides

This is very similar to the test for sulphates. Make a solution of your suspected chloride, bromide or iodide and add enough dilute *nitric* acid to make it acidic. Then add some silver *nitrate* solution.

The acid is added to react with and remove other substances which might also produce precipitates with silver nitrate solution.

A white precipitate (of silver chloride) shows the presence of chloride ions.

$$Ag^+(aq) + Cl^-(aq) \rightarrow AgCl(s)$$

A pale cream precipitate (of silver bromide) shows the presence of bromide ions.

$$Ag^+(aq) + Br^-(aq) \rightarrow AgBr(s)$$

A yellow precipitate (of silver iodide) shows the presence of iodide ions.

$$Ag^+(aq) + I^-(aq) \rightarrow AgI(s)$$

All of these precipitates tend to discolour towards greys and pale purples on exposure to light. The bromide and iodide colours are quite difficult to distinguish between in practice. There is a follow-on test involving ammonia solution which helps to sort them out.

## Activity 1: Identification test

**Investigative skills**

P

Plan an experiment to distinguish between sodium chloride and potassium nitrate. You may use flame tests or any solution tests.

## Testing for hydrogen ions or hydroxide ions

### Hydrogen ions, $H^+$

The presence of hydrogen ions makes a solution acidic. You can test this with indicators or by adding a reactive metal like magnesium (looking for hydrogen evolved) or a carbonate like sodium carbonate (looking for carbon dioxide).

### Hydroxide ions, $OH^-$

Hydroxide ions in a solution make it alkaline. You can test this with indicators. Alternatively, you can make use of the fact that hydroxide ions react with ammonium ions to produce ammonia gas.

$$NH_4^+(s) + OH^-(aq) \rightarrow NH_3(g) + H_2O(l)$$

Add some solid ammonium chloride to a solution which might contain hydroxide ions, warm it and test for ammonia with moist red litmus paper.

## Summary tables

Tests on gases

| Gas | Soluble/ insoluble | Odour | Moist red/ blue litmus | Special test |
|---|---|---|---|---|
| hydrogen | insoluble | none | no change | pop test – explodes with a burning splint |
| oxygen | slightly soluble | none | no change | relights a glowing splint |
| carbon dioxide | fairly soluble | none | blue → reddish | lime water goes cloudy |
| chlorine | soluble | strong | blue → red then bleaches | bleaching action with indicator paper |
| sulphur dioxide | soluble | strong | blue → red | i) acidified dichromate changes from orange to green ii) acidified potassium manganate(VII) changes from purple to colourless |
| hydrogen chloride | soluble | strong | blue → red | i) fumes in moist air ii) white smoke with ammonia |
| ammonia | soluble | strong | red → blue | white smoke with HCl gas |

Tests on ions using dilute ammonia solution

| Precipitate | With excess alkali | Ion indicated |
|---|---|---|
| white | insoluble | $Al^{3+}$, $Pb^{2+}$, temporarily hard water |
| white | soluble | $Zn^{2+}$ |
| white, rapidly turning green, then slowly brown | insoluble | $Fe^{2+}$ |
| red-brown | insoluble | $Fe^{3+}$ |
| pale blue | soluble (deep blue) | $Cu^{2+}$ |
| no precipitate | – | $Na^+$, $K^+$, $NH_4^+$, $Ca^{2+}$ |

Tests on ions using sodium hydroxide

| Precipitate | With excess alkali | Ion indicated |
|---|---|---|
| white | insoluble | $Mg^{2+}$, $Ca^{2+}$ |
| white | soluble | $Zn^{2+}$, $Pb^{2+}$, $Al^{3+}$, $Sn^{2+}$, $Sn^{4+}$ |
| white, rapidly turning green, then slowly brown | insoluble | $Fe^{2+}$ |
| white, slowly turning pale brown | insoluble | $Mn^{2+}$ |
| red-brown | insoluble | $Fe^{3+}$ |
| dark brown | insoluble | $Ag^+$ |
| green | insoluble | $Ni^{2+}$ |
| blue | insoluble | $Cu^{2+}$, $Co^{2+}$ |
| no precipitate | – | $Na^+$, $K^+$, $NH_4^+$, $Ba^{2+}$ |

# End of Chapter Checklist

**In this chapter you have learnt that:**

- gases must be collected in different ways according to their solubility

- there are special tests to identify many gases

- water changes the colour of cobalt chloride paper from blue to pink and anhydrous copper sulphate from white to blue

- flame tests identify metal ions

- precipitates of metal compounds can be identified by their colours and solubilities

- acid radicals, such as carbonate or nitrate, have special tests to identify them.

# Questions

1. A colourless gas that explodes with a pop is:

   A  oxygen
   B  nitrogen
   C  chlorine
   D  hydrogen                                (1)

2. A green gas that bleaches indicator paper is:

   A  nitrogen dioxide
   B  carbon dioxide
   C  sulphur dioxide
   D  chlorine                                (1)

3. A blue-green flame colour indicates:

   A  zinc ions
   B  copper ions
   C  barium ions
   D  calcium ions                            (1)

4. A liquid that turns anhydrous copper sulphate blue could be:

   A  lemonade
   B  petrol
   C  hexane
   D  sunflower oil                           (1)

5. Which of the following statements is/are true?

I    carbon dioxide is colourless
II   carbon dioxide puts out flames
III  carbon dioxide is lighter than air
IV  carbon dioxide turns lime water cloudy

A   I and II
B   I, II and III
C   I, II and IV
D   III and IV                           (1)

6. a) Draw and describe *two* ways to collect hydrogen gas. (5)

b) What is the test for hydrogen? (1)

c) Write a word equation for burning the gas in air. (1)

d) How could you distinguish between hydrogen and oxygen? (3)

7. a) Describe *two* different tests for water and explain what you would see. (4)

b) How could you reverse one of the tests? (2)

c) Would either test work with a cola drink? Explain your answer. (3)

d) Do either of the tests show how pure a water sample is? (1)

8. a) A substance releases a strong-smelling, colourless gas when heated. The gas produces white smoke when combined with fumes of hydrogen chloride gas. Explain these changes. (5)

b) How can a solution of sodium hydroxide be used to identify metal ions? Give *two* examples. (4)

c) Why is sodium hydroxide hazardous to use? (1)

9. a) Describe the reaction of carbonates with acids and give a word equation for *one* example. (5)

b) i)   Which gas is released in the reaction?

   ii)  What test could you carry out to identify it? (3)

c) How could you show whether this gas is lighter or heavier than air? (2)

10. Four common methods of collecting gases are:

A   downwards into a gas jar
B   upwards into a gas jar
C   over water
D   in a gas syringe

Which method would you use in each of the following cases?

a) To collect a dry sample of carbon dioxide in order to do a reaction with it. (2)

b) To measure the amount of hydrogen produced in 10 seconds during the reaction between dilute sulphuric acid and magnesium. (2)

c) To collect a sample of carbon monoxide – a colourless, odourless, poisonous gas, insoluble in water and with approximately the same density as air. (2)

d) To collect a sample of ammonia in order to do a reaction with it. (2)

11. Name the gas being described in each of the following cases.

a) A green gas which bleaches damp litmus paper. (1)

b) A gas which dissolves readily in water to produce a solution with a pH of about 11. (1)

c) A gas which turns blue litmus paper red and potassium dichromate(VI) paper from orange to green. (1)

d) A gas which produces a white precipitate with calcium hydroxide solution. (1)

e) A gas which pops when a lighted splint is placed in it. (1)

f) A steamy gas which turns blue litmus paper red. (1)

g) A gas which relights a glowing splint. (1)

12. Describe fully how you would carry out the following tests. In each case, describe what you would expect to happen.

a) A flame test for lithium ions in lithium chloride. (2)

b) A test for iron(II) ions in iron(II) sulphate. (2)

c) A test for sulphate ions in iron(II) sulphate. (2)

d) A test for the presence of water. (2)

e) A test for the carbonate ion in sodium carbonate. (2)

f) A test for the hydroxide ion in sodium hydroxide. (2)

13. **A** is an orange solid which dissolves in water to give an orange solution. When sodium hydroxide solution is added to a solution of **A**, an orange-brown precipitate, **B**, is formed. Adding dilute nitric acid and silver nitrate solution to a solution of **A** gives a white precipitate, **C**.

a) Identify **A**, **B** and **C**. (6)

b) Write equations (full or ionic) for the reactions producing **B** and **C**. (6)

14. **D** is a white crystalline solid which dissolves in water to give a colourless solution. Addition of sodium hydroxide solution to a solution of **D** produces a white precipitate, **E**, insoluble in excess sodium hydroxide solution. A flame test on **D** proves negative. Addition of dilute hydrochloric acid and barium chloride solution to a solution of **D** gives a white precipitate, **F**.

a) Identify **D**, **E** and **F**. (6)

b) Write equations (full or ionic) for the reactions producing **E** and **F**. (6)

15. **G** is a colourless crystalline solid which reacts with dilute nitric acid to give a colourless solution, **H**, and a colourless, odourless gas, **I**, which turns lime water milky. **G** has a bright orange flame colour.

a) Identify **G**, **H** and **I**. (6)

b) Write an equation (full or ionic) for the reaction between **G** and dilute nitric acid. (6)

**16.** *a)* Identify the substances lettered from **A** to **L** in the flow scheme below. (Don't worry if you don't know anything about electrolysis. This uses electricity to split compounds up. The cathode is the negative electrode and the anode is the positive one. In this case, the electrodes would be made of carbon and attached to a 6 volt battery.)

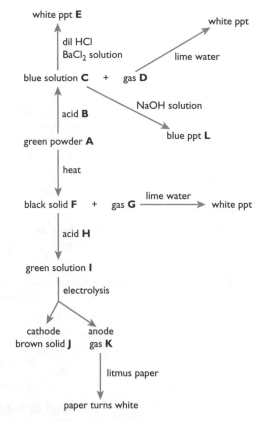

(12 × 2)

*b)* Write equations (full or ionic) for the reactions

| | | |
|---|---|---|
| *i)* | between **A** and **B** | (2) |
| *ii)* | solution **C** and barium chloride solution | (2) |
| *iii)* | solution **C** and sodium hydroxide solution | (2) |
| *iv)* | heating **A** | (2) |
| *v)* | the reaction between **F** and **H** | (2) |

# End of Section Questions

**1 a)** Chlorine was bubbled through a solution of potassium iodide.

    **i)** Describe, with a reason, any precaution that you would have to take in using the chlorine.

    **ii)** Balance the ionic equation for the reaction involved.

$$Cl_2(g) + I^-(aq) \rightarrow Cl^-(aq) + I_2(s)$$

    **iii)** Describe what you would expect to see happen in the solution.

    **iv)** Describe the function of the chlorine in the reaction with the iodide ions.

*(7 marks)*

**b)** Samples of a very pale green solution, **G**, were tested as follows:

| Test | Observation |
| --- | --- |
| A sample of solution was acidified with dilute nitric acid and silver nitrate solution was added | A white precipitate (**H**) was formed |
| A small amount of sodium hydroxide solution was added to a sample of **G** | A dark green precipitate (**I**) was formed |
| Chlorine was bubbled through a sample of **G** | The pale green solution turned yellow (solution **J**) |
| A small amount of sodium hydroxide solution was added to solution **J** | An orange-brown precipitate (**K**) was formed |

    **i)** Use the results from the first two tests to identify solution **G**.

    **ii)** Identify precipitates **I** and **K**.

    **iii)** Suggest the identity of solution **J**.

    **iv)** Write an ionic equation for the formation of **I**.

*(6 marks)*

**Total 13 marks**

**2** Mendeleev produced his Periodic Table by arranging the elements in order of their atomic masses. When argon was discovered, its atomic mass turned out to be slightly higher than potassium's. In this instance, Mendeleev reversed the usual order in the Periodic Table.

| Mendeleev's order | | | Atomic mass order | | |
| --- | --- | --- | --- | --- | --- |
| Group 0 | Group 1 | Group 2 | Group 0 | Group 1 | Group 2 |
| Ne | Na | Mg | Ne | Na | Mg |
| $A_r$ | K | Ca | K | $A_r$ | Ca |

**a)** State *one* physical property of potassium which suggests that it should be in the same Group as sodium rather than with neon. *(1 mark)*

**b)** Give any *one* chemical property of potassium that is similar to that of sodium. Say what the potassium reacts with and what is formed. Write the balanced equation for the reaction. *(4 marks)*

**c) i)** Draw dot-and-cross diagrams to show the electronic structures of sodium and potassium atoms.

    **ii)** What happens to these structures when sodium or potassium react to form compounds?

    **iii)** Explain why potassium is more reactive than sodium. *(6 marks)*

**d)** Argon is chemically unreactive and its molecules are monatomic. What is a monatomic molecule? Explain why argon's molecules are monatomic. *(3 marks)*

**e)** Give a use for argon. *(1 mark)*

**Total 15 marks**

**3** Iron is produced in a blast furnace by the reduction ore, of its haematite, $Fe_2O_3$.

**a)** What do you understand by the term *reduction*? *(1 mark)*

**b)** Give the proper chemical name for haematite. *(1 mark)*

**c)** The main heat source in the furnace is provided by burning coke in air.

$$C(s) + O_2(g) \rightarrow CO_2(g)$$

What name is given to a reaction which produces heat? *(1 mark)*

**d)** The main reducing agent in the furnace is carbon monoxide. Write an equation to show its formation. *(1 mark)*

**e)** Balance the equation:

$$Fe_2O_3(s) + CO(g) \rightarrow Fe(l) + CO_2(g) \quad \text{(1 mark)}$$

**f)** Limestone is added to the furnace to help in the removal of impurities in the ore like silicon dioxide, $SiO_2$. Explain the chemistry of this. *(3 marks)* *g)* Give *one* use for slag. (*1 mark*)

*h)* The impure iron from the blast furnace can be used to make cast iron, but most is converted into various steels.

|  | cast iron | mild steel | high carbon steel | stainless steel |
|---|---|---|---|---|
| contains | 4% C | 0.25% C | 1.5% C | 18% Cr 8% Ni |

*i)* Give *one* use in each case for cast iron, mild steel, high carbon steel and stainless steel. (*4 marks*)

*ii)* Give *two* effects of increasing the proportion of carbon mixed with the iron. (*2 marks*)

**Total: 15 marks**

> **When you have completed this chapter, you will be able to:**
> - understand the physical properties of hydrocarbons
> - explain fractional distillation
> - list the uses of the fractions
> - explain why some fractions are cracked
> - explain the properties of alkanes and alkenes
> - distinguish between saturated and unsaturated molecules
> - explain the difference between substitution and addition
> - write an organic formula
> - name organic compounds
> - draw isomers and explain their differences.

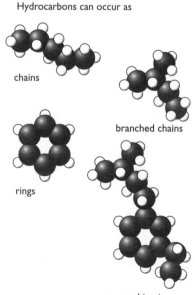

*This sticky black liquid underpins modern life.*

Hydrocarbons can occur as

chains

branched chains

rings

or a combination

## What is crude oil (petroleum)?

### The origin of crude oil

Millions of years ago plants and animals living in the sea died and fell to the bottom. Layers of sediment formed on top of them. Their shells and skeletons formed limestone. The soft tissue was gradually changed by high temperatures and pressures into crude oil. Crude oil is a **finite, non-renewable resource**. Once all the existing supplies have been used they won't be replaced – or at least, not for many millions of years.

### Crude oil contains hydrocarbons

Crude oil is a mixture of **hydrocarbons** – compounds containing carbon and hydrogen only. Hydrocarbons are **organic** molecules. The term *organic* was originally used because it was believed that organic compounds could only come from living things. Now it is used for any carbon compound except for the very simplest (like carbon dioxide and the carbonates).

Hydrocarbons can exist as chains, branched chains or rings of carbon atoms with hydrogens attached.

### How the properties of hydrocarbons change with size of molecule

As the number of carbon atoms in the molecules increases, several properties of the compounds change in a regular pattern. Most of these changes are the result of increasing attractions between neighbouring molecules. As the molecules get bigger, these **intermolecular attractions** increase and it gets more difficult to pull one molecule away from its neighbours.

As the molecules get bigger:

- Boiling point increases – the larger the molecule, the higher the boiling point. This is because large molecules are attracted to each other more strongly than smaller ones. More heat is needed to break these stronger attractions to produce the widely separated molecules in the gas.

- The liquids become less volatile. The bigger the hydrocarbon, the more slowly it evaporates at room temperature. This is again because the bigger molecules are more strongly attracted to their neighbours and so don't turn to a gas so easily.

- The liquids flow less easily (they become more viscous). Liquids containing small hydrocarbon molecules are runny. Those containing large molecules are much stickier because of the greater attractions between their molecules.

- Bigger hydrocarbons do not burn as easily as smaller ones. This limits the use of the bigger ones as fuels.

> We usually count a substance as being volatile if it turns to a vapour easily at room temperature. This means that it will evaporate quickly at that temperature.

## Types of formula for organic molecules

### Molecular formulae

A molecular formula simply counts the numbers of each sort of atom present in the molecule, but tells you nothing about the way they are joined together. For example, the molecular formula of propane is $C_3H_8$, and the molecular formula of ethene is $C_2H_4$.

Molecular formulae are very rarely used in organic chemistry, because they don't give any useful information about the bonding in the molecule. You might use them in equations for the combustion of simple hydrocarbons where the structure of the molecule doesn't matter. For example:

$$C_3H_8(g) + 5O_2(g) \rightarrow 3CO_2(g) + 4H_2O(g)$$

### Three-dimensional formulae

It is possible to draw 3-D molecular shapes on a flat piece of paper. The wedge shape is a bond that points towards you. A dotted line is a bond pointing away from you. The ordinary lines are bonds in the plane of the paper.

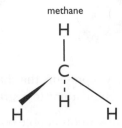

methane

3-D structure of methane is tetrahedral.

The bond angle in methane, and in other similar molecules, is about 109°. This is the angle in tetrahedral molecules and groups.

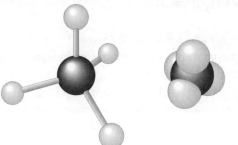

*Bond angles.*

## Structural formulae

A structural formula shows how the atoms are joined up. There are two ways of representing structural formulae – they can be drawn as a displayed formula or they can be written out as, for example, $CH_3CH_2CH_3$. You need to be confident about doing it either way.

### Displayed formulae

A displayed formula shows all the bonds in the molecule as individual lines. You need to remember that each line represents a pair of shared electrons.

The diagram shows a model of butane, together with its displayed formula.

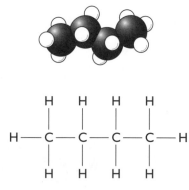

Notice that the way the displayed formula is drawn bears no resemblance to the shape of the actual molecule. Displayed formulae are always drawn with the molecule straightened out and flattened. They do, however, show exactly how all the atoms are joined together.

### The normal way to draw a structural formula

For anything other than the smallest molecules, drawing a fully displayed formula is very time-consuming. You can simplify the formula by writing, for example, $CH_3$ or $CH_2$ instead of showing all the carbon–hydrogen bonds.

Butane could be written quite quickly as $CH_3CH_2CH_2CH_3$ – and this shows all the necessary detail. But you have to be very careful. For example, all of these structures represent butane, even though they look different:

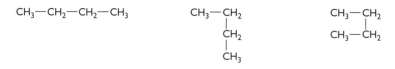

Each structure shows four carbon atoms joined up in a chain, but the chain has simply twisted. This happens in real molecules as well.

Not one of the structural formulae accurately represents the shape of butane. The convention is to write it with all the carbon atoms in a straight line.

A molecule like propene, $C_3H_6$, has a carbon–carbon double bond. That is shown by drawing two lines between the carbon atoms to show the two pairs of shared electrons. You would normally write this in a simplified structural formula as $CH_3CH{=}CH_2$.

> If you are asked to draw the structure for a molecule, always draw it in the fully displayed form. You can't lose any marks by doing this, whereas you might if you use the simplified form. If, on the other hand, you are just writing a structure in an equation, you can use whichever version you prefer.

> The best way to understand this is to make some models. If you don't have access to atomic models, use blobs of Plasticine joined together with bits of match sticks to represent the bonds You will find that you can change the shape of the model by rotating the bonds. That's what happens in real molecules.

1. What is crude oil and how was it formed?
2. Use an example to explain the difference between a molecular formula and a structural formula.
3. Draw the displayed formula for butane, $C_4H_{10}$.

# Naming organic compounds

Names for organic compounds can look quite complicated, but they are no more than a code which describes the molecule. Each part of a name tells you something specific about the molecule. One part of a name tells you how many carbon atoms there are in the longest chain, another part tells you whether there are any carbon–carbon double bonds, and so on.

## Coding the chain length

Look for the following letters in the name:

| Code letters | No. of carbons in chain |
|---|---|
| meth | 1, e.g. methane |
| eth | 2, e.g. ethane |
| prop | 3, e.g. propane |
| but | 4, e.g. butane |
| pent | 5, e.g. pentane |
| hex | 6, e.g. hexane |

> You have to learn these! The first four are the tricky ones because there isn't any pattern; 'pent' means 5 (as in **pent**agon), and 'hex' means 6 (as in **hex**agon).

**But**ane has a chain of four carbon atoms; **pro**pane has a chain of three carbon atoms.

## Coding for the type of compound

### *Alkanes*

Alkanes are hydrocarbons in which all the carbons are joined to each other with single covalent bonds. Compounds like this are coded with the ending '**ane**'. For example, eth**ane** is a 2-carbon chain (because of 'eth') with a carbon–carbon single bond, $CH_3CH_3$.

> In more complicated molecules, the presence of the code 'an' in the name again shows that the carbons are joined by single bonds. For example, you can tell that propan-1-ol contains 3 carbon atoms ('prop') joined together by carbon–carbon single bonds ('an'). The coding on the end gives you more information about the molecule. This is explained in the next chapter.

*Names of some straight-chain alkanes.*

| Number of carbons | Molecular formula | Name | Number of carbons | Molecular formula | Name |
|---|---|---|---|---|---|
| 1 | $CH_4$ | methane | 11 | $C_{11}H_{24}$ | undecane |
| 2 | $C_2H_6$ | ethane | 12 | $C_{12}H_{26}$ | dodecane |
| 3 | $C_3H_8$ | propane | 13 | $C_{13}H_{28}$ | tridecane |
| 4 | $C_4H_{10}$ | butane | 14 | $C_{14}H_{30}$ | tetradecane |
| 5 | $C_5H_{12}$ | pentane | 15 | $C_{15}H_{32}$ | pentadecane |
| 6 | $C_6H_{14}$ | hexane | 18 | $C_{18}H_{38}$ | octadecane |
| 7 | $C_7H_{16}$ | heptane | 20 | $C_{20}H_{42}$ | eicosane |
| 8 | $C_8H_{18}$ | octane | 21 | $C_{21}H_{44}$ | heneicosane |
| 9 | $C_9H_{20}$ | nonane | 30 | $C_{30}H_{62}$ | triacontane |
| 10 | $C_{10}H_{22}$ | decane | 100 | $C_{100}H_{202}$ | hectane |

## Alkenes

Alkenes contain a carbon–carbon double bond. This is shown in their name by the ending '**ene**'. For example, eth**ene** is a 2-carbon chain containing a carbon–carbon double bond, $CH_2\text{=}CH_2$. With longer chains, the position of the double bond could vary in the chain. This is shown by numbering the chain and noting which carbon atom the double bond *starts* from.

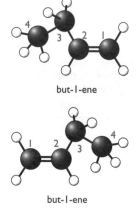

but-1-ene

but-1-ene

| | | |
|---|---|---|
| $CH_2\text{=}CHCH_2CH_3$ | but-1-ene | a 4-carbon chain with a double bond starting on the first carbon |
| $CH_3CH\text{=}CHCH_3$ | but-2-ene | a 4-carbon chain with a double bond starting on the second carbon |
| $CH_3CH_2CH\text{=}CHCH_2CH_3$ | hex-3-ene | a 6-carbon chain with a double bond starting on the third carbon |

How do you know which end of the chain to number from? The rule is that you number from the end which produces the smaller numbers in the name.

The diagrams both show the same molecule, but one has been flipped over so that what was originally on the left is now on the right, and vice versa.

It would be silly to change the name every time the molecule moved! Both of them are called but-1-ene.

### Coding for branched chains

Hydrocarbon chains can have side branches on them. You are only likely to come across two small side chains:

| Side chain | Coded |
|---|---|
| $CH_3^-$ | methyl |
| $CH_3CH_2^-$ or $C_2H_5^-$ | ethyl |

> Notice that the count of the number of carbons in the side chain is coded exactly as before. 'meth' shows a 1-carbon side chain; 'eth' shows 2 carbons.

The name of a molecule is always based on the *longest* chain you can find in it. The position of the chain is shown by numbering exactly as before.

The longest chain in the molecule in the next diagram has four carbon atoms ('**but**') with no double bonds ('**ane**'). The name is based on butane. There is a **methyl** group branching off the **number 2** carbon. (Remember to number from the end which produces the smaller number.)

The compound is called 2-methylbutane.

$$\overset{4}{C}H_3\text{—}\overset{3}{C}H_2\text{—}\overset{2}{C}H\text{—}\overset{1}{C}H_3$$
$$|$$
$$CH_3$$

Where there is more than one side chain, you describe the position of each of them.

$$CH_3$$
$$\overset{1}{C}H_3\text{—}\overset{2}{C}\text{—}\overset{3}{C}H_3$$
$$|$$
$$CH_3$$

The longest chain in the molecule in the diagram has three carbon atoms and no double bonds. Therefore the name is based on propane.

There are two methyl groups attached to the second carbon. The compound is 2,2-dimethylpropane. The 'di' in the name shows the presence of the two methyl groups. '2,2-' shows that they are both on the second carbon atom.

You can reverse the process and draw a structural formula from a name. All you have to do is decode the name. For example, what is the structural formula for **2,3-dimethylbut-2-ene**?

Start by looking for the code for the longest chain length. '**but**' shows a 4-carbon chain. '**ene**' shows that it contains a carbon–carbon double bond starting on the second carbon atom ('-**2-ene**').

There are two methyl groups ('**dimethyl**') attached to the second and third carbon atoms in the chain ('**2,3-**'). All you have to do now is to fit all this together into a structure.

$$\overset{\displaystyle CH_3}{\underset{\displaystyle CH_3}{\overset{1\quad 2\ |\ \ 3\quad 4}{C-C=C-C}}}$$

Start by drawing the structure without any hydrogen atoms on the main chain. It doesn't matter whether you draw the $CH_3$ groups pointing up or down. Then add enough hydrogens so that each carbon atom is forming four bonds.

The final structure is:

$$\overset{\displaystyle CH_3}{\underset{\displaystyle CH_3}{CH_3-\overset{|}{C}=\overset{|}{C}-CH_3}}$$

> **4.** How many carbon atoms does pentane have?
> **5.** What does the 'ene' in the name of an alkene tell you?
> **6.** In the compound 2,2-dimethylpropane, what does the 'di' tell you?

## Structural isomerism

Structural isomers are molecules with the same molecular formula, but with different structural formulae. Examples will make this clear.

### Structural isomerism in the alkanes

#### Isomers of butane, $C_4H_{10}$

If you had some atomic models and picked out four carbon atoms and ten hydrogen atoms, you would find that it was possible to fit them together in more than one way. The two different molecules formed are known as **isomers**. Both have the molecular formula $C_4H_{10}$, but they have different structures. **Structural isomerism** is the existence of two or more different structures with the same molecular formula.

If you look carefully at the models in the diagram, you can see that you couldn't change one into the other just by bending or twisting the molecule. You would have to take the model to pieces and rebuild it. That's a simple way of telling that you have got isomers.

The 'straight chain' isomer is called butane. The branched chain has a 3-carbon chain with no carbon–carbon double bond ('propane') with a methyl group on the second carbon. The name is 2-methylpropane.

> It is actually much more important to be able to decode names to give structures than the other way around. If you don't know what a teacher or an examiner is talking about you are completely lost!

> **Warning!** Don't confuse the word **isomer** with **isotope**. Isotopes are atoms with the same atomic number but different mass numbers.

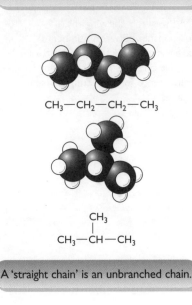

$$CH_3-CH_2-CH_2-CH_3$$

$$\underset{\displaystyle CH_3-CH-CH_3}{\overset{\displaystyle CH_3}{\overset{|}{\ }}}$$

A 'straight chain' is an unbranched chain.

### Isomers of pentane, $C_5H_{12}$

There are three isomers of pentane:

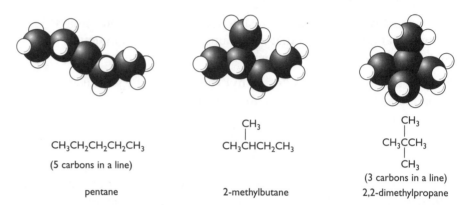

$CH_3CH_2CH_2CH_2CH_3$

(5 carbons in a line)

pentane

$CH_3CHCH_2CH_3$
|
$CH_3$

2-methylbutane

$CH_3CCH_3$
|
$CH_3$ (top) $CH_3$ (bottom)

(3 carbons in a line)

2,2-dimethylpropane

$CH_3CHCH_3$
|
$CH_2$
|
$CH_3$

Students frequently think that they can find another isomer as well. If you look closely at this 'fourth' isomer, you will see that it is just 2-methylbutane rotated in space.

To avoid this sort of problem, always draw your isomers so that the longest carbon chain is drawn horizontally. Check each isomer after you have drawn it to be sure that you have done that.

### Physical differences between the isomers

The various isomers will have slightly different physical properties because they will experience slightly different intermolecular forces.

Branched chains have weaker intermolecular attractions than straight ones. Intermolecular forces are only effective over very short distances. The more branching there is in a chain, the more difficult it is for the molecules to get close to each other.

You can see the effect of this on the boiling points of the isomers of pentane:

On the graph showing the relationship between number of carbons and boiling point for the alkanes on page 269, the boiling points were for the straight chain alkanes.

|  | Boiling point/°C |
|---|---|
| pentane | 36.3 |
| 2-methylbutane | 27.9 |
| 2,2-dimethylpropane | 9.5 |

As the amount of branching increases, boiling point falls.

### Structural isomerism in the alkenes

### Ethene and propene

Ethene, $CH_2{=}CH_2$, doesn't have any isomers. Propene, $CH_3CH{=}CH_2$, doesn't have a structural isomer which is still an alkene. (You can find a structural isomer of $C_3H_6$ which doesn't have a carbon–carbon double bond by joining the carbon atoms in a ring.)

## Butene, $C_4H_8$

Butene has three structural isomers containing a carbon–carbon double bond.

$CH_3CH_2CH=CH_2$ but-1-ene

$CH_3CH=CHCH_3$ but-2-ene

$$CH_3\underset{\underset{CH_3}{|}}{C}=CH_3$$
2-methylpropene

Notice the way that you can vary the position of the double bond as well as branch the chain.

## Pentene, $C_5H_{10}$

These are all structural isomers of pentene:

$CH_3CH_2CH_2CH=CH_2$ pent-1-ene

$CH_3CH_2CH=CHCH_3$ pent-2-ene

$$CH_3CH_2\underset{\underset{CH_3}{|}}{C}=CH_2$$
2-methylbut-1-ene

$$CH_3\underset{\underset{CH_3}{|}}{C}HCH=CH_2$$
3-methylbut-1-ene

$$CH_3CH=\underset{\underset{CH_3}{|}}{C}CH_3$$
2-methylbut-2-ene

> At first sight this looks really worrying! There are two issues. First, do you understand the names? If not, get that sorted out before you do anything else.
>
> Second, could you draw at least a few of these isomers? Use a scrap of paper and see how many isomers of $C_5H_{10}$ you can find. Draw fully displayed formulae showing all the bonds. Remember that you must have a carbon–carbon double bond. Better still – use some models.

# Separating crude oil

## Fractional distillation

The crude oil is heated and passed into a **fractionating column** which is cooler at the top and hotter at the bottom. The crude oil is split into various **fractions**.

How far up the column a particular hydrocarbon gets depends on its boiling point. Suppose a hydrocarbon boils at 120°C. At the bottom of the column, the temperature is much higher than 120°C and so the hydrocarbon remains as a gas. As it travels up through the column, the temperature gets lower. When the temperature falls to 120°C, that hydrocarbon will start to turn to a liquid. It condenses and can be tapped off.

The hydrocarbons in the petroleum gases have boiling points which are so low that the temperature of the column never falls low enough for them to condense to liquids.

The temperature of the column isn't hot enough to boil the large hydrocarbons found in the fuel oil and this remains as a liquid. Some of the fuel oil is fractionally distilled under reduced pressure. The residue at the end of all this is **bitumen**, which is used in road making.

> Things boil at lower temperatures if you reduce the pressure. Distillation under reduced pressure prevents the large molecules breaking up as a result of the high temperatures.

## Uses of the fractions

All hydrocarbons burn in air (oxygen) to form carbon dioxide and water and release lots of heat in the process. They can therefore be used as **fuels**.

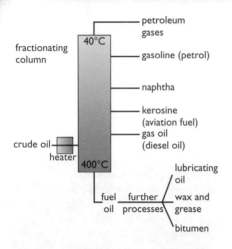

petroleum gases
40°C
fractionating column
gasoline (petrol)
naphtha
kerosine (aviation fuel)
gas oil (diesel oil)
crude oil
heater
400°C
lubricating oil
fuel oil — further processes — wax and grease
bitumen

Don't try to learn these equations – there are too many possible hydrocarbons you could be asked about. Provided you know (or are told) the formula, they are easy to balance.

*Kerosene is used as aviation fuel.*

*Bitumen is used in road construction.*

For example, burning methane (the major constituent of natural gas):

$$CH_4(g) + 2O_2(g) \rightarrow CO_2(g) + 2H_2O(l)$$

...or burning octane (one of the hydrocarbons present in petrol):

$$2C_8H_{18}(l) + 25O_2(g) \rightarrow 16CO_2(g) + 18H_2O(l)$$

If there isn't enough air (or oxygen) you get **incomplete combustion**. This leads to the formation of carbon (soot) or carbon monoxide instead of carbon dioxide. Carbon monoxide is colourless and odourless and very poisonous.

Carbon monoxide is poisonous because it combines with **haemoglobin** (the molecule which carries oxygen in the bloodstream), preventing it from carrying the oxygen. You are made ill, or even die, because not enough oxygen gets to the cells in your body.

### Petroleum gases

Petroleum gases are a mixture of methane, ethane, propane and butane which can be separated into individual gases if required. These gases are commonly used as LPG (liquefied petroleum gas) for domestic heating and cooking.

### Gasoline (petrol)

As with all the other fractions, petrol is a mixture of hydrocarbons with similar boiling points. Its use is fairly obvious!

### Naphtha

Naphtha is used as a source of organic chemicals for industry as well as a constituent of petrol. Useful molecules like ethene and propene can be made by **cracking** the naphtha.

### Kerosene

Kerosene is used as fuel for jet aircraft, as domestic heating oil and as 'paraffin' for small heaters and lamps.

### Gas oil (diesel oil)

This is used for buses, lorries, some cars, and railway engines where the line hasn't been electrified. Some is also cracked to make other organic chemicals and produce more petrol.

### Fuel oil

This is used for ships' boilers and for industrial heating.

Some of the fuel oil is also distilled again, under reduced pressure, to make lubricating oil, grease, wax (for candles) and bitumen.

### Bitumen

Bitumen is a thick black material which is melted and mixed with rock chippings to make the top surfaces of roads.

# Cracking

Although the fractions from crude oil distillation are useful fuels, there are two problems:

- The amounts of each fraction you get will depend on the proportions of the various hydrocarbons in the original crude oil, not the amounts in which they are needed. Far more petrol is needed, for example, than is found in crude oil.

- Apart from burning, the hydrocarbons in crude oil are fairly unreactive. To make other organic chemicals from them they must first be converted into something more reactive.

Cracking is a useful process in which large hydrocarbon molecules are broken into smaller ones. The big hydrocarbon molecules in gas oil, for example, can be broken down into the smaller ones needed for petrol.

The majority of the hydrocarbons found in crude oil have single bonds between the carbon atoms. During the cracking process, molecules are also formed which have double bonds between carbon atoms. These new molecules are much more reactive and can be used to make lots of other things.

### How cracking works

#### *The conditions*

The naphtha or gas oil fraction is heated to give a gas and then passed over a catalyst of mixed silicon dioxide and aluminium oxide at about 500°C. Cracking can also be carried out at higher temperatures without a catalyst.

#### *The reactions*

Cracking is just an example of thermal decomposition – a big molecule splitting into smaller ones. The molecules are broken up in a fairly random way. This is just one possibility:

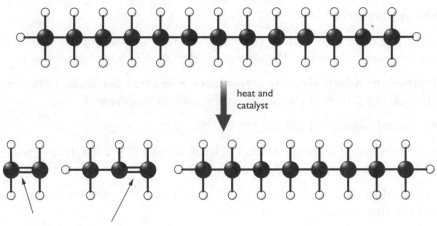

heat and catalyst

double bonds formed

## Activity 1: Comparing viscosity

**Investigative skills**

P

Plan an experiment to compare the viscosity of engine oil at different temperatures. Explain what safety precautions would be necessary. How could you make it a fair test?

---

The molecules have been drawn to show the various covalent bonds. They have also been 'straightened out'. The real molecules are much more 'worm-like'! Organic molecules are almost always drawn in this simplified form.

As an equation, this would read:

$$C_{13}H_{28}(l) \rightarrow C_2H_4(g) + C_3H_6(g) + C_8H_{18}(l)$$

Cracking produces a mixture of **alkanes** and **alkenes**. Alkanes are hydrocarbons in which all the carbon atoms are joined by single bonds – like $C_{13}H_{28}$ and $C_8H_{18}$. An alkene contains at least one carbon–carbon double bond – like $C_2H_4$ and $C_3H_6$.

The molecule might have split quite differently. All sorts of reactions are going on in a catalytic cracker. Two other possibilities might be:

$$C_{13}H_{28}(l) \rightarrow 2C_2H_4(g) + C_9H_{20}(l)$$
$$C_{13}H_{28}(l) \rightarrow 2C_2H_4(g) + C_3H_6(g) + C_6H_{14}(l)$$

Some reactions might even produce a small percentage of free hydrogen. For example:

$$C_{13}H_{28}(l) \rightarrow 2C_2H_4(g) + C_3H_6(g) + C_6H_{12}(l) + H_2(g)$$

In this case, all the hydrocarbons formed will have double bonds. They are all alkenes.

10. Give two uses for kerosene.
11. What is meant by cracking?
12. Give an example of thermal decomposition.

## Alkanes and alkenes

### The alkanes

The alkanes are a family of simple hydrocarbons. They contain carbon–carbon single bonds. Alkanes are described as **saturated** hydrocarbons in the sense that they contain the maximum possible number of hydrogen atoms for a given number of carbons.

The three smallest alkanes are shown in the diagram.

Methane is the major component of natural gas. Ethane and propane are also present in small quantities in natural gas, and are important constituents of the petroleum gases from crude oil distillation.

methane, $CH_4$        ethane, $C_2H_6$

propane, $C_3H_8$

### *Homologous series*

A homologous series is a family of compounds with similar properties because they have similar bonding. The alkanes are the simplest homologous series.

**Members of a homologous series have a general formula**. In the case of the alkanes, if there are 'n' carbons, there are '2n+2' hydrogens.

The general formula for the alkanes is $C_nH_{2n+2}$.

So, for example, if there are 3 carbons, there are $(2 \times 3) + 2 = 8$ hydrogens. The formula for propane is $C_3H_8$.

If you wanted the formula for an alkane with 15 carbons, you could easily work out that it was $C_{15}H_{32}$ – and so on.

*Physical properties of some alkanes.*

| Name | Formula | Melting point/°C | Boiling point/°C | Density/ g cm$^{-3}$ | Physical state at room temperature |
|---|---|---|---|---|---|
| methane | $CH_4$ | –182 | –164 | – | gas |
| ethane | $C_2H_6$ | –183 | –89 | – | gas |
| propane | $C_3H_8$ | –188 | –42 | – | gas |
| butane | $C_4H_{10}$ | –138 | –1 | – | gas |
| pentane | $C_6H_{12}$ | –129 | 36 | 0.63 | liquid |
| hexane | $C_6H_{14}$ | –95 | 69 | 0.66 | liquid |
| heptane | $C_7H_{16}$ | –91 | 98 | 0.68 | liquid |
| pentadecane | $C_{15}H_{32}$ | 10 | 270 | 0.77 | 'liquid' |
| hexadecane | $C_{16}H_{34}$ | 22 | 287 | 0.78 | solid |
| heptadecane | $C_{17}H_{36}$ | 22 | 302 | 0.78 | solid |

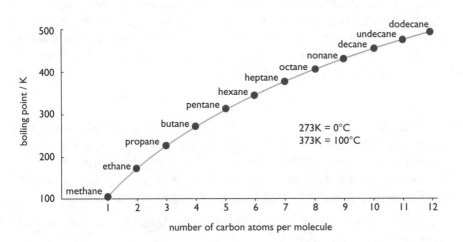

*Boiling temperatures of straight-chain alkanes.*

**Members of a homologous series have trends in physical properties**. The molecules of the members of a homologous series increase in size in a regular way. There is always a difference of $CH_2$ between one member and the next.

As the molecules get bigger, the intermolecular forces between them increase. This means that more energy has to be put in to break the attractions between one molecule and its neighbours.

One effect of this is that the boiling points increase in a regular way.

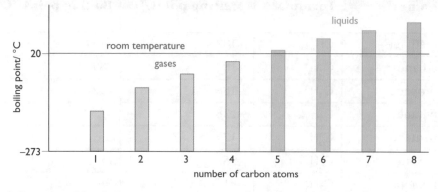

*Boiling points of the alkanes.*

Notice that the first four alkanes are gases at room temperature. All the other ones you are likely to come across are liquids. Solids start to appear at about $C_{17}H_{36}$.

**Members of a homologous series have similar chemical properties**. Chemical properties are dependent on bonding. Because alkanes only contain carbon–carbon single bonds and carbon–hydrogen bonds, they are all going to behave in the same way. These are strong bonds and the alkanes are fairly unreactive. All you will need to know for your examination is that they burn and that they can be cracked.

You might argue that methane doesn't have a carbon–carbon bond, and so might have different properties. It does – to the extent that it can't be cracked to produce smaller hydrocarbons!

### Burning of alkanes

The alkanes provide us with a very useful range of fuels. Natural gas contains methane and many gas cookers use either propane or butane gases. When hydrocarbons burn completely in air, the products are always the same – carbon dioxide and water vapour. If there is too little air (oxygen), poisonous carbon monoxide may form. This is called **incomplete combustion**.

If there is plenty of oxygen and the alkane molecule has few carbon atoms, combustion will be complete:

$$C_nH_{(2n + 2)} + \tfrac{1}{2}(3n = 1)O_2 \rightarrow nCO_2 + (n + 1)H_2O$$

For example, if n = 6:

| hexane | + | oxygen | → | carbon dioxide | + | water |
|--------|---|--------|---|----------------|---|-------|
| $2C_6H_{14}(l)$ | + | $19O_2(g)$ | → | $12CO_2(g)$ | + | $14H_2O(g)$ |

In general:

| hydrocarbon | + | oxygen | complete<br>→<br>combustion | carbon dioxide | + | water |
|-------------|---|--------|------|----------------|---|-------|

### Halogenation

Carbon is in Group 4 of the Periodic Table and can form a maximum of four bonds. When alkanes react with halogens, such as chlorine, the halogen must take the place of a hydrogen atom. We call this a **substitution reaction**. These reactions use light energy to start, either daylight or ultraviolet light.

methane   +   chlorine   →   chloromethane   +   hydrogen chloride

$$CH_4 \quad + \quad Cl_2 \quad → \quad CH_3Cl \quad + \quad HCl$$

This can continue with further substitutions like this:

$$CH_3Cl \quad + \quad Cl_2 \quad → \quad CH_2Cl_2 \quad + \quad HCl$$
dichloromethane

These are the last two substitutions of hydrogen by chlorine atoms.

$$CH_2Cl_2 \quad + \quad Cl_2 \quad → \quad CHCl_3 \quad + \quad HCl$$
trichloromethane
*or* chloroform

$$CHCl_3 \quad + \quad Cl_2 \quad → \quad CCl_4 \quad + \quad HCl$$
tetrachloromethane
*or* carbon tetrachloride,
a solvent

In sunlight, chlorine reacts so fast with alkanes that it can explode. Bromine is slower to react than chlorine. The typical reaction of the alkanes is **substitution**.

## The alkenes

The alkenes are another family (homologous series) of hydrocarbons. They all contain a carbon–carbon double bond. Alkenes are **unsaturated** hydrocarbons. The presence of the double bond means that they don't contain as many hydrogen atoms as the corresponding alkane.

### The general formula

Alkenes have a general formula $C_nH_{2n}$. There isn't an alkene with just one carbon atom. The two smallest alkenes are ethene and propene.

### Shapes of alkene molecules

In alkenes the bonds are arranged symmetrically around the carbon. The bond angle is 120° and the bonds point to the corners of an equilateral triangle.

### Physical properties

These are very similar to the alkanes. Remember that the small alkanes with up to four carbon atoms are gases. The same thing is true for the alkenes. They are gases up to $C_4H_8$, and the next dozen or so are liquids.

### Chemical reactions of the alkenes

In common with all hydrocarbons, alkenes burn in air or oxygen to give carbon dioxide and water. For example:

$$C_2H_4(g) + 3O_2(g) → 2CO_2(g) + 2H_2O(l)$$

More importantly, alkenes undergo **addition reactions**. Part of the double bond breaks and the electrons are used to join other atoms onto the two carbon atoms.

Alkenes are unsaturated compounds and their typical reaction is **addition**, unlike the substitution reactions of the alkanes. After undergoing an addition reaction, the unsaturated alkene becomes a saturated alkane.

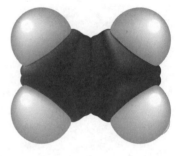

*Alkene bond angles.*

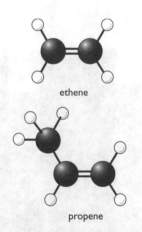

ethene

propene

When ethene is reacted with an alkaline solution of the oxidising agent potassium manganate(VII) it produces ethane-1,2-diol. This is known commercially as ethylene glycol and is used as an antifreeze in car engines in some countries.

Saturated compounds have single bonds; unsaturated compounds can have double or triple bonds.

**Hydrogenation** is the addition of hydrogen to a compound. If a mixture of ethene and hydrogen is passed over a nickel catalyst at about 150°C, ethane is formed.

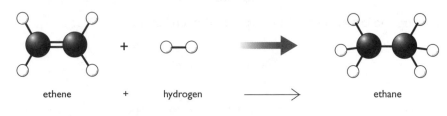

ethene    +    hydrogen    $\longrightarrow$    ethane

$$CH_2{=}CH_2(g) + H_2(g) \rightarrow CH_3CH_3(g)$$

> Notice the way that ethene is written in the equation to show the double bond. The ethane formed is also written in a way that shows exactly how the hydrogen has been added.

This isn't an important reaction for ethene itself. This reaction happens with any molecule containing a carbon–carbon double bond, and ethene is taken as a simple example. Ethene is too useful to turn into ethane. Hydrogenation of a double bond is important industrially in the manufacture of margarine.

Bromine adds to alkenes without any need for heat or a catalyst. The reaction is often carried out using bromine solution ('bromine water'). For example, with ethene:

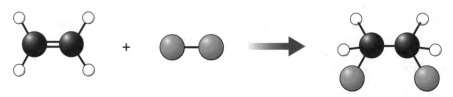

The product is called 1,2-dibromoethane and is a colourless liquid.

You can write this as an equation in two ways. The first is very close to the way the reaction is shown using the models:

The other method takes up rather less space:

$$CH_2{=}CH_2(g) + Br_2(aq) \rightarrow CH_2BrCH_2Br(l)$$

Any compound with a carbon–carbon double bond will react with bromine in a similar way. This is used to **test for a carbon–carbon double bond**.

If you shake an unknown organic compound with bromine water and the orange bromine water is decolorised, the compound contains a carbon–carbon double bond.

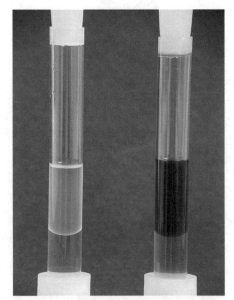

*The results of shaking a liquid alkene (left-hand tube) or alkane (right-hand tube) with bromine water.*

The left-hand tube in the photograph shows the effect of shaking a liquid alkene with bromine water. The organic layer is on top. You can see that the bromine water has been completely decolorised – showing the presence of the carbon–carbon double bond.

The right-hand tube in the photograph shows what happens if you use a liquid alkane – which doesn't have a carbon–carbon double bond. The colour of the bromine is still there. Interestingly, most of the colour is now in the top organic layer. That is because the covalent bromine is more soluble in the organic compound than it is in water.

13. What is the main difference between saturated and unsaturated compounds?
14. What is the test for a double bond?
15. What is the typical reaction of alkenes?

## The down-side of the oil industry

Simply because it is so big and important, the oil industry is also responsible for many of the world's environmental problems. Extraction and transport of oil results in obvious problems like oil spills, but there are all sorts of more important secondary effects from oil use.

There are many natural sources of hydrocarbon pollution. Methane is released by ruminant animals such as cows, and by termites and bacteria that cause decay. Some trees also release hydrocarbons such as isoprene. Hydrocarbons in the atmosphere act as greenhouse gases, like carbon dioxide. In major cities, the main hydrocarbon pollutant is unburned fuel from car and truck engines. These materials can produce smog (smoke with fog) in cities when the low air quality will be a hazard to the health of those living there. This is one of the reasons for the development of alternative power sources such as hydrogen-powered fuel cells and electric cars.

*Obvious pollution from an oil spill.*

Global warming is the result of carbon dioxide produced from burning fuels. Acid rain is produced by sulphur dioxide caused by the presence of sulphur in fuels, or by nitrogen oxides produced when they burn in air. Acid rain damages trees and makes lakes acidic, so killing the fish. Buildings are also attacked, especially those made of limestone (calcium carbonate). The holes in the ozone layer are caused by chlorinated solvents – again a product of the oil industry.

The majority of plastics in use are **non-biodegradable**. This means that they aren't broken down by living organisms and so they last in the environment almost for ever. Some of this problem can be overcome by recycling, but there are so many different sorts of plastics with different characteristics that this isn't easy.

# End of Chapter Checklist

**In this chapter you have learnt that:**

- the physical properties of hydrocarbons depend on the size and shape of the molecule
- fractional distillation is used to separate compounds in crude oil
- some fractions containing large molecules are cracked to give more useful, smaller molecules
- alkanes are saturated and react by substitution
- alkenes are unsaturated and react by addition
- there are rules to follow when naming organic compounds
- isomers have the same atoms arranged in a different structure.

# Questions

1. In general, the larger the molecule the larger the:

   A   reactivity
   B   boiling point
   C   transparency
   D   evaporation rate                                    (1)

2. The different compounds in crude oil are separated by:

   A   crystallisation
   B   burning
   C   fractional distillation
   D   filtering                                           (1)

3. When large molecules are broken up it is called:

   A   substitution
   B   addition
   C   cracking
   D   distillation                                        (1)

4. When bromine is added to an alkene the colour changes from brown to:

   A   green
   B   colourless
   C   purple
   D   red                                                 (1)

5. Which of the following statements is/are true?

   I    alkenes contain a double bond
   II   alkenes react by addition
   III  alkenes are less reactive than alkanes
   IV   alkenes are all gases

   A   I and II
   B   II and III
   C   III and IV
   D   I and IV                                            (1)

6. a) Explain how petroleum is separated by fractional distillation. Include a diagram in your answer.   (6)

   b) Name *two* fractions from crude oil and give a use for each.                                        (2)

   c) What is the link between the boiling point and size of most alkane molecules?                       (2)

7. a) Write word and balanced symbol equations for the complete combustion of octane.                    (4)

   b) What are the tests for the products of this reaction?                                               (4)

   c) What is meant by incomplete combustion?             (2)

8. a) Write displayed formulae for butane and hexane.    (4)

   b) What are isomers?                                   (2)

   c) Draw and name the isomers of pentane.              (4)

9. a) How do alkanes and alkenes react with halogens? Give balanced symbol equations to explain your answers.   (6)

   b) Which reacts faster, chlorine or bromine?          (2)

   c) What could you predict about the reaction of methane with fluorine?                                 (2)

10. a) Alkanes are *saturated* hydrocarbons. What do you understand by the term *saturated*?              (2)

    b) Undecane is an alkane with 11 carbon atoms.

       i)   Write down the molecular formula for undecane. (1)

       ii)  What physical state (solid, liquid or gas) would you expect undecane to be in at room temperature?   (2)

       iii) Write an equation for the complete combustion of undecane.                                    (2)

11. A gaseous hydrocarbon with three carbon atoms decolorised bromine water.

   a) Write the displayed formula for the hydrocarbon. (1)

   b) Write the balanced symbol equation for the reaction between the hydrocarbon and bromine. (2)

   c) Write the balanced symbol equation for the complete combustion of the hydrocarbon in oxygen. (2)

   d) The hydrocarbon was mixed with hydrogen and passed over a hot nickel catalyst. Write an equation for the reaction, and name the product. (3)

12. a) Write down the names of the following hydrocarbons:
   i) $CH_4$, ii) $CH_3CH_2CH_3$, iii) $C_5H_{12}$, iv) $CH_3CH=CH_2$, v) $C_2H_4$, vi) $CH_2=CHCH_2CH_3$. (6)

   b) Write fully displayed formulae (showing all the bonds) for: i) butane, ii) ethane, iii) but-2-ene, iv) 2-methylbut-2-ene. (4)

13. a) What do you understand by the term *structural isomerism*? (2)

   b) There are two structural isomers of $C_4H_{10}$.

   i) Draw their structures and name them. (2)

   ii) Which of these isomers has the lower boiling point? Explain your reasoning. (2)

   c) There are five structural isomers of $C_6H_{14}$. Draw their structures and name them. (5)

   d) How many structural isomers can you find with a molecular formula $C_4H_8$? You don't need to restrict yourself to alkenes. Draw their structures and name as many as you can. (4)

## Chapter 26: Polymers and Polymerisation

> **When you have completed this chapter, you will be able to:**
> - explain addition polymerisation
> - distinguish between monomers and polymers
> - know the names and uses of polymers
> - describe the properties and uses of polyesters and polyamides.

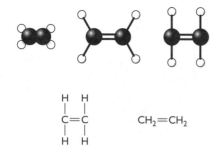

### The polymerisation of ethene

Ethene is one of the alkenes produced by cracking. It is the smallest hydrocarbon containing a carbon–carbon double bond. These are all different ways of writing or drawing an ethene molecule.

Under the right conditions, molecules containing carbon–carbon double bonds can join together to produce very long chains. Part of the double bond is broken, and the electrons in it are used to join to neighbouring molecules. This is called **addition polymerisation**.

Polymerisation is the joining up of lots of little molecules (the **monomers**) to make one big molecule (the **polymer**). In the case of ethene, lots of ethene molecules join together to make **poly(ethene)** – more usually called polythene.

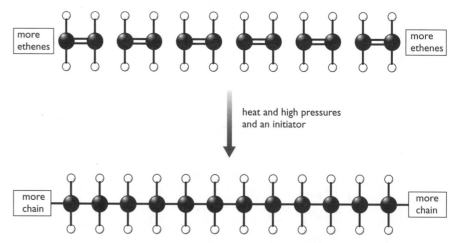

heat and high pressures and an initiator

The chain length can vary from about 4000 to 40 000 carbon atoms.

For normal purposes, this is written using **displayed formulae**. A displayed formula is rather like the models drawn here, but with symbols for the atoms rather than circles.

For exam purposes, the acceptable form is:

#### Monomers and polymers

A **monomer** is a simple chemical compound that can join with lots of other identical molecules. The result is a long chain compound known as a polymer. The prefix mono- means one; the prefix poly- means many. For example, the monomer ethene joins up to give a polymer called poly(ethene) or polythene.

An **initiator** is used to get the process started. You mustn't call it a catalyst, because it gets used up in the reaction. People occasionally wonder what happens at the ends of the chains. They don't end tidily! Bits of the initiator are bonded on at either end.

In this structure for poly(ethene), 'n' represents a large but variable number. It simply means that the structure in the brackets repeats itself lots and lots of times in the molecule.

## Uses of poly(ethene)

Poly(ethene) comes in two types – low density poly(ethene) (LDPE) and high density poly(ethene) (HDPE).

Low density poly(ethene) is made by compressing ethene and heating to 200°C in the presence of traces of oxygen. The molar mass varies from 50 000 up to 300 000.

Low density poly(ethene) is mainly used as a thin film to make polythene bags. It is very flexible and not very strong.

High density poly(ethene) is made in a different way. Ethene gas is passed into a solvent containing a Ziegler catalyst. The catalyst contains a titanium compound, a transition metal compound. The catalyst is removed at the end of the reaction. The molar mass is higher than with low density poly(ethene): it is about 50 000 up to 3 million.

High density poly(ethene) is used where greater strength and rigidity is needed – for example, to make plastic milk bottles. If you look underneath a plastic milk bottle, you will probably find the letters HDPE next to a recycling symbol.

## The polymerisation of propene

Propene is another alkene – this time with three carbon atoms in each molecule. Its formula is normally written as $CH_3CH=CH_2$. Think of it as a modified ethene molecule – with a $CH_3$ group attached in place of one of the hydrogen atoms.

When propene is polymerised you get **poly(propene)**. This used to be called polypropylene.

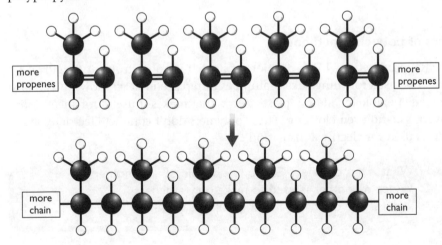

Write this as:

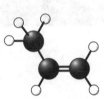

This is the real shape...

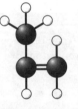

...but you'll make life much easier for yourself if you think of it like this

You will find that it is much easier to work out the structure of its polymer if you keep the $CH_3$ group tucked up out of the way when you draw it.

*Poly(propene) is used to make crates...*

*...and ropes.*

## Uses of poly(propene)

Poly(propene) is somewhat stronger than poly(ethene). It is used to make ropes and crates (among many other things). If an item has a recycling mark with PP inside it or near it, it is made of poly(propene).

## The polymerisation of chloroethene

Chloroethene is an ethene molecule in which one of the hydrogen atoms is replaced by a chlorine. Its formula is $CH_2{=}CHCl$. It used to be called vinyl chloride. Polymerising chloroethene gives you poly(chloroethene). This is usually known by its old name, polyvinylchloride or PVC.

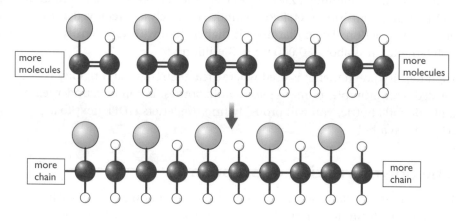

You would write it as:

It doesn't matter when you draw this whether you put the chlorine atom on the left-hand carbon atom or the right-hand one.

## Uses of poly(chloroethene)

Poly(chloroethene) – PVC – has lots of uses. It is quite strong and rigid and so can be used for things like drainpipes or replacement windows. It can also be made flexible by adding 'plasticisers'. That makes it useful for sheet floor coverings, and even clothing. These polymers don't conduct electricity and PVC is used for electrical insulation.

*PVC is used to insulate electric cables.*

> 1. What is a monomer? Give one example of a monomer.
> 2. Which monomer is used to make poly(ethene)?
> 3. What is the old name for poly(chloroethene)?

## Thermosoftening and thermosetting plastics

Plastics made of polymer chains like poly(ethene), poly(propene) and PVC are described as **thermosoftening plastics**. Although the polymer chains are held together by strong covalent bonds, the intermolecular forces between the chains are much weaker. If you heat the plastic gently, there is enough energy to break the intermolecular forces, and the plastic melts. On cooling, the plastic becomes solid again.

In a **thermosetting plastic**, cross-links are set up between the individual chains during the polymerisation process. This means that the whole piece of plastic is essentially one huge molecule.

This won't melt on gentle heating, because you would have to break strong covalent bonds. Stronger heating causes the plastic to char, but it won't melt.

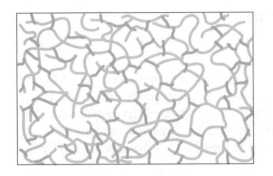

*In a thermosetting plastic individual chains (pale) are cross linked (dark).*

Melamine (used as a plastic coating on some furniture) and certain glues are thermosetting plastics.

### Dangers from burning plastics

When plastics burn they can produce a range of poisonous gases. Because of their high carbon content, they are likely to produce carbon monoxide as well as carbon dioxide, unless there is a very plentiful air supply. Carbon monoxide is poisonous. The thick black smoke produced by burning plastics is due to carbon – again produced by incomplete combustion.

Plastics containing chlorine (such as PVC) produce hydrogen chloride when they burn. Plastics containing nitrogen (such as polyurethane) produce hydrogen cyanide. Both hydrogen chloride and hydrogen cyanide are extremely poisonous.

*Burning plastics produce poisonous gases as well as heat and flames.*

### The oil industry

*The oil industry is BIG business!*

### Benefits of the oil industry

It is hard to imagine what life would be like if the oil industry suddenly stopped producing oil and gas. Transport – land, sea and air – would collapse. No more plastics. No more artificial fibres like polyester or nylon. No heating oil. Major power cuts as oil and gas fired power stations stopped producing electricity. Modern detergents would disappear. The pharmaceutical industry couldn't get its essential raw materials, so medicines would run out. Disease would flourish.

## Polyesters and polyamides

### Polyamides

Some polymers form in a different way to poly(ethene) and other addition polymers. If one monomer contains an –OH group and the other an –H, then they add by forming a molecule of water. When they polymerise, the water molecule is split out between them. This is called a condensation reaction. Polyamides occur naturally in proteins and they are formed by condensation reactions.

We can make polyamides artificially. The best known example is **nylon**. To make an amide link (like a peptide link) you need both a carboxylic acid and an amino group. If each of these molecules is double-ended, the monomers link up to give a new polymer. This is how nylon is formed.

$$HOOC^- —\square—COOH + H_2N^- —\square—NH_2 \rightarrow$$
$$HOOC —\square—CONH—\square—NH_2 + H_2O$$

If each of the monomers (represented above by a block) contains six carbon atoms, we get nylon 66. There are other types of nylon.

#### Uses of polyamides

Nylon is used to make toothbrushes, ropes that do not rot, tights, carpets and clothing. Nylon fibres are stronger than natural fibres such as cotton.

### Polyesters

When alcohols react with carboxylic acids they produce an ester and water. If the monomers are double-ended, containing two functional groups each, a polyester forms.

$$HOOC —\square—COOH + HO^-—\square—OH \rightarrow HOOC^- —\square—COO—\square—OH$$
'carboxylic acid'　　　'alcohol'　　　'ester link'

This is another example of **condensation polymerisation**. Two large molecules join by splitting out a small molecule between them. In this case the small molecule is water.

#### Uses of polyesters

Polyester fibres are very strong and are used to make clothing. An example is Terylene or Dacron. The material can be given a permanent crease, useful for trousers.

*Nylon finds applications in many different products.*

# End of Chapter Checklist

## In this chapter you have learnt that:

- addition polymerisation is used to form polythene and PVC (poly(chloroethene))
- polymers contain many small units called monomers, joined together
- each polymer has its own special properties and uses
- polyamides and polyesters are made by condensation polymerisation.

## Questions

1. The small units that join up to make a polymer are called:

   A peptides
   B dimmers
   C monomers
   D plastics                                                    (1)

2. The old name for poly(chloroethene) is:

   A polythene
   B PVC
   C polypropene
   D polyurethane                                                (1)

3. Which of the following is a well known polyamide?

   A polystyrene
   B nylon
   C plastic
   D bakelite                                                    (1)

4. a) Draw a diagram of monomers to show how poly(ethene) is formed.                                    (4)

   b) Is this addition or condensation polymerisation? Explain your answer.                              (2)

   c) Give two uses for poly(ethene).                            (2)

   d) What raw material is used to make this polymer?           (2)

5. a) Draw the structure of part of a molecule of poly(chloroethene).                                  (3)

   b) Describe two uses of this polymer and explain why it is chosen.                                    (4)

   c) Why is poly(chloroethene) dangerous to burn?             (2)

   d) State one method of waste disposal for plastics, other than burning.                              (1)

6. a) Draw structural diagrams to explain condensation polymerisation.                                  (4)

   b) Give two examples of polymers formed in this way.   (2)

7. Propene, $C_3H_6$, can be polymerised to make poly(propene).

   a) What do you understand by the term polymerisation? (2)

   b) Draw a displayed formula (showing all the bonds) for propene.                                      (1)

   c) Draw a diagram to show the structure of a poly(propene) chain. Restrict yourself to showing three repeating units.                             (2)

   d) Give one use for poly(propene) and explain the properties which make it suitable for that use.     (2)

   e) Styrene has the formula $C_6H_5CH\!=\!CH_2$. Write an equation to show what happens when styrene is polymerised to make polystyrene. Your equation should clearly show the structure of the polystyrene. (Show the $C_6H_5$ group as a whole without worrying about its structure.)                          (3)

   f) Both poly(propene) and polystyrene are thermosoftening plastics. Describe a simple experiment that you could do to show this.     (3)

**When you have completed this chapter, you will be able to:**

- describe the physical and chemical properties of alcohols
- understand how roadside alcohol meters work
- name and draw the structures of alcohols, acids and esters
- describe the reactions of carboxylic acids.

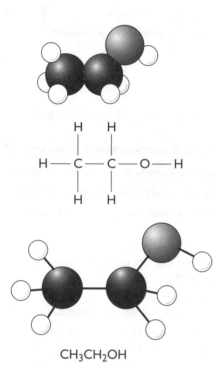

CH₃CH₂OH

*Various ways of drawing ethanol.*

## Alcohols

What everybody knows as 'alcohol' is just one member of a large family (homologous series) of similar compounds. Alcohols all contain an –OH group covalently bonded onto a carbon chain. Their general formula is $C_nH_{2n+1}OH$.

The familiar alcohol in drinks is $C_2H_5OH$ (or, better, $CH_3CH_2OH$), and should properly be called **ethanol**.

### Alcohol in drinks

Alcohol depresses the activity of some of the higher parts of the brain. As a result, it reduces inhibitions. Even small quantities of alcohol reduce concentration.

The alcohol content of different drinks varies enormously. By law, bottles or cans have to tell you the percentage of alcohol by volume they contain. This means that you can work out the volume of pure alcohol that the drink contains.

Wine typically contains about 12% alcohol by volume. A reasonably sized wine glass holds about 125 cm³. 12% of 125 cm³ is 15 cm³. Drinking a glass of wine is therefore the equivalent of drinking 15 cm³ of pure alcohol.

In some countries, 10 cm³ of alcohol is defined as '1 unit of alcohol'. The maximum recommended weekly consumption for adult men was 21–28 units, and for women was 14–21 units in 2001. Recommendations change from time to time. It is easy to find out the current recommendations by doing an internet search. Spirits, such as rum, are distilled to increase the alcohol percentage. The same is true of whisky and vodka.

Drinking excessive amounts of alcohol is the cause of all sorts of social problems – both domestic and public violence, for example. More than half of road traffic accidents involving injury are alcohol related.

### *The roadside alcohol test for drivers*

Since drinking alcohol has such a bad effect on the skill and reaction times of drivers, a test was needed to identify drunk drivers. The breathalyser is a portable test kit. The driver breathes into the meter and the tube changes colour if there is too much alcohol in the exhaled air.

# Activity 1: The alcohol meter colour change

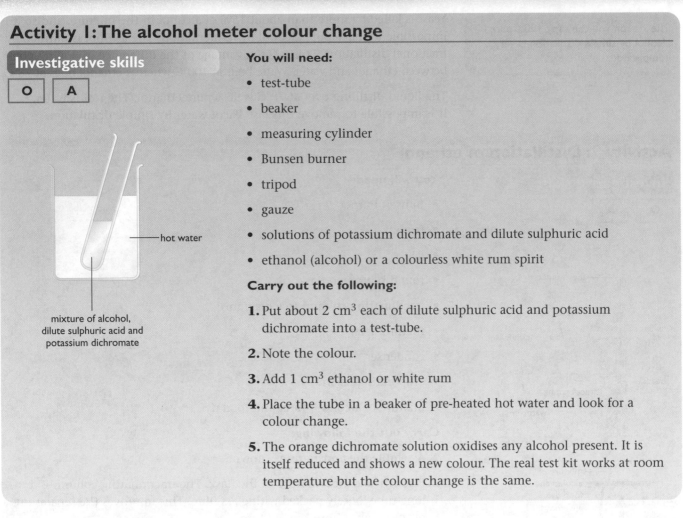

hot water

mixture of alcohol, dilute sulphuric acid and potassium dichromate

**You will need:**

- test-tube
- beaker
- measuring cylinder
- Bunsen burner
- tripod
- gauze
- solutions of potassium dichromate and dilute sulphuric acid
- ethanol (alcohol) or a colourless white rum spirit

**Carry out the following:**

1. Put about 2 cm$^3$ each of dilute sulphuric acid and potassium dichromate into a test-tube.

2. Note the colour.

3. Add 1 cm$^3$ ethanol or white rum

4. Place the tube in a beaker of pre-heated hot water and look for a colour change.

5. The orange dichromate solution oxidises any alcohol present. It is itself reduced and shows a new colour. The real test kit works at room temperature but the colour change is the same.

Long term consumption of alcohol beyond the recommended limits can cause serious liver damage.

## The production of alcohol (ethanol – C$_2$H$_5$OH)

### Making ethanol by fermentation

Yeast is added to a sugar solution and left in the warm for several days in the absence of air ('**anaerobic**' conditions). **Enzymes** in the yeast convert the sugar into ethanol and carbon dioxide. The process is known as **fermentation**.

The biochemistry is very complicated. First, the sugar (sucrose) is split into two smaller sugars, glucose and fructose. Glucose and fructose have the same molecular formulae, but different structures. They are isomers.

Enzymes in the yeast convert these sugars into ethanol and water in a multitude of small steps. All we normally write are the overall equations for the reactions.

$$C_{12}H_{22}O_{11}(aq) + H_2O(l) \rightarrow C_6H_{12}O_6(aq) + C_6H_{12}O_6(aq)$$

sucrose                                           glucose              fructose

$$C_6H_{12}O_6(aq) \rightarrow 2C_2H_5OH(aq) + 2CO_2(g)$$

glucose              ethanol              carbon dioxide

*Sugar cane – the raw material for making ethanol.*

Yeast is killed by more than about 15% of alcohol in the mixture, and so it is impossible to make pure alcohol by fermentation. The alcohol is purified by fractional distillation. This takes advantage of the difference in boiling point between ethanol and water. Water boils at 100°C whereas ethanol boils at 78°C.

The liquid distilling over at 78°C is 96% pure ethanol. The rest is water. It is impossible to remove this last 4% of water by simple distillation.

## Activity 2: Distillation of ethanol

### Investigative skills

| O | A |
|---|---|

thermometer
water out
condenser
glass beads
fractionating column
water in
mixture of alcohol and food dye, and water
ethanol
heat

**You will need:**

- Bunsen burner
- tripod
- gauze
- round-bottomed flask
- fractionating column (see diagram)
- thermometer
- condenser
- evaporating basin
- mixture of alcohol, water and food dye

**Carry out the following:**

1. Assemble the apparatus as shown.

2. Place the alcohol mixture in the flask. The fractionating column is a long tube loosely packed with glass fibre. This improves the separation of the alcohol from the mixture.

3. Heat the mixture until it boils and observe the temperature.

4. Collect the distillate in the range 75–85°C. (The boiling point of ethanol is 78°C.) (**Warning:** Flammable)

5. Pour 2 cm$^3$ of the distillate into an evaporating basin. Ignite it using a splint. If it is alcohol it should burn with a clean blue flame.

6. How much evidence is there that the distillate is alcohol and not water?

7. Would this experiment work with wine or rum?

### Making ethanol by the hydration of ethene

Ethanol is also made by reacting ethene with water – a process known as hydration.

$$CH_2{=}CH_2(g) + H_2O(g) \rightarrow CH_3CH_2OH(g)$$

Starting materials:    ethene and steam
Temperature:    300°C
Pressure:    60–70 atmospheres
Catalyst:    phosphoric acid

Only a small proportion of the ethene reacts. The ethanol produced is condensed as a liquid and the unreacted ethene is recycled through the process.

### Comparing the two methods of producing ethanol

At the moment, countries which have easy access to crude oil produce ethanol mainly from ethene, but one day the oil will start to run out. At that point, the production of ethanol from sugar will provide an alternative route to many of the organic chemicals we need.

Those countries whose climate is too cold to grow sugar cane may be able to grow an alternative sugar crop. Sugar beet can also be used to produce sugar, from which alcohol can be made.

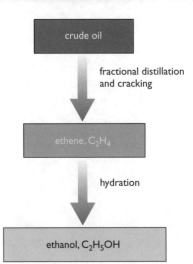

From crude oil to ethanol.

| | Fermentation | Hydration of ethene |
|---|---|---|
| Use of resources | Uses renewable resources – sugar beet or sugar cane. | Uses finite resources. Once all the oil has been used up there won't be any more. |
| Type of process | A batch process. Everything is mixed together in a reaction vessel and then left for several days. That batch is then removed and a new reaction is set up. This is inefficient. | A continuous flow process. A stream of reactants is constantly passed over the catalyst. This is more efficient than a batch process. |
| Rate of reaction | Slow, taking several days for each batch. | Rapid. |
| Quality of product | Produces very impure ethanol which needs further processing. | Produces much purer ethanol. |
| Reaction conditions | Uses gentle temperatures and ordinary pressure. | Uses high temperatures and pressures, needing a high input of energy. |

1. What is a breathalyser?
2. What is fermentation?
3. Why are gin or rum distilled?

### Uses of ethanol

Ethanol is sold as 'industrial methylated spirit'. This is ethanol with a small amount of another alcohol, methanol, added to it. Methanol is poisonous, and makes the industrial methylated spirit unfit to drink, and so avoids the high taxes on alcoholic drinks.

Ethanol is widely used as a solvent – for example, for cosmetics and perfumes. It is relatively safe and is a good solvent for the complex organic molecules which don't dissolve in water.

Ethanol is also a useful fuel. It burns to form carbon dioxide and water, producing about two-thirds as much energy per litre as petrol.

$$C_2H_5OH(l) + 3O_2(g) \rightarrow 2CO_2(g) + 3H_2O(l)$$

Mixtures of petrol with 10–20% ethanol – known as **gasohol** – are increasingly used in some countries, such as Brazil. These are countries which have little or no oil industry to produce their own petrol. On the other hand, they often have a climate which is good for growing sugar cane. Ethanol can be produced by fermenting the sugar, and then mixed with imported petrol. This saves money on imports.

### Some reactions of the alcohols

#### *Alcohols burn*

All alcohols burn to form carbon dioxide and water. For example, with methanol:

$$2CH_3OH(l) + 3O_2(g) \rightarrow 2CO_2(g) + 4H_2O(l)$$

Methanol is being tried as an alternative fuel for cars. To find out the current state of research into methanol as an alternative fuel, try an internet search on **methanol fuel**.

The use of ethanol in fuel has already been mentioned.

#### *Dehydrating alcohols*

Dehydration refers to the removal of water from a compound. The dehydration of ethanol produces ethene. Ethanol vapour is passed over hot aluminium oxide acting as a catalyst.

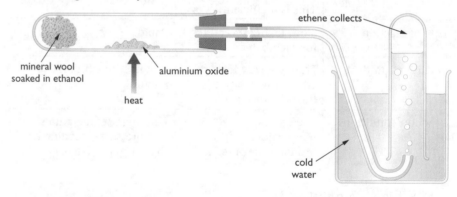

$$CH_3CH_2OH(g) \rightarrow CH_2{=}CH_2(g) + H_2O(l)$$

Notice that the –OH group is removed, together with a hydrogen from the neighbouring carbon atom.

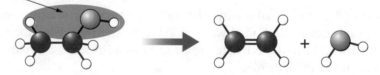

Other alcohols would dehydrate in a similar way. For example:

$$CH_3CH_2CH_2OH(g) \rightarrow CH_3CH{=}CH_2(g) + H_2O(l)$$

### Alcohols react with sodium

Alcohols react gently with sodium to produce hydrogen. For example, with ethanol:

$$2CH_3CH_2OH(l) + 2Na(s) \rightarrow 2CH_3CH_2ONa(\text{solution in ethanol}) + H_2(g)$$

Comparison of the reaction of sodium metal with water and with ethanol

| Reaction with water | Reaction with ethanol |
|---|---|
| sodium floats | sodium sinks |
| lots of heat produced | gets warm |
| hydrogen gas produced | hydrogen gas produced |
| sodium hydroxide, a strong alkali, also produced in solution | sodium ethoxide, a salt, also produced in solution |

The organic product is called sodium ethoxide. This is a useful way of disposing of small amounts of unwanted sodium, or of treating small sodium spills on the bench in the lab.

> This is similar to the reaction between sodium and water to produce sodium hydroxide and hydrogen, but is much less vigorous.

### Alcohols can be oxidised

A bottle of wine left open to the air turns sour. The French for 'sour wine' is *vin aigre*, which has been distorted into **vinegar**. The ethanol in the wine is oxidised by air with the help of bacteria to form ethanoic acid, $CH_3COOH$. The old name for ethanoic acid was acetic acid.

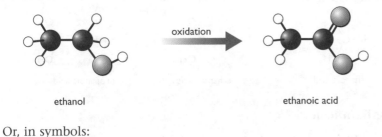

ethanol                    ethanoic acid

Or, in symbols:

$$CH_3CH_2OH \xrightarrow{\text{oxidation}} CH_3COOH$$

Ethanol can also be oxidised by a mixture of potassium dichromate and sulphuric acid. This is the same reaction as in the breathalyser. The acidic solution of potassium dichromate is orange in colour. As it oxidises the alcohol to a carboxylic acid, the dichromate is itself reduced. The final colour is green. This colour change of the transition metal ion shows that the reaction has happened. You will find more about acids like ethanoic acid in the next few pages – including their reactions with alcohols to make **esters**.

---

4. What is formed when alcohols burn?
5. How can you turn ethanol into ethane?
6. What would you see when sodium metal reacts with ethanol?

Here is a comparison of four main types of organic compound:

| Name | Formula | Example |
|------|---------|---------|
| alkane | $C_nH_{(2n+2)}$ | ethane $C_2H_6$ |
| alcohol | $C_nH_{(2n+1)}OH$ or ROH | ethanol $C_2H_5OH$ |
| acid | RCOOH (R is an alkyl group) | ethanoic acid $CH_3COOH$ |
| ester | RCOOR' (where R and R' can be different) | methyl propanoate $C_2H_5COOCH_3$ |

# Carboxylic acids

Acids such as ethanoic acid are known as **carboxylic acids**. The term 'carboxylic' refers to the presence of the –COOH group. These acids form another homologous series.

## Names and structures

The carboxylic acid group in a molecule is coded by the ending '**oic acid**'. The code for the number of carbon atoms ('meth', 'eth', etc.) includes the one in the –COOH group. For example, ethanoic acid has a total of two carbon atoms and so is $CH_3COOH$.

Methanoic acid, ethanoic acid and propanoic acid don't have any structural isomers which are also acids.

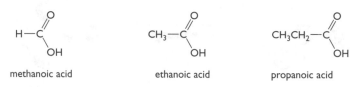

methanoic acid        ethanoic acid        propanoic acid

## Uses of ethanoic acid

Vinegar is a dilute solution of ethanoic acid. The familiar smell of vinegar is due to the acid. Naturally produced vinegars will have their origin described on the label – for example, 'wine vinegar' or 'cider vinegar'. Cheap vinegar may well be a product of the chemical industry and now has to be called 'non-brewed condiment'.

Vinegar is used as a flavouring and a preservative.

## Reactions of the carboxylic acids

Ethanoic acid is taken as typical of the carboxylic acids. It is the one most commonly used in the lab because it is the cheapest.

### Acid properties

Ethanoic acid is a weak acid with a pH about 2 to 3, depending on the concentration of the solution. It will turn blue litmus paper red, and reacts with all the things you expect acids to react with.

*Pepper sauce is acid-based so can be used in preserving food.*

*Reactions between carboxylic acids and alcohols*

Heating a mixture of ethanoic acid and ethanol with a few drops of concentrated sulphuric acid produces a sweet smelling liquid called **ethyl ethanoate**. This is one member of a family (homologous series) of compounds called **esters**.

$$CH_3COOH(l) + CH_3CH_2OH(l) \rightleftharpoons CH_3COOCH_2CH_3(l) + H_2O(l)$$
$$\text{ethanoic acid} \qquad \text{ethanol} \qquad \text{ethyl ethanoate}$$

$$\text{alcohol} \quad + \quad \text{acid} \quad \rightleftharpoons \quad \text{ester} \quad + \text{ water}$$

The concentrated sulphuric acid isn't written into the equation because it is a catalyst and isn't used up in the reaction.

The ethyl ethanoate has the lowest boiling point of any of the substances present, and is distilled off as soon as it is formed.

# Esters

### Names and structures

Students often have more trouble with the names and structures of esters than with any other organic compounds. Look again at the formation of ethyl ethanoate:

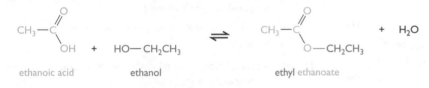
ethanoic acid + ethanol ⇌ ethyl ethanoate + H₂O

The ethanol has been written back-to-front so that you can more easily see how everything fits together. The confusing thing about the name of the ethyl ethanoate is that the two halves are written the other way around from the way they appear in the formula.

Here is another example:

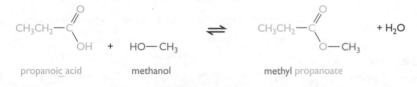
propanoic acid + methanol ⇌ methyl propanoate + H₂O

To be absolutely accurate, the oxygen attached to the ethyl group in the ethyl ethanoate comes from the ethanol and ought really to be coloured red.

You won't need to worry about this unless you do a Chemistry degree!

### Uses for esters

The small esters like ethyl ethanoate are commonly used as solvents, but most esters are used in flavourings and perfumes. The typical smell of bananas, raspberries, pears or any other fruit is due in part to naturally occurring esters. Food chemists create artificial food flavourings using mixtures of esters and other organic compounds.

You have to read labels *very* carefully. For example, 'strawberry **flavoured**' means that the flavouring has to come from real strawberries. 'Strawberry **flavour**' means that the flavouring can be entirely artificial. Chemists have produced as many as 200 different versions of strawberry flavour designed for different products.

# End of Chapter Checklist

## In this chapter you have learnt that:

- alcohols can be oxidised and this reaction is used to test suspected drunk drivers
- alcohols can be converted into carboxylic acids
- acids and alcohols react to form esters
- natural oils and fats contain esters.

## Questions

1. Organic compounds containing the group –OH are called:

   A   esters
   B   alkalis
   C   alcohols
   D   salts                                            (1)

2. Alcohols can be oxidised to form:

   A   acids
   B   bases
   C   soap
   D   esters                                           (1)

3. a) Given some sugar and some yeast, describe briefly how you would produce an impure solution of ethanol (alcohol).                                   (4)

   b) How would you produce a reasonably pure sample of ethanol from the impure mixture?              (2)

   c) Industrially, much alcohol is made by the *hydration* of ethene. Explain, with the help of an equation, what you understand by the term *hydration*.           (2)

   d) Give the conditions for the manufacture of ethanol from ethene.                                (2)

4. a) Write the equation for the combustion of ethanol, $C_2H_5OH$.                                      (2)

   b) Carbon dioxide is a 'greenhouse gas', and increased amounts of it in the atmosphere are a cause of global warming. It could be argued that burning ethanol made by fermentation doesn't add to global warming, but burning ethanol made from the hydration of ethene does. How could you justify this strange statement?  (6)

   c) What other environmental advantages are there in producing ethanol by fermentation?             (2)

5. Some studies have suggested that the reason soft drinks are so popular with children is that they contain large amounts of refined sugar. Many young children suffer more dental decay than was the case in previous generations when these drinks were unknown. There is no agreement as to whether these two issues are connected. By doing an internet (or other) search, find out the most recent scientific view of

this. Write a short article (maximum 300 words) for a lifestyle magazine, summarising what you find. Add diagrams or pictures if you think they help.             (10)

6. Draw structures for:

   a) propan-1-ol                                       (2)

   b) the compound formed when propan-1-ol is dehydrated by passing its vapour over hot aluminium oxide   (2)

   c) the carboxylic acid which would be obtained if propan-1-ol was oxidised                          (2)

   d) the ester formed when propan-1-ol reacts with ethanoic acid.                                    (2)

7. a) A unit of alcohol is 10 cm$^3$ of pure alcohol. How many units are there in:

   i)   half of a 750 cm$^3$ bottle of white wine containing 12% alcohol by volume                      (4)

   ii)  3 pints of beer containing 4% alcohol by volume (1 pint = 568 cm$^3$)                           (2)

   iii) a small (350 cm$^3$) bottle of brandy containing 36% alcohol by volume?                        (4)

8. Identify substances A to F in the following reactions.

   a)        hydration                sodium metal
       A      →      ethanol      →      B + C          (3)

   b)      ethanoic acid            sodium hydroxide
       D      →    ethyl ethanoate    →    E + F        (3)

   c) Write word and balanced symbol equations for each of the reactions in a) and b).                  (4)

9. Identify substances G to M in the following reactions.

   a) ethanol + G → ethyl propanoate                   (2)

   b)                                heat
       calcium hydrogen carbonate    →    H + I + J     (3)

   c)            yeast
       sugar    →    K + L,
       then L + M → solution M goes cloudy              (3)

   d) Write word and balanced symbol equations for the reaction in c).                                 (2)

# End of Section Questions

1. Crude oil (petroleum) is a complex mixture of *hydrocarbons*. The crude oil can be separated into simpler mixtures (called *fractions*) by taking advantage of differences in boiling points between the various hydrocarbons.

   **a)** What do you understand by the term *hydrocarbon*? (2 marks)

   **b)** What is the relationship between the number of carbon atoms in a hydrocarbon and its boiling point? (1 mark)

   **c)** What name is given to the process of producing the simpler mixtures from the crude oil? (1 mark)

   **d)** One of the fractions produced from crude oil is called *kerosine*. Give *one* use for kerosine. (1 mark)

   **e)** One of the hydrocarbons present in kerosine is an alkane containing 10 carbon atoms, called decane. Write the molecular formula for decane. (1 mark)

   **f)** The hydrocarbon $C_{15}H_{32}$ (also present in kerosene) burns to form carbon dioxide and water. Write the equation for the reaction. (2 marks)

   **g)** How would you test the products when $C_{15}H_{32}$ burns to show that carbon dioxide has been formed? (2 marks)

   **h)** Name the environmental problem that is caused by the formation of carbon dioxide during the combustion of hydrocarbons. (1 mark)

   **i)** Liquefied petroleum gas (LPG), used for domestic heating and cooking, is propane, $C_3H_8$. Burning propane in badly maintained appliances or in poorly ventilated rooms can cause death. Explain why that is. (4 marks)

   **Total 15 marks**

2. Some of the gas oil fraction from crude oil is broken into smaller molecules by heating it in the presence of a catalyst. A mixture of *saturated* and *unsaturated* hydrocarbons is formed.

   **a)** Explain the difference between a saturated and an unsaturated hydrocarbon. (2 marks)

   **b)** What name is given to the process of breaking up the gas oil fraction in this way? (1 mark)

   **c)** When a molecule $C_{17}H_{36}$ was heated in the presence of a catalyst, it broke up to give two molecules of ethene, $C_2H_4$, one molecule of propene, $C_3H_6$, and another molecule, X.

   **i)** Write a balanced symbol equation for the reaction. (2 marks)

   **ii)** Is molecule X an alkane or an alkene? (1 mark)

   **d)** Some propene is converted into propenenitrile which is polymerised to make a fibre used for textiles. 'Orlon', 'Acrilan' and 'Courtelle' are all poly(propenenitrile). The structure of poly(propenenitrile) is:

   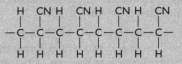

   **i)** What do you understand by the term *polymerisation*? (2 marks)

   **ii)** By looking at the structure of the polymer, suggest a structural formula for the monomer, propenenitrile. (2 marks)

   **e)** Poly(chloroethene) (also called PVC) is a polymer produced from chloroethene.

   **i)** State *one* use for poly(chloroethene). (1 mark)

   **ii)** Poly(chloroethene) is non-biodegradable and produces hydrogen chloride gas when it burns. What are the implications of this for disposal of PVC? (4 marks)

   **Total 15 marks**

3. **a)** Draw fully displayed structural formulae for i) propane, ii) propene. (2 marks)

   **b)** Propane and propene are both gases. Describe a simple test which would enable you to distinguish between them. You should describe what you would do and what you would see in each case. Write an equation for any reaction(s) you describe. (4 marks)

   **c)** Propene can be *hydrogenated* by passing it, together with hydrogen, over a catalyst.

   **i)** Name the catalyst used. (1 mark)

   **ii)** Write an equation for the reaction. (1 mark)

   **iii)** Give an example of a large-scale process which involves the hydrogenation of a carbon–carbon double bond. (1 mark)

   **Total 9 marks**

4. This question is about *structural isomerism*.

   **a)** What are *structural isomers*? (2 marks)

   **b)** The diagram shows the structure of one of the isomers of $C_5H_{12}$.

   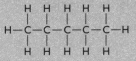

*i)* Draw the structures of the other two isomers of $C_5H_{12}$. *(2 marks)*

*ii)* Which of the three isomers has the highest boiling point? Explain your answer. *(3 marks)*

**c)** Draw the structures of the two alcohols which have the molecular formula $C_3H_8O$. *(2 marks)*

**d)** The molecular formula $C_3H_6O_2$ is shared by one carboxylic acid and two esters. Draw the structures of the acid and one of the esters. Name the acid and the ester that you have drawn. *(4 marks)*

**Total 13 marks**

5. Ethanol, $CH_3CH_2OH$, can be made by *fermentation* followed by *fractional distillation*, or by the *hydration of ethene*.

**a)** Describe briefly how an impure dilute solution of ethanol is made by fermentation. *(5 marks)*

**b)** Fermentation is controlled by *enzymes*. What are enzymes? *(2 marks)*

**c)** State the different boiling points of ethanol and water which enable them to be separated by fractional distillation. *(2 marks)*

**d)** Write an equation to show the hydration of ethene. *(1 mark)*

**e)** State any *two* conditions used during the hydration reaction. *(2 marks)*

**f)** Explain which of the two processes has the advantage in terms of:

*i)* the use of resources *(2 marks)*

*ii)* the conditions used *(2 marks)*

*iii)* the speed of production. *(2 marks)*

**Total 18 marks**

6. This question is about ethanol, $CH_3CH_2OH$, and its reactions.

**a)** Ethanol can be used as a fuel. Write the equation for the complete combustion of ethanol. *(2 marks)*

**b)** Ethanol is likely to be found among the ingredients of some perfumes and cosmetics. Explain why it is used in these products. *(2 marks)*

**c)** If ethanol vapour is passed over hot aluminium oxide, it dehydrates and a gas is produced.

*i)* Name the gas produced and write an equation for its formation. *(2 marks)*

*ii)* Draw a simple diagram of the apparatus you could use to produce and collect a few test tubes of the gas using this reaction. *(4 marks)*

**d)** If ethanol is warmed with ethanoic acid in the presence of some concentrated sulphuric acid, ethyl ethanoate is formed.

*i)* What type of compound is ethyl ethanoate? *(1 mark)*

*ii)* What is the function of the concentrated sulphuric acid? *(1 mark)*

*iii)* Write an equation for the reaction, clearly showing the structure of the ethyl ethanoate. *(2 marks)*

**Total 14 marks**

7. **a)** When wine is exposed to bacteria in the air, it turns sour as the ethanol is oxidised to ethanoic acid, $CH_3COOH$.

*i)* What common name is given to the solution of ethanoic acid formed in this way? *(1 mark)*

*ii)* Give *two* different uses for the solution. *(2 marks)*

**b)** Ethanoic acid is a *weak acid*.

*i)* What do you understand by the term *weak* as applied to acids? *(1 mark)*

*ii)* Describe a simple experiment by which you could show that dilute ethanoic acid was acidic, without using either indicators or a pH meter. *(2 marks)*

**c)** This part of the question concerns the compound:

*i)* Name this compound. *(1 mark)*

*ii)* Suggest a possible commercial use for the compound. *(1 mark)*

*iii)* What reagents would you need to make this compound from ethanoic acid? *(2 marks)*

**Total 10 marks**

# Appendix A: Safety Symbols

## Hazard Warning Symbols

**Toxic**
These substances can cause death. They may be poisonous when swallowed, breathed in, or absorbed through the skin.

**Harmful**
These are similar to toxic substances but less dangerous.

**Corrosive**
These substances attack and destroy material like wood, and living tissue including skin and eyes.

**Irritant**
These aren't as dangerous as corrosive substances, but still cause reddening or blistering of the skin.

**Highly flammable**
These substances catch fire easily.

**Oxidising**
These substances provide oxygen and cause other things to burn more fiercely.

## Hazchem Codes

These are used when dangerous substances are transported. The symbols give important information to the emergency services so that they can respond safely and quickly.

code to identify the hazardous substance

code for the type of substance needed to neutralise the risk — in this case, foam

code for the type of breathing apparatus and protective clothing

type of hazard

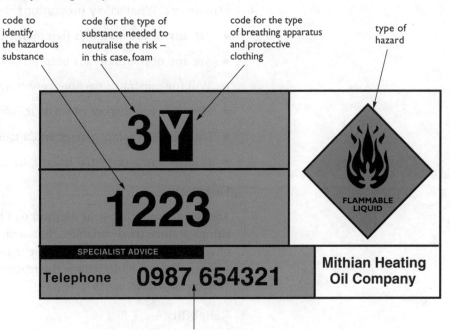

source of further information

# Appendix B: Experimental and Investigative Skills

## Experimental skills

Chemistry is an experimental subject and you should be given lots of opportunities to carry out practical work. This practical work, or the alternative written paper, will form 20 percent of your final mark for the examination.

You will see that the experiments in this book have code letters: P, O, A and E which refer to the different experimental skills. These skills are explained in detail later on.

Here is a list of some of the most important skills you need to develop during the chemistry course.

- Predict what is likely to happen in an experiment.
- Write a detailed method, rather like a recipe in cooking.
- Make some relevant measurements making it clear what units you are using.
- Record your results, for example, in a table.
- Display your results, for example as a line graph.
- Analyse your results, looking out for any unexpected (anomalous) results.
- Evaluate the data: is there enough of it to reach a conclusion? Did you choose a sensible range?
- Suggest ideas for extra work that would extend your investigation and produce more data.

There are some other points that you must consider before you start any experiment. What safety precautions should you take? For example:

- Are any of the materials flammable or toxic?
- Are any of the materials harmful if you touch them?
- Will the apparatus become very hot?
- How can you dispose of waste materials safely at the end?
- Do you need safety glasses and a laboratory coat?
- Is loose hair a potential hazard, for example with gas flames?

### Fair testing

You must check that your method will be a fair test. If you change two things at once (two variables), how will you know which of them caused a change? For example, if you raise the temperature and also stir faster, how will you be sure which made the reaction speed up? Always change one variable at a time.

### Reliability

If you carry out an experiment just once, your results may be unusual or even wrong. You cannot be sure because there is nothing with which to compare the results. For reliable results you need to repeat your experiments to check that the findings are similar.

## The experimental skills P, O, A and E

### P is for planning

You need a plan before you can carry out any experiments.

You need to consider:

- the apparatus and materials you need
- what to change (variable) and what to keep the same (the other variables)
- which scientific ideas support your choices
- predict what you think may happen and why
- mention any earlier work that is useful for your new plan.

### O is for obtaining evidence

This is the experimental part of your work.

You need to consider:

- the accuracy of your measurements and the units
- how many repeats are needed for reliable results
- how to record your observations
- if there is enough evidence to draw a valid conclusion.

### A is for analysing evidence

You need to look for patterns in your results and for unusual results (anomalies).

You need to consider:

- how best to display your results, for example a graph or pie chart
- whether you can draw a best-fit line on the graph
- what your results tell you – the conclusion
- the scientific reasons for your results.

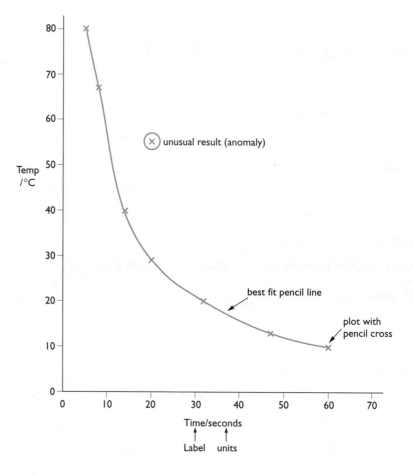

## E is for evaluating

A good scientist will review the experiments and the data and consider what improvements could be made.

You need to consider:

- the reliability of your evidence
- what were the anomalies and what might have caused them
- if you could improve your method next time
- if you have enough evidence to support your conclusions
- if you can summarise your findings.

# Appendix C: The Periodic Table

Group

| 1 | 2 | | | | | | | | | | | 3 | 4 | 5 | 6 | 7 | 0 |
|---|---|---|---|---|---|---|---|---|---|---|---|---|---|---|---|---|---|
| | | | | | | | | | | | | | | | | | 4 **He** Helium 2 |

| | | | | | | | | | | | | |
|---|---|---|---|---|---|---|---|---|---|---|---|---|

Period

| Period | | | | | | | | | | | | | | | | | |
|---|---|---|---|---|---|---|---|---|---|---|---|---|---|---|---|---|---|
| 1 | | | | | | | 1 **H** Hydrogen | | | | | | | | | | |
| 2 | 7 **Li** Lithium 3 | 9 **Be** Beryllium 4 | | | | | | | | | | | 11 **B** Boron 5 | 12 **C** Carbon 6 | 14 **N** Nitrogen 7 | 16 **O** Oxygen 8 | 19 **F** Fluorine 9 | 20 **Ne** Neon 10 |
| 3 | 23 **Na** Sodium 11 | 24 **Mg** Magnesium 12 | | | | | | | | | | | 27 **Al** Aluminium 13 | 28 **Si** Silicon 14 | 31 **P** Phosphorus 15 | 32 **S** Sulphur 16 | 35.5 **Cl** Chlorine 17 | 40 **Ar** Argon 18 |
| 4 | 39 **K** Potassium 19 | 40 **Ca** Calcium 20 | 45 **Sc** Scandium 21 | 48 **Ti** Titanium 22 | 51 **V** Vanadium 23 | 52 **Cr** Chromium 24 | 55 **Mn** Manganese 25 | 56 **Fe** Iron 26 | 59 **Co** Cobalt 27 | 59 **Ni** Nickel 28 | 63.5 **Cu** Copper 29 | 65 **Zn** Zinc 30 | 70 **Ga** Gallium 31 | 73 **Ge** Germanium 32 | 75 **As** Arsenic 33 | 79 **Se** Selenium 34 | 80 **Br** Bromine 35 | 84 **Kr** Krypton 36 |
| 5 | 86 **Rb** Rubidium 37 | 88 **Sr** Strontium 38 | 89 **Y** Yttrium 39 | 91 **Zr** Zirconium 40 | 93 **Nb** Niobium 41 | 96 **Mo** Molybdenum 42 | 99 **Tc** Technetium 43 | 101 **Ru** Ruthenium 44 | 103 **Rh** Rhodium 45 | 106 **Pd** Palladium 46 | 108 **Ag** Silver 47 | 112 **Cd** Cadmium 48 | 115 **In** Indium 49 | 119 **Sn** Tin 50 | 122 **Sb** Antimony 51 | 128 **Te** Tellurium 52 | 127 **I** Iodine 53 | 131 **Xe** Xenon 54 |
| 6 | 133 **Cs** Caesium 55 | 137 **Ba** Barium 56 | 139 **La** Lanthanum 57 | 179 **Hf** Hafnium 72 | 181 **Ta** Tantalum 73 | 184 **W** Tungsten 74 | 186 **Re** Rhenium 75 | 190 **Os** Osmium 76 | 192 **Ir** Iridium 77 | 195 **Pt** Platinum 78 | 197 **Au** Gold 79 | 201 **Hg** Mercury 80 | 204 **Tl** Thallium 81 | 207 **Pb** Lead 82 | 209 **Bi** Bismuth 83 | 210 **Po** Polonium 84 | 210 **At** Astatine 85 | 222 **Rn** Radon 86 |
| 7 | 223 **Fr** Francium 87 | 226 **Ra** Radium 88 | 227 **Ac** Actinium 89 | | | | | | | | | | | | | | | |

Key

| |
|---|
| Relative atomic mass |
| Symbol |
| Name |
| Atomic number |

# Index